Ontario Hydro
Has Monopoly's

Ontario Hydro at the Millennium

Has Monopoly's Moment Passed?

EDITED BY
RONALD J. DANIELS

Published for
University of Toronto Faculty of Law
by
McGill-Queen's University Press
Montreal & Kingston • London • Buffalo

© University of Toronto Faculty of Law 1996
ISBN 0-7735-1426-0 (cloth)
ISBN 0-7735-1430-9 (paper)

Legal deposit second quarter 1996
Bibliothèque nationale du Québec

Printed in Canada on acid-free paper

Canadian Cataloguing in Publication Program

Main entry under title:
Ontario Hydro at the millennium: has monopoly's moment passed?
Collection of papers presented at University of Toronto Electric Power Project Conference held June 5-6, 1995 at the University of Toronto.
Includes bibliographical references.
ISBN 0-7735-1426-0 (bound)
ISBN 0-7735-1430-9 (pbk.)
1. Ontario Hydro – Congresses. 2. Ontario Hydro – Reorganization – Congresses. 3. Electric utilities – Ontario – Congresses. I. Daniels, Ron (Ronald Joel), 1959- II. University of Toronto Power Project Conference (1995 : University of Toronto).
HD9685.C34058 1996 333.79'32'060713 C96-900078-2

Contents

Introduction

This monograph is the culmination of an extraordinary collaborative project between the Faculty of Law, University of Toronto and Ontario Hydro – the University of Toronto Electric Power Project. The project was established in 1994, and was designed to involve several of Canada's most distinguished academic commentators from a wide range of different disciplines in an intense year-long investigation of the possibilities for, and challenges of, significant restructuring of the electric power industry in Ontario.

The rationale for the project was rooted in the dizzying pace of change confronting the electric power industry throughout the world. By and large, the most important catalyst for change in the industry is the growth of new forms of technology that undermine the received rationale for the industry's vertically integrated structure. In the generation segment, reliance on combined-cycle gas turbine generation allows electricity producers to achieve low average production cost at low levels of scale, thereby undermining the case for the achievement of scale economies through large, monopoly suppliers. In the distribution segment, the growth of smart metering based on the new information technologies allows for the development of demand side management systems that can be more creatively supplied by specialized providers.

In Ontario, the industry's woes are exacerbated by several other problems, including: the creation of significant cost overruns experienced by Ontario Hydro in bringing new capacity on stream in the last decade (particularly the nuclear assets); significant price increases imposed by Ontario Hydro in the last several years (above increases in the consumer price index); significant excess capacity (again related to the acquisition of the nuclear assets); poor financial performance of Ontario Hydro (high leverage, significant writedowns, reliance on a government financing guaranty); and a dysfunctional regulatory structure that limits effective public oversight and paves the way for backdoor micromanagement by the provincial government, particularly insofar as provincial industrial policy goals are concerned.

In tandem, these factors have resulted in calls by large and small consumers alike for significant industry restructuring aimed at increasing the industry's overall competitiveness. Indeed, this concern has been made all the more urgent by the willingness of large industrial consumers of electricity to exit the system, either through self-generation or through physical relocation to jurisdictions in which electricity is supplied at lower cost. The latter scenario is, of course, particularly vexing given the loss of high-paying manufacturing jobs that follow in train.

The studies prepared for this volume explore a number of different aspects of industry restructuring from a broad public policy perspective. That is, in devising a research program for the project, we deliberately sought to examine the impact of different restructuring scenarios on the public weal. This perspective reflects the long-standing and distinctive role played by Ontario Hydro in the province's economic development, in the public health and safety concerns surrounding various aspects of electricity generation and transmission, and in the significant industrial relations challenges entailed by any non-trivial changes to the industry.

As several of the studies included in the volume indicate, rational industry restructuring motivated by efficiency goals will invariably involve some degree of vertical de-integration of the industry (i.e., separation of the generation and transmission segments of the industry) and, following from that, some degree of privatization of Hydro's assets. In the main, the case for de-integration and privatization derives from the incompatibility of the existing industry and regulatory structure with open, robust market competition. Given Ontario Hydro's dominance of the industry's generation segment (the company possesses 90% of the province's generating capacity), monopoly control over the industry's transmission segment, and regulatory control over the distribution segment (Ontario Hydro, not the Ontario Energy Board, sets rates for the 300 or so downstream municipal utilities), it is simply inconceivable that private capital would make significant sunk investments in generation capacity given the strong innate incentives that exist within Hydro (as currently constituted) to favour the electricity supplied by its own upstream suppliers. Currently, such favouritism can be achieved easily through largely non-transparent decisions regarding the operation of, or investment in, the grid. Thus, de-coupling of the generation and transmission segments appears a necessary antecedent to enhanced industry competition.

In this vein, it should be noted that privatization without de-integration enjoyed scant support from project participants. Although such a reform would maximize the size of government's take from the sale of industry assets, it would do little to improve the efficient operation of the industry. Indeed, it would most likely have the opposite effect. Were privatization to proceed without restructuring, the government would most likely be called on to provide credible long-term assurances respecting the value of private investment by, for instance, restricting the scope for competitive entry (which would have pernicious efficiency effects and also require the introduction of obtrusive regulation).

Yet, despite general agreement on the need for some industry de-integration and privatization, agreement on the precise nature of such measures proved far more elusive. As the work presented in this volume indicates, opinion was divided on several crucial design issues that will impact the restructuring enterprise. For instance, what should the logical endpoint of industry restructuring be: wholesale or retail competition? If Hydro's generation and transmission assets are unbundled, what portion of Hydro's generating assets need to be sold off to private interests to ensure competitive provision in that segment? Can and should Hydro's nuclear assets be privatized? If so, what kind of modifications are necessary to the federal regulatory regime to ensure appropriate operation and maintenance of nuclear assets by private operators? How large are the costs from stranded investment in the nuclear assets, and who should bear them: Ontario taxpayers or ratepayers? Is there a meaningful difference between the two groups?

More generally, when privatized, how should the generation assets be organized (i.e., number of generation companies, composition of assets)? What is the magnitude of the efficiency gains from rationalization in the retail segment of the industry? How should these gains be realized (through market forces, through regulatory or political fiat)? More broadly, what is the role for the downstream distributing companies in industry restructuring? Should bargains involving certain consumer classes (i.e., embedded transfers from urban to rural consumers) continue to be protected? What kinds of general modifications should be made to the regulatory regime to promote responsive, efficient regulation? More particularly, how should various environmental externalities related to electricity generation be regulated (i.e., through plant or industry specific regulation)? What is the magnitude of the job losses likely to be suffered by Hydro employees from privatization, and how should such losses be redressed?

While agreement proved elusive on several of these issues, the studies and accompanying commentary presented in this volume constitute a first step in the identification and analysis of the various design issues involved in the restructuring enterprise. For that, I am extremely grateful to Ontario Hydro, particularly to Ms. Eleanor Clitheroe, Executive Vice-President, Chief Financial Officer and Managing Director, Corporate Business Group of Ontario Hydro, for providing us with the financial and moral support to undertake this project. Without her inspiration and commitment this project would have never been undertaken. I also want to express my appreciation to Ms. Pia Bruni of the Faculty of Law, University of Toronto, who cheerfully and efficiently coordinated all aspects of the Project's activities. A debt is also owed to Ms. Wendy Cuthbertson, Special Advisor at Ontario Hydro, for her administrative support of the Project.

Preparation of this volume commenced shortly after I was appointed Dean of the Faculty of Law, University of Toronto. I was fortunate in being able to rely on the very competent editorial staff of McGill-Queen's University Press to fill the editorial void created by my new responsibilities. In particular, I want to thank the able and patient leadership of Mr. Philip Cercone of McGill-Queen's. Of course, for all the people mentioned above, the standard disclaimer applies. Responsibility for all errors and omissions is that of the authors and commentators alone. As this was an independent research project, Ontario Hydro and the University of Toronto are in no way responsible for the positions advanced herein.

Ronald J. Daniels
Dean
Faculty of Law
University of Toronto

Contributors

THOMAS ADAMS, Energy Probe

STEPHEN R. ALLEN, Energy Research Group

LAURENCE BOOTH, Faculty of Management, University of Toronto

JAKE BROOKS, Independent Power Producers' Society of Ontario

DAVID CAMERON, Department of Political Science, University of Toronto

RONALD J. DANIELS, Faculty of Law, University of Toronto

DONALD N. DEWEES, Department of Economics, University of Toronto

KEITH DINNIE, Ontario Hydro

KENT L. EDWARDS, Windsor Utilities Commission

WILLIAM A. FARLINGER, Ontario Hydro

NEIL FREEMAN, public policy consultant and adjunct professor of Political Science, University of Toronto

MICHAEL GILLESPIE, Ontario Hydro

DAVID S. GOLDSMITH, Association of Major Power Consumers in Ontario

MYRON J. GORDON, Faculty of Management, University of Toronto

DAVID GOULDING, Ontario Hydro

ERIK F. HAITES, Margaree Consultants Inc.

PAUL HALPERN, Faculty of Management, University of Toronto

GERALD HECKMAN, LL.B., Clerk to the Federal Court of Canada, Trial Division, B.Sc. (Eng. Sci.)

ROBERT HOWSE, Faculty of Law, University of Toronto

H.N. JANISCH, Faculty of Law, University of Toronto

I. H. (TONY) JENNINGS, Municipal Electric Association

ARTHUR KROEGER, Carleton University

I. BRUCE MACODRUM, Toronto Hydro

FRANK MATHEWSON, Department of Economics, University of Toronto

YVES MÉNARD, Hydro-Québec

ANDREW MULLER, Department of Economics, McMaster University

JOHN D. MURPHY, Power Workers' Union, CUPE Local 1000-CLC

DUANE PENDERGAST, Atomic Energy of Canada

DAVID POCH, environmental lawyer

ERIC PRESTON, Ontario Hydro

A. STEPHEN PROBYN, Probyn & Company; Independent Power Producers' Society of Ontario

ANDREW J. ROMAN, Miller Thomson

STEPHEN F. SHERWIN, Foster Associates Inc.

MICHAEL J. TREBILCOCK, Faculty of Law, University of Toronto

KARL WAHL, Mississauga Hydro-Electric Commission (Roland Herman presented)

PETER WARRIAN, Centre for International Studies, University of Toronto

ADONIS YATCHEW, Department of Economics, University of Toronto

The Future of Ontario Hydro: A Review of Structural and Regulatory Options

RONALD J. DANIELS
MICHAEL J. TREBILCOCK

I. INTRODUCTION

As Ontario Hydro approaches both the millennium and its own centenary, what does the future hold for one of the dominant institutions in the economic and political life of Ontario over the past century? The genesis of Ontario Hydro lies in the creation by the government of Ontario in 1906 of a permanent Ontario Hydro Electric Commission, initially to construct and operate a provincial transmission grid which would deliver power from privately owned hydro electric generators on the Niagara River to various municipally owned distribution systems in Southwestern Ontario. In a history that has now been well documented, under the dominating leadership of Sir Adam Beck over its first 20 years, Ontario Hydro broadened its vision to embrace a province-wide transmission grid and the progressive acquisition of most privately owned generating facilities in the province, as well as the construction of massive new generating facilities of its own. By the 1930s the essential structure of Ontario Hydro that has persisted to the present day was firmly established.[1] The public power movement, which Adam Beck spearheaded and which was vigorously supported by a large coalition of municipalities, decried the real and alleged abuses of private monopolies in the generation and local distribution segments of the industry; the unacceptibility of foreign ownership of major generating facilities and the exportation of power by these facilities to U.S. consumers; emphasized that electric generating resources (especially hydro electricity) were community resources ("people's power") that the public should not have to pay a rate of return on to private owners; and that rapid public development of an electricity industry in Ontario was a key ingredient in a long-term development or industrial strategy for Ontario, where low-cost and widely accessible electricity would become a critical element in Ontario's comparative advantage in attracting and promoting industry. While these issues have taken on a different complexion over the years, and are likely to be cast in different relief in the future, they are also likely to remain at the centre of debates over the future objectives, structure, regulation, and public accountability of Ontario Hydro.

A brief profile of the electricity industry in Ontario at the present time is a necessary prelude to an evaluation of future policy options.[2] Ontario Hydro is a corporation without share capital, whose Board of Directors is appointed by the provincial government. The issue of the ownership of the company is debated. Some of the municipal electric utilities (MEUs) claim that they are the residual owners of Ontario Hydro, due to the contributions that they have paid over the years in excess of direct power costs, which are reflected in Ontario Hydro's net equity. On the other hand, the provincial government has guaranteed all of Ontario Hydro's debt, which currently amounts to about $35 billion, or about 30% of total provincial public indebtedness. Ontario Hydro generates about 90% of the electric power sold in the province. Its generating facilities comprise 79 generating stations: 68 hydraulic, 5 nuclear, and 6 fossil fuel. Nuclear represents 62% of the total electricity supplied in the province; hydro electric 24%; fossil 10%; non-utility generation 3.7%; and purchases 0.3%.[3]

As Table 1 indicates, fossil and nuclear generation entail considerably larger total costs than hydro electricity.

Ontario Hydro is by far the largest user of nuclear power in Canada and among the largest in North America. Grid interconnections between the province and neighbouring states and provinces could enable three U.S. states – New York, Michigan, and Minnesota – to provide 12 % of total installed capacity, and Quebec and Manitoba over 5%, figures that would be much higher if expressed as a percentage of the electricity that has actually been supplied in the province over the last several years.[4] Recent plans (now suspended) by Ontario Hydro envisaged that local NUG generating capacity would account for about 10% of total system capacity by the year 2014.[5] Currently, the provision of electricity by NUGs to the grid is governed by purchase contracts concluded between Hydro and the independent generators. Typically, these contracts incorporate off-peak curtailment provisions which allow Hydro to curtail the NUG's outputs during off-peak periods up to a specified number of hours per year. Further, some of the purchase contracts contain caps that limit Hydro's obligation to pay for electricity generated above stipulated thresholds.

Table 1[6]
Comparison of Ontario Hydro Generation Costs By Fuel Type (1993)

		Operating Cost		*Capital Cost*	
	Energy Generated (TWh)	*Fuel (¢/kWh)*	*OM&A (¢/kWh)*	*Int. & Depn. (¢/kWh)*	*Total Cost (¢/kWh)*
Hydroelectric	36.80	0.33	0.28	0.49	1.10
Fossil	18.10	2.52	1.30	3.01	6.83
Nuclear	77.50	0.51	1.02	3.91	5.44
Weighted Average	132.40	10.74	0.85	2.84	4.42

The province-wide transmission grid owned and operated by Ontario Hydro delivers about 70% of the total power generated in the province to 309 MEUs, which own and operate the local distribution systems, with the balance being split evenly between direct delivery by Ontario Hydro to about 950,000 rural retail customers outside the jurisdictions of any of the MEUs and direct sales by Hydro to about 100 large industrial customers. As with Ontario Hydro itself, there are some ambiguities as to the ownership of the MEUs. Some municipal councils take the view that they own the municipal electric utilities in their jurisdictions. On the other hand, other MEUs believe they are owned by the citizens at large who elect the Electric Commissioners of the local utility. Under the provisions of the Power Corporation Act, under which Ontario Hydro is constituted, Hydro is required to provide "power at cost." While there are severe ambiguities as to what this entails in the pricing of power to different classes of customers, it has not precluded substantial cross-subsidies from urban to rural consumers (in excess of $100 million per year), nor apparently has it precluded some recent discounting of bulk power purchases by some large industrial users threatening to exit from the system.

Under the terms of the Power Corporation Act, the provincial government, through the Minister of Energy, can issue Policy Directives after consultation with the Board of Directors of Ontario Hydro, with which the Corporation must comply. In addition, the Act provides for a Memorandum of Understanding between the Corporation and the Minister, which must be renewed at least once every three years and sets out the accountability and reporting requirements governing the Corporation's relationship with the Minister and the government, and matters of government policy that the Corporation will respect in the conduct of its affairs. The external regulation of Ontario Hydro is sui generis, and does not follow conventional modes of regulation in electricity or other utilities. Ontario Hydro sets its own rates for those customers it serves directly, sets wholesale rates for the supply of electricity to the MEUs, and regulates retail rates that may be charged by the MEUs. These rates are subject to review by the Ontario Energy Board (OEB), but the OEB's powers are limited to making recommendations to the Corporation and the Minister of Energy and are not binding on the Corporation, at least without Ministerial directive. Capital expenditure plans have been subject to various ad hoc forms of review by Select Committees of the Ontario Legislature, Task Forces, Royal Commissions and, most recently, by the Environmental Assessment Board. The number and extent of these ad hoc reviews since 1960 are reflected in the following table (taken from Waverman and Yatchew)[7], although from its early history Ontario Hydro has rarely been far from the political spotlight. There were major cost overruns in building radial railway systems and its own large hydro generating plant at Queenston in the period during and after World War I, and corporate expenditures vastly in excess of legislative appropriations precipitated external audits and the first of many Royal Commissions or like inquiries in 1920. As early as 1923, debts incurred on behalf of Ontario Hydro amounted to one-half of the entire provincial debt.[8]

The question that this paper addresses is whether there are any reasons to suppose that Ontario Hydro's existing structure will be any less adequate in providing low-cost, reliable electricity to Ontario consumers in the next century than it has in the last?

II. AN INDUSTRY UNDER STRESS

The Sources of Stress

A number of factors have conspired to render contestable a number of the assumptions upon which the present structure, regulation, and performance of the electricity industry in Ontario are based. These were thrown into recent political relief when Ontario Hydro's customers faced average rate increases of 8.6% in 1991, 11.8% in 1992, and 7.9% in 1993, at a time of severe recession in the province. In the face of public outcries provoked by these rate increases, Ontario Hydro has committed itself for the rest of the decade to holding rate increases to no more than increases in the CPI. In 1993, the Corporation incurred a net loss after restructuring charges of over $3.6 billion – the largest corporate loss in Canadian history. Restructuring charges related to severance and redundancy costs; write-downs related to assets recorded at values in excess of market value; and plant closure costs. In addition, in

Table 2
List of Major Special Inquiry and Legislative Committee Reports Related to Ontario Hydro

Year	*Name of Committee and Title Report*
1960	Royal Commission; Report on the Purchase of Lands by Hydro-Electric Power Commission of Ontario
1973	Select committee on the Hydro-Electric Power Commission of Ontario Hydro New Head Office Building (Hearings)
1973	Task Force Hydro (Committee on Government Productivity); Hydro in Ontario
1975	Ontario Environmental Hearing Board; Public Hearing on Ontario Hydro Bradley Georgetown 500 kv Transmission Line Right-of-Way between Point 33 near Colbeck and Point 95 near Limehouse
1976	Select Committee on Inquiry into Hydro's Proposed Bulk Power Rates; A New Public Policy Direction for Ontario Hydro
1976-80	Reports of the Royal Commission on Electricity Power Planning, Volumes 1-9
1980	Select Committee on Ontario Hydro Affairs; Final Report on the Safety of Ontario's Nuclear Reactors
1980	Select Committee on Ontario Hydro Affairs; Report on the Management of Nuclear Fuel Waste
1980	Select Committee on Ontario Hydro Affairs; Report on Proposed Uranium Contracts
1980	Select Committee on Ontario Hydro Affairs; Special Report on the Need for Electricity Capacity
1980	Select Committee on Ontario Hydro Affairs; Final Report on the Mining, Milling and Refining of Uranium in Ontario
1985	Select Committee on Energy; Report on Darlington Nuclear Generating Station
1986	Select Committee on Energy; Final Report on Toward a Balanced Electricity System
1987	Select Committee on Environment; Report on Acid Rain in Ontario
1989	Select Committee on Energy; Report on Ontario Hydro Draft Demand and Supply Planning Strategy
1990	Select Committee on Energy; Interim Report on Climate Change

1993 Ontario Hydro reduced its full-time work force by 24% from 29,600 to 22,600. Ontario Hydro's current debt to equity ratio is about 92%. In the absence of provincial guarantees on this debt, a recent report by RBC Dominion Securities[9] concludes that Ontario Hydro would have had difficulty meeting the requirement for a BB rating by Standards and Poors based on its most recent financial results. Ontario Hydro now also possesses substantial excess capacity. Ontario Hydro's reserve margin between installed capacity and maximum peak demand has risen sharply from 28.1% in 1989 to 64.8% in 1993. This margin compares with the Canadian average of 38% and 19% for selected U.S. utilities in 1993.[10] Compounding Hydro's current problems is its relatively low level of equity capitalization (even for a publicly held utility); in 1993, its equity capitalization ratio stood at 7.9% versus a 15.45% average for crown-owned utilities in Canada and 41.67% for selected investor-owned utilities in the U.S.[11]

Most of Ontario Hydro's immediate difficulties can be attributed to the following factors: (1) over-estimation of future demand; (2) over-expansion and related borrowings with respect to its nuclear facilities in the 1970s and 1980s; (3) substantial cost over-runs and disappointing operating performance of a number of these facilities, in part itself a function of a federal-provincial industrial strategy designed to promote the atomic energy sector in Ontario through Atomic Energy of Canada Ltd. and the development of the CANDU reactor, which was seen as having a major domestic and export potential, with nuclear energy being touted as "too cheap to meter"; (4) wide-spread adoption of energy conservation measures on the demand side; (5) the severe recession of the early 1990s, which had the immediate impact of reducing demand for electricity (particularly by industrial and commercial users), and a significant impact on Ontario Hydro's revenues; (6) a changing longer-term industrial structure for the Ontario economy, with less dependence on heavy manufacturing and greater reliance on the service sector and knowledge-based industries, which are less energy intensive; (7) declining prices for substitute sources of energy, particularly natural gas, and to some extent oil, which can be readily used as an alternative to electricity for heating purposes and many industrial applications; and (8) an anachronistic regulatory structure characterized by dispersed and fragmented authority that has at times subverted public transparency and fostered government micromanagement.

However, this set of factors, in themselves, might not justify a fundamental rethinking of the organization and accountability of the electricity industry in Ontario. After all, some of them can be regarded as simply reflecting one-time misjudgments of demand and supply requirements; others can be regarded as reflecting temporary conditions in the broader provincial economy; yet others (i.e., regulatory distortions) could be redressed with targeted legal and institutional changes alone. However, in addition to long-term changes in the structure of demand noted above, dramatic rates of technological innovation on the supply side are undermining traditional assumptions about minimum efficient levels of scale in electricity generation. Combined cycle gas turbines (CCGTs) that entail a fraction of the capital costs of existing generating facilities now hold out immediate prospects of power generation at significantly lower average total cost than that of established generating technology (along with dramatically lower environmental externalities).[12] Not only does CCGT technology facilitate competitive entry by relatively small-scale plants (i.e., generating capacities in the order of 200-600 MW), but, further, the implementation of these plants can be achieved much more quickly than alternative generators, given their short construction times (technology for these plants is widely available) and the generally less obtrusive level of regulatory review given the lower levels of environmental risk. The effect of the widespread availability of CCGT technology is that a competitive generating sector characterized by significant levels of risk assumed by private investors is immediately feasible, rather than a natural monopoly sector that has required either external regulation or public ownership, or both, in the past. In addition, much more localized power generation technology is rapidly developing, such as wind farms, fuel cells, photo-voltaic solar cells etc. Much of this new technology can be installed by end-users to serve their own power needs and in some cases also to supply power to other end users through the local distribution system (so-called distributed generation or retail wheeling).[13] The implications of this new generation technology for Ontario Hydro are reflected in existing cost differentials. In 1993, the average cost of power to Hydro's large industrial customers was about 5.1 cents per kilowatt hour and the average cost to the MEUs was about 6.2 cents per kilowatt hour compared to the estimated costs of power generated by new gas turbine facilities of around 4 to 4.5 cents per kilowatt hour at current natural gas prices.[14] On a comparative basis, Hydro's average revenue per kilowatt hour (6.31 cents) was one of the highest in Canada (rates for Ontario cities were higher than most other cities in Canada, though lower than many cities in the United States). Furthermore, if industrial rates are expressed as a percentage of residential rates, consumers in Toronto and Ottawa pay 82.6% and 86.7%, respectively, compared to a North American average of 74.9%.[15] These costs and hence price differentials raise the prospect of large industrial consumers and larger MEUs embarking upon policies of self-generation or purchase from local non-utility generators and exiting the present integrated electricity system in Ontario, casting an increasing proportion of the burden of recovery of sunk investment costs onto remaining customers, who are likely to be disproportionately urban residential and rural customers. These customers, in turn and to the extent feasible, are likely to substitute away from electricity and towards natural gas, oil, or propane for heating and other uses,[16] thus setting in motion an unravelling of the existing integrated system, or as it is sometimes more dramatically put – a "death spiral." An additional factor driving in the same direction is the amount of excess capacity possessed by many existing electric utilities in North America, which makes it profitable for them to sell power outside their traditional jurisdictions at any price above short-run marginal costs. Many electric utilities in the Northeast of the U.S. presently possess excess capacity and, according to the report of Ontario Hydro's Financial Restructuring Group,[17] about 20% of Ontario Hydro's peak demand could be displaced by U.S.-based suppliers. The report notes that the threat from U.S. suppliers on a purely price-driven basis through sale of bulk power at heavily discounted rates to selected customers could develop rapidly and could potentially be very significant.[18]

Another set of technological innovations affecting the performance of the industry is the development of information systems that measure the amount of electricity used by consumers and the available sources and prices of supply on a continuous basis.[19] On the basis of this information, consumers are able (again with the assistance of smart metering systems) to adopt consumption patterns that render the private benefits of their electricity usage convergent with the social costs that are thereby occasioned. Whereas the new market-based demand side management systems vest consumption and planning decisions in the hands of consumers with local knowledge, the traditional demand side management systems sought to correct cognitive infirmities, perceived to have been suffered by consumers, by the utility specifying and then subsidizing desirable conservation measures, which were, in turn, predicated on the existence of cross-subsidies associated with a highly protected electricity market.[20] The key implication of these innovative demand-based technologies is that because they do not appear to be associated with scale economies nor are they based on centralized planning, the need to place singular reliance on the interventions of large, vertically integrated utilities to achieve environmental goals is attenuated, thereby removing another rationale for the continuing existence of the current industry structure.[21] Moreover, the provision of innovative demand side management systems could constitute an important and competitively robust segment of the industry itself if current regulatory barriers that limit entry in the distribution segment were relaxed.

Comparative Reform Experience

It is important to emphasize at this juncture that the pressures facing Ontario Hydro are not sui generis. In many jurisdictions throughout both the developed and developing world the structure and regulation of the domestic electricity industry is being fundamentally reevaluated – for many of the same reasons. Table 3 from the Report of the Ontario Hydro Financial Restructuring Group, summarizes many of these initiatives.[22]

CANADA

In Canada, the Nova Scotia Power Corporation has recently been privatized and the Newfoundland-Labrador Hydro Commission is in the process of being privatized. The Alberta Department of Energy in a recent report[23] proposes the creation of an independent power pool, which generators would compete to supply, with prices being averaged for existing generation facilities to consumers in the province. However, the proposals envisage that the cost of future generation would not be averaged. Rather, each local distribution utility would be responsible for obtaining the new generating capacity needed to meet the needs of its customers. The report proposes further study of the option of allowing customers of distribution utilities the right to make their own purchase arrangements for new power supply. The proposals envisage that existing generating units would continue to be regulated by the Public Utilities Board under some form of incentive regulation (e.g., lagged cost-of-service/rate-of-return regulation, or price caps/RPI-X). Transmission charges would also be regulated under some form of incentive regulation. However, local distribution companies would pay the same rate for transmission out of the pool, meaning that all distributors throughout Alberta would pay the same price for transmission regardless of how far they are from sources of generation. To supply power into the pool, generators and importers would pay charges or receive credit based on the location of supply, and therefore the extent of their use of the transmission system. This would encourage suppliers to locate facilities so as to maximize the efficiency of the system. Transmission access and rates, as well as the planning of new transmission facilities, would be monitored by a new Electric Transmission Council comprising representatives of consumer groups, municipalities, distribution companies, generators, and exporters. The Council would be responsible for ensuring that generators, importers, distributors, and exporters have open access to the power pool on a non-discriminatory basis. With respect to distribution charges, distribution costs for investor owned

Table 3
Financial Restructuring, Experience in Other Jurisdictions

Component	*United States*	*Canada*	*England and Wales*	*Italy*
Objectives/Drivers of Restructuring	To foster a competitive market	- In Nova Scotia: recapitalization to reduce indebtedness - To facilitate 1993 and 1994 planned rate increase - Province received a part of proceeds from the offering	- Broaden share ownership - Increase efficiency	- Reduce budget deficit
Nature of Restructuring 2A. Structure Before - Generation - Transmission - Distribution - Energy Services - Regulatory & Market Framework	- Fully vertically integrated utilities services discreet areas through monopolistic control - Independent producers can sell to distributors but not directly to customers - No competition at transmission or distribution level - Return on capital (debt + equity) regulated by state utility commissions	Nova Scotia Power Inc., wholly owned by the province - Regulated by PUB (Nova Scotia Board of Commissioners of Public Utilities)	State-owned CEGB broken into several independent units in preparation for privatization	- Generation, transmission, and distribution by state-owned monopoly, ENEL - Municipal utilities seeking to break ENEL's monopoly in cross-border power sales - ENEL produces 83% of total electric output, while the remainder is produced by auto makers and municipal utilities
2B. Structure After - Generation - Transmission - Distribution - Energy Services	- Fully integrated - Trend towards open access transmission, retail wheeling, and deregulated and competitive generation	- Nova Scotia Power Inc., inclusive of generation, transmission, and distribution assets was sold to private sector	- 3 separate generating companies (National Power, PowerGen, Nuclear Electric) - 12 separate distribution companies (RECs), which	N/A

- Regulatory & Market Framework			jointly own the National Grid Company - Competitive pool pricing	
2c. Associated Financial Restructuring - Asset sales/values - Allocation of debt - Share oferings - Other (Taxation, etc.)	N/A	- Common share offerings	- Share offerings of National Power, PowerGen, and 12 RECS	- ENEL, along with ENI, INA, and IRI, were transformed into joint stock companies as a step toward privatization
3. Results of Restructuring - Rates - Costs - Service - Efficiency gains - Customer Acceptance - Customer Choices	N/A	- Improved capital structure with lower leverage - Improved EBIT margin	- Employment reduced from 47,000 to 26,000 and a 20% employment reduction in the RECS - Rate decreases of up to 10.15%	- Improved financial structure, profitablility, and quality of service
4. Restructuring Status	- Some movement towards further privatization or sale of municipal systems - 80% of electric industry already privately owned	- Privatization completed for Nova Scotia Power - Privatizations being considered in Newfoundland, Northwest Territories, and City of Calgary.	- Privatization completed	- Privatization in process

Table 3 continued
Financial Restructuring, Experience in Other Jurisdictions

Component	*Norway*	*Argentina*	*Brazil*	*Chile*
Objectives/Drivers of Restructuring	- Create efficiencies and cost reductions	- Liberalize economy and reduce country's debt	- Reduce government deficit created by state-owned enterprises	- Finance public sector deficit and broaden share ownership
Nature of Restructuring 2A. Structure Before - Generation - Transmission - Distribution - Energy Services - Regulatory & Market Framework	- One-third of generation by independent state-owned power company (Statkraft) and two-thirds by regional producers – both municipalities and private producers - Statkraft owns 85% of main transmission grid and has monopoly control of import and export of electricity - 250 local distribution utilities - Clearinghouse auctions surplus power at spot prices	- Attract foreign investment - Integrated state-owned entities, Agua y Energia Electrica, Servicios Electricos de Gran Buenos Aires, and the Compania Italo-Argentino de Electricidad - Edusur – power distribution	- Electrobras, national holding company, produces power and supplies it to regional distributors - LIGHT (controlled by Electrobras), Electropaulo, and Celesc are the distribution companies - Unified tariff system (i.e., one price for all) - Federal government forced utilities with ROA in excess of 12% to transfer profits to utilities with ROA less than 10% (CRC policy)	- 2 major interconnected grids, SIC and SING - State-owned utilities
2B. Structure After - Generation - Transmission - Distribution	- Reforms implemented in 1991 intended to reduce number of distribution utilities	- Significant asset pools now owned by private international power entities	- Ended the CRC policy - Excess returns are to be used to pay down debt financed by government	- National Energy Commission proposes distribution tariffs

- Energy Services - Regulatory & Market Framework	- Create concession system to regulate trading power		- Electrobras controls 72% of generating capacity and is responsible for technical and strategic orientation	- ENERSIS – a public company – acquires state-owned utilities to be privatized, i.e., Chilectra and Rio Maipo, ENDESA
2c. Associated Financial Restructuring - Asset sales/values - Allocation of debt - Share oferings - Other (Taxation, etc.)	N/A	- Asset sales of SEBGA and Agua y Energia in Buenos Aires - Asset sales of Edusur to a consortium of ENERSIS, ENDESA, and Chilectra	-Regulatory reform -LIGHT to be privatized in 1994	- 4 employee-related companies formed ENERSIS - ENERSIS acquired other state-owned utilities - ENERSIS was sold to private market in an equity offering and continued acquiring state-owned utilities
3. Results of Restructuring - Rates - Costs - Service - Efficiency gains - Customer Acceptance - Customer Choices	- Rate decreases of up to 20%	- Better maintained power generation system	-Improved efficiency -Reduced company losses due to technical faults or energy thefts	- ENERSIS as a holding company acquired and turned around state-owned companies in Chile and Argentina
4. Restructuring Status	- Not privatized, only restructured	- Privatization partially completed	-Privatization still in process	- In process

Table 3 continued
Financial Restructuring, Experience in Other Jurisdictions

Component	*Australia*	*Sweden*	*Spain*
Objectives/Drivers of Restructuring	- Implement market reform in utility industry	- Competition through competing generators and open transmission	- Broaden shareholder base of ENDESA
Nature of Restructuring 2A. Structure Before - Generation - Transmission - Distribution - Energy Services - Regulatory & Market Framework	- State-owned utilities – Integrated, i.e.: Electricity Commission of NSW, State Electricity Commission of Victoria	- 2 major generating companies (1 state-owned) which also transmit and distribute electricity - 50% of total electricity produced by state-owned Vattanfall - The private sector and municipalities produce 25% and 19% of total electricity, respectively - Deregulated industry except for the main grid which is still controlled by government	- INI state-owned holding company -owned 96% of ENDESA - Integrated utilities
2B. Structure After - Generation - Transmission - Distribution - Energy Services	- Formed National Grid among different States and opened to private sector	- Bill pending on proposed deregulation; if passed, it will take effect on January 1, 1995 with provisions including: - Government ownership of grid - Non-discriminatory transmission	-INI's holding decreased to 75.6% through initial share offering of ENDESA and to 66% in 1994 - ENDESA –generating system company sold to the private sector through share offerings

- Regulatory & Market Framework		- Utilities required to supply customers unwilling to switch	- ENDESA Group – the holding company owns ENDESA and several distribution subsidiaries - Red Electrica – the national transmission system
2c. Associated Financial Restructuring - Asset sales/values - Allocation of debt - Share oferings - Other (Taxation, etc.)	- Corporatization/Commercialization - Restructured into SBUs (Strategic Business Units)-Implemented market based transfer pricing - Invited private sector into grid developments	N/A	- Privatization through international public offering
3. Results of Restructuring - Rates - Costs - Service - Efficiency gains - Customer Acceptance - Customer Choices	- Profitable Government-owned utilities - Leaner organization structure - Improved efficiencies - Competitive cost of generation through balancing among different states' generators	- Enhanced competive environment	- New regulatory system was implemented to allow for recovery of depreciation and a guaranteed return on assets based on pre-determined "standard value" of assets - Post-restructuring, ENDESA Group acquires interests of other Spanish utilities, i.e., Sevillana, FECSA, and ERZ and currently produces 35% of total Spanish electricity
4. Restructuring Status	- Not privatized, only corporatized	- Status quo, state-owned utility not privatized	- Privatization was completed, INI's holding in ENDESA is reduced further by a recent secondary share offering

utilities would continue to be regulated by the Public Utilities Board, but not those for municipally owned distribution companies.

The new structure of the Alberta electricity industry under the foregoing proposals is depicted as follows:

Figure 1
Open Competition for Generation[24]

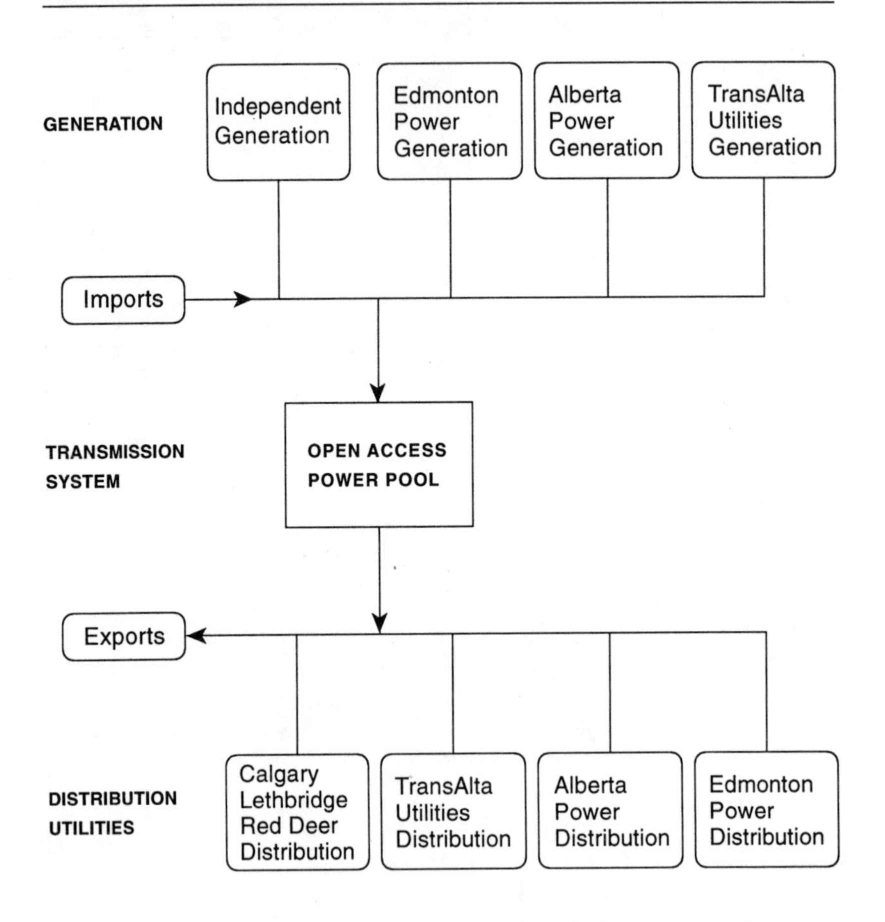

Note: Medicine Hat could sell energy to the power pool as a generator or purchase as a distribution utility.

THE UNITED KINGDOM

In the U.K., as of March 1990, the state-owned fully vertically integrated Central Electricity Generating Board was privatized through both horizontal and vertical de-integration. With respect to generation, the nuclear power plants (accounting for about 20% of generating capacity in England and Wales) were retained as a state-owned enterprise, while other generating facilities were privatized through two new corporations, National Power and PowerGen, and entry permitted to new non-utility generators and generators in Scotland and France.[25] These generators now compete to supply a power pool or exchange that operates, in effect, a half-hourly spot market for bulk power. The grid, which operates the exchange, in turn became vertically integrated with the local distribution companies. Twelve regional electricity companies (RECs) – successors to the previous 12 area boards – now jointly own the National Grid Company. The RECs in turn were privatized. Currently the RECs hold monopoly franchises for the supply of smaller customers. However, competition is possible for larger customers of whom there are currently about 4,000 and who account for about 40% of the total market volume. By early 1993, more than 40% of these users had chosen a supplier other than their REC. In April 1994, the franchise limit was reduced, creating about 40,000 more non-captive customers (around 50% of the market) and four years after this date all customers will be free to choose their own supplier. The bilateral contracts concluded in the system between upstream suppliers and downstream consumers (or their proxies) are typically contracts for differences whereby either buyer or seller compensates the other for differences between the contract price and the spot price generated by the pool at the time of delivery. The transmission charges and local distribution charges are subject to regulation by the Director-General of Electricity Supply and the Office of Electricity Regulation (OFFER), subject to appeal to the Monopolies and Mergers Commission. Initially, generation prices were not regulated at all, but public concerns over excessive price increases, widely attributed to insufficient competition amongst generators, have recently led the Director-General of Electricity Supply to secure undertakings from National Power and PowerGen to divest some of their generating facilities and as an interim measure to abide by an average pool price over the next two years that does not exceed a price cap that is about 7% lower than average price levels in 1993-94.[26]

The structure of the electricity industry in England and Wales before and after the 1990 reforms is depicted in the following diagram:

Figure 2
Electricity Industry Structure[27]

Old structure of the industry in England and Wales

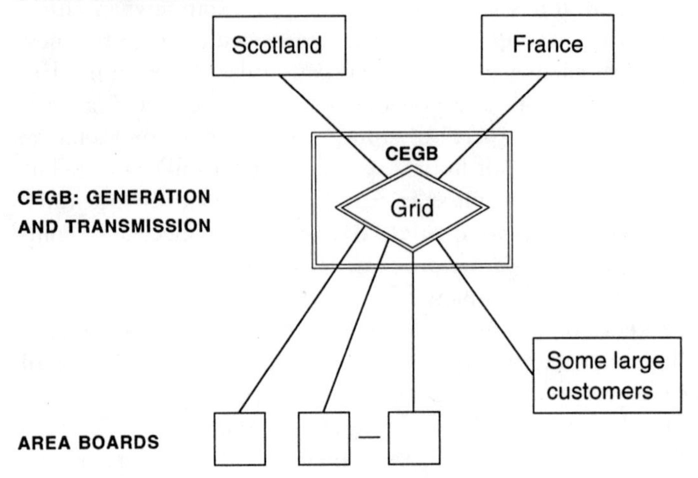

New structure of the industry in England and Wales

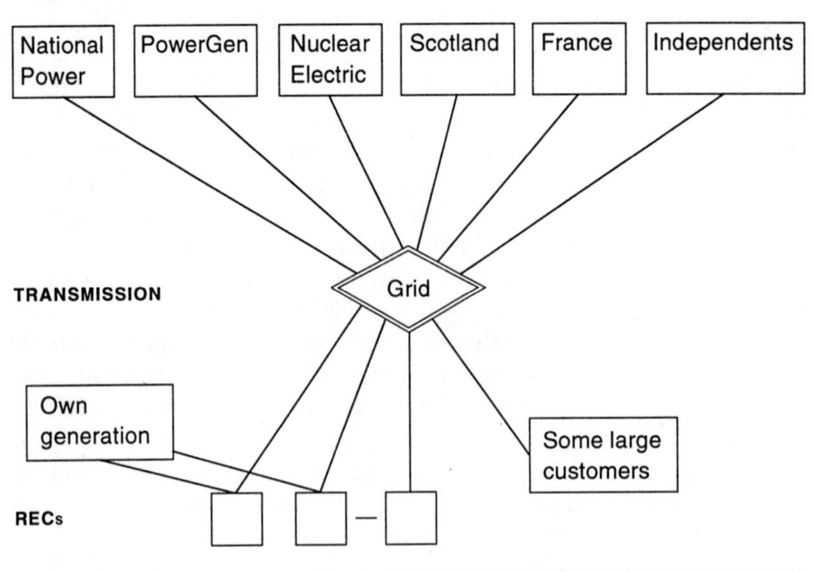

The experience with the British reforms has been mixed. On the one hand, electricity restructuring has clearly generated improvements in the short-run operating efficiency of generators and distribution companies as measured by significant reductions in staffing combined with stable or increasing output.[28] Even state-owned Nuclear Electric has achieved very substantial productivity improvements due in large measure to reduced employment.[29] As well, most consumers have experienced reduced prices for electricity when measured against inflation, and for those who have not, at least part of the explanation resides in the removal of subsidies. On the other hand, the reforms have been plagued by several problems that are mainly the result of the government's decision to create a duopolistic market in generation and to maintain distortions in input pricing (artificially high prices for coal). The duopolistic market structure has enabled generators to secure prices for electricity that exceed marginal costs, which, in turn, have required repeated interventions by the regulator to curtail market power. This market power has extended to strategic manipulation through electricity bidding of both the uplift charge (designed to address generators who are either constrained on or off the system because of transmission congestion problems) and the capacity or value of lost load charge (designed to bring short- and long-run marginal costs into equilibrium).[30] The exercise of a duopolistic market may have also induced inefficient entry by firms wishing to exploit the rents accruing to generators.

THE UNITED STATES

In the U.S., the Federal Public Utility Regulatory Policies Act of 1978 (PURPA) precipitated major changes in wholesale power markets and in public policy toward competition and vertical integration in the industry. Before PURPA there was effectively no non-integrated independent generating sector in the U.S. The primary purpose of this change in policy was to encourage improvements in energy efficiency through expanded use of cogeneration and by creating a market for electricity produced from renewable fuels and fuel waste. Essentially, under PURPA vertically integrated utilities were required to service incremental generation needs through independent suppliers at the utility's avoided cost. Although it was not explicitly intended to restructure the whole industry, an important legacy of PURPA has been its effect of dissipating the previously ubiquitous vertical integration between generation, transmission, and distribution. The supply of independent generating capacity proved to be larger than anticipated, the need for new generating capacity significantly less, and unexpectedly low natural gas prices and the associated heavy reliance on CCGTs took the competitive balance away from new large coal and nuclear units and towards independent generators. The Federal Energy Policy

Act of 1992 made further changes, including an expansion of the Federal Energy Regulatory Commission's (FERC) power to order utilities to provide wheeling services to support wholesale power transactions between large industrial users and local distribution companies on the one hand, and generators on the other. The Act did not itself promote retail wheeling, which would entail customers buying power directly from generators, and precludes FERC from ordering it.[31] However, the Blue Book proposal by the California Public Utilities Commission would go this further step and provide, as under the new British regime, for phased-in retail contracting.

Proposals by the Wisconsin Electric Power Company[32] provide an indication of how full-scale retail competition might work. Under these proposals, the generation pool or Poolco coordinates the movement of power from where it is generated to where it is needed. The three primary functions of Poolco are to create a spot market or pool for selling and buying electricity; to provide access to the transmission system; and to dispatch pool members' power plants on a least-cost (or merit-order) basis. The generation pool operates in the short term to facilitate system interactions and to allow for longer-term bilateral contracts between sellers and buyers of electricity. Companies that generate electricity (Gencos) are competitive entities operating and maintaining existing or new power plants. They bid into the short-term power pool for least-cost dispatch and they form contracts directly with buyers. Customer service companies (Retailcos) sell electricity and perhaps other forms of energy to customers along with value-added services. These companies buy electricity from brokers, the spot market, or directly from Gencos. Energy merchants provide marketing and brokering services for Gencos and Retailcos as well as municipal utilities and large customers. The transmission grid (Gridco) remains a natural monopoly that would require some form of regulation, as would distribution entities (Linecos) that construct, operate, and maintain the local distribution wires. In both cases, it is envisaged that incentive-based regulation such as price caps would be adopted.

The existence of a Poolco in this model assumes that virtually all investment decisions for new generation and transmission facilities occur through commercial bilateral contracts. Each half hour, the Poolco determines the market-clearing price and settles spot transactions with users of the pool in a method similar to the way utility system control centres in the U.S. make decisions today on whether to run their own marginal generating units or to buy energy from a neighbouring utility. A spot market allows contracting parties to buy and sell incremental amounts of electricity in the market, which means bilateral contracts do not necessarily have to match physical operations perfectly. The parties might act independently in the spot market with monetary payments between them based on the difference between the spot market price and the contractually defined price. Such contracts for differences can take many different forms, providing a flexible vehicle for allocating risks. Spot market prices depend on only a small fraction of the money flows between customers and generators. With respect to potentially stranded assets of incumbent generators, the Wisconsin Electric proposals envisage that these would be recovered in a non-avoidable access fee on the distribution system and billed as a separate unbundled charge during a transition period. This charge, which would consist of over-market asset value, would decrease through an accelerated depreciation of assets – book value to market value – through the transition period. At the end of the transition asset values would decrease to market value. Any costs associated with social programmes would become part of a permanent access fee.

The Wisconsin Electric proposals claim that divesting parts of Wisconsin Electric's business may be neither necessary nor desirable: generation can be deregulated and direct access to wholesale and retail customers provided without divestiture.[33] Formation of a Poolco to create a spot market is claimed to eliminate the need for divestiture. In order to avoid or mitigate problems of self-dealing or conflict of interest between the transmission arm of the utility and its generating arms, the proposals envisage the creation of a Regional Transmission Group comprising representatives of entities with an interest in transmission and operation – marketers, generators, Gridcos, state regulatory commissions, and others, which would govern all Gridcos within the Regional Transmission Group and set operational rules and system reliability for all network operations, including the Poolco. The industry structure envisaged by Wisconsin Electric's proposals is depicted in the following diagram:

Figure 3
Conceptual Framework for a Restructured Industry

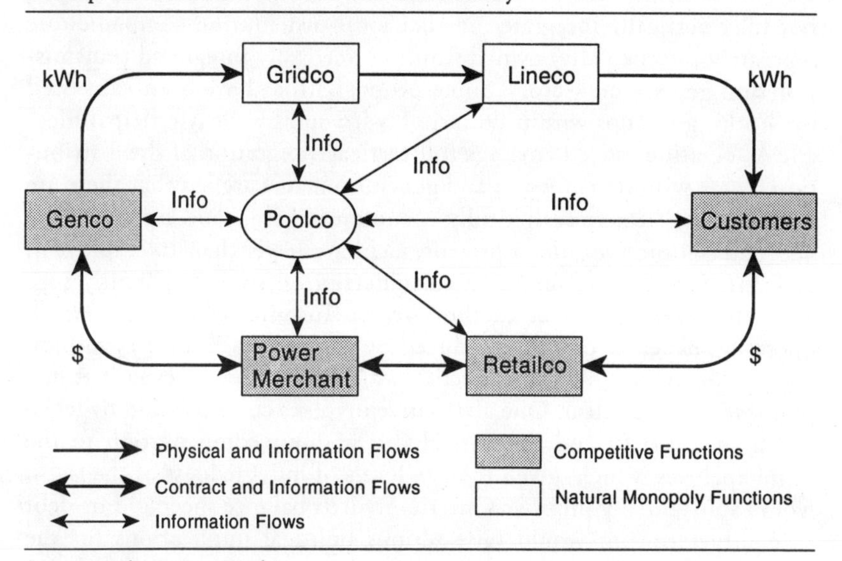

See Appendix in Jere Jacobi.

The Three Basic Reform Models

From these brief descriptions of reforms or reform proposals in other jurisdictions, three basic models (each with a range of structural and regulatory sub-options) emerge. Model 1 (competitive access to a Power Pool) entails electricity generators competing on a merit-order basis for access to a power pool that operates the spot market, provides access to the transmission system, and dispatches pool members' power plants on a least-cost basis to the grid. Model 2 (wholesale competition) involves competition among generators for wholesale customers, entailing bilateral contracting between generators and local distribution companies and large industrial company customers, probably supported by or supplemented by a pool or exchange operating a spot market. Model 3 (retail competition) entails competition among generators for retail customers where industrial, commercial, and residential customers, through various kinds of intermediaries, would contract with generators for the supply of electricity, probably again supported or supplemented by a spot market, and paying regulated transmission and local distribution charges for use of the grid and local wire system on a common carrier basis. One feature that these three models share in common is some measure of de-integration of the vertically integrated natural monopoly model of the electricity industry. Here it needs to be noted that the structure of the electricity in Ontario differs from that traditionally observed in other jurisdictions in at least two respects: first, almost the entire electricity system in Ontario is publicly owned (generation, transmission, and local distribution); second, the industry is not fully vertically integrated in that local distribution companies are separately (municipally) owned from the vertically integrated transmission and generation sectors. Some proposals that have been canvassed by Working Groups within Ontario Hydro and by the Municipal Electric Association indeed envisage full vertical integration of the distribution sector with transmission and generation. It is argued that there are far too many MEUs to realize fully economies of scale and scope. Of the 300-odd MEUs in Ontario at present, 224 have fewer than 10 employees, and 131 have fewer than 1,000 customers. One study apparently suggests that if all the MEUs in the Greater Toronto area were merged, operating expenses could be reduced by roughly 25%, or $95 million per year.[34] As to how this objective would be achieved, even if it had policy merit, is unclear. One alternative involves consolidation by legislation of the MEUs and Ontario Hydro without compensation to the municipalities, which, given the low levels of indebtedness of the MEUs, would substantially improve Ontario Hydro's balance sheet and its debt to equity ratio but would raise serious political implications for the province's relationships with the municipalities. It is hard to imagine that the municipalities would voluntarily surrender their unencumbered local distribution assets in exchange for an equity interest in a wholly integrated utility that contained several generating assets whose value is extremely uncertain, given growing levels of generation competition. This is especially so when the diluted voice of the municipalities in an integrated company is considered (the municipalities would likely only possess a minority of shares in the integrated company). Another option is the purchase of the MEUs in exchange for, or financed by, leveraged capital (bonds) of Ontario Hydro. While no additional common equity would be brought into Ontario Hydro's balance sheet, from the municipalities' perspective one advantage of the receipt of leveraged capital (bonds) rather than an interest in Hydro would be the ability to sell or redeem them at any time, while an exchange of assets for shares would only be attractive if some form of dividend was contemplated or there was some ability to sell the shares into the market or back to Hydro or to the province for cash at some point in time.[35]

Basic Premises in Reviewing the Options

ECONOMIES IN VERTICAL COORDINATION SHOULD NOT BE UNNECESSARILY ENDANGERED

Assuming that full vertical integration is not feasible, even if it were desirable (which we will question below), an over-arching feature of all three models described above moves in the opposite direction, i.e., toward more vertical de-integration, and is strongly under-scored by the experiences of history in this and other utility sectors. As eloquently described by Armstrong and Nelles in *Monopoly's Moment,*[36] and as further analyzed by Baldwin in a Canadian context[37] and Priest in a U.S. context,[38] the evolution toward vertical integration in many utility sectors in the late 19th and early 20th centuries – in the U.S. typically parallelled by the creation of independent state Public Utility Commissions to regulate the vertically integrated monopolies, and in Canada by either public ownership or independent regulation, or both – was the severity of the contracting problems engendered by what were typically bilateral monopoly relationships at different stages of the production and distribution processes in these industries. For example, in both Canada and the U.S. in the last century, exclusive contractual franchises were typically granted by cities or municipalities to local lighting, water, or tramway companies on a long-term basis with prices and terms of service stipulated in the franchise, but these relationships were often characterized by bitter conflicts typically at their most intense at contract renewal time – over contractual terms and performance thereof.

Hold-ups and other forms of reciprocal opportunism were pervasive. This historical experience suggests a salutary caution in evaluating radical proposals for the de-integration of the electricity industry in Ontario and elsewhere. Economies of scale and coordination between what are technologically intimately interrelated segments of the industry, i.e., generation, transmission, and distribution, could be forfeited without careful attention to the institutional and market arrangements that will address these interfaces. Here, we emphasize the importance of devising institutional mechanisms that will ensure effective coordination of different segments of the industry in a restructured world if these currently integrated segments are unbundled from one another. In other words, we should be careful to avoid any form of political or policy "cycling" by ignoring the lessons of history and uncritically adopting reforms that commit us to a "back to the future" strategy.[39]

THE ELECTRICITY INDUSTRY SHOULD NO LONGER BE UTILIZED AS A MAJOR INSTRUMENT OF INDUSTRIAL POLICY

Another general issue that is likely to shape the choice of future policy options for Ontario Hydro is the role of Ontario Hydro as an instrument of state-orchestrated industrial policy. Clearly, industrial policy considerations were central to the creation of Ontario Hydro and its early pattern of expansion. Equally, the heavy investment in nuclear power generation in the 1960s and 1970s was a central aspect of federal/provincial policies to promote a nuclear reactor industry in Canada – the latter a significant source of Ontario Hydro's current financial difficulties. At least from a contemporary vantage point, we are sceptical of the wisdom of assigning Ontario Hydro a major industrial policy role, given the now complete elaboration of a province-wide electricity grid. Moreover, future uncertainties about demand forecasts, the general evolution of the Ontario economy, future technologies, etc. suggest that in adopting policies that will shape the future structure and regulation of the electricity industry in Ontario, a significant premium should be placed on incrementalism, pluralism, and decentralization in terms of the loci of decision making, rather than attempting to make some once-and-for-all, system-wide set of collective decisions as to the future of the industry. This would be to risk similar forms of systemic errors to those made in the past. In this respect, we emphasize the contribution that private capital markets can make to efficient long-run investment planning for this industry. In the absence of public financial assistance (including implicit guarantees), capital markets can draw on and aggregate the dispersed knowledge and expectations of investors in the electrical industry to generate a more rational assessment of the alleged merits of a given project.

MANAGING THE INHERENT TENSION BETWEEN INSTITUTIONAL AUTONOMY AND INSTITUTIONAL ACCOUNTABILITY NEEDS BETTER MECHANISMS

Another general consideration suggested by a reading of the history of Ontario Hydro is that the choice of institutional arrangements for the non-competitive, natural monopoly elements of the industry, whether publicly or privately owned, will pose difficult institutional challenges in fashioning an appropriate balance between institutional autonomy and institutional accountability. The diffused, ad hoc, politically driven and crisis-oriented oversight mechanisms that have been adopted in the past have failed to provide the institutional expertise, continuity, and memory for effective external oversight. In arguing for the virtues of a more consistent and integrated oversight focus, we believe that it is crucial that the oversight functions be as precisely defined as possible and that the mandate of the oversight agency be as narrow, rather than as broad as possible. In particular, it should not be charged with the primary responsibility for fashioning essentially political trade-offs, and the regulatory functions of the oversight agency should not be structured in such a way that it constitutes a surrogate second tier of managerial decision makers constantly second-guessing, or micromanaging, the strategic and managerial decisions of firms in the industry, which would entail the serious risk of undermining any efficiency gains that might be realized by restructuring the Ontario electricity industry along more flexible, dynamic, and decentralized lines. In achieving an appropriate balance between institutional autonomy and institutional accountability, ideally appropriate performance benchmarks (e.g., as to prices and capacity) would be stipulated ex ante, and the external oversight agency would focus its attention primarily on ensuring that these output targets are met, while giving maximum latitude to firms in the industry to organize their inputs in such fashion as they judge to be most efficient. In this vein, we believe it essential that the rules which are to govern this industry be drawn precisely and clearly ex ante so that private investors will be able to make long-term investments on the basis of relatively well understood expectations of how their investments will be treated.

INDUSTRY STRUCTURE AND INDUSTRY REGULATION ARE SOMETIMES SUBSTITUTES AND SOMETIMES COMPLEMENTS

In terms of general considerations that orient our analysis of the reform options, it is important to stress that structural arrangements are often a substitute for regulation, and that to the extent that competitive structures can be promoted in various segments of the industry, to that extent the case for extensive regulation, or in some cases any regulation at all, is obviated. Thus, in reviewing reform options, we begin in every case with a consideration of structural arrangements and then attempt to identify the minimum and most targeted forms of regulation required to ensure that the performance of the industry does not deviate substantially from appropriate performance benchmarks. In some cases, regulation can be conceived as of a complement to competition; in other cases, a substitute for it. Canada's competition (or anti-trust) laws (the federal Competition Act), administered by the Federal Competition Policy Bureau and enforced primarily by the Competition Tribunal, can be thought of as a form of regulation designed to complement competition by discouraging collusive, exclusionary or predatory practices, discouraging anti-competitive mergers, and preventing abuse of dominant market positions. Similarly, in an industry characterized by some elements of natural monopoly and some areas of potential competition, a major role of regulation should be to complement the role of competition by ensuring a level playing field, particularly where the competitive and natural monopoly elements of the industry interface (as they inevitably will). With respect to those elements of the industry where competition is both unlikely and indeed inefficient, i.e., the natural monopoly elements of the industry, regulation serves the function of a substitute for competition by attempting to ensure that pricing and output decisions as closely as possible replicate those that would prevail under competitive conditions.

REFORM PROPOSALS SHOULD AVOID EMBODYING MULTIPLE AND IRRECONCILABLE GOALS IN THE SAME POLICY INSTRUMENTS

Reform of regulated industries, particularly when it involves scope for privatization of state-owned industries, will implicate a range of different public policy goals, including: enhanced efficiency (allocative, productive, and administrative), government revenue raising, political decentralization, and equity.[40] However, owing to the conflicts engendered by attempts at simultaneous realization of these divergent goals, policy makers will, of necessity, have to make choices among these goals, adopting some to guide the reform enterprise and remitting others to other more finely honed instruments. This tension is most apparent when the goals of enhanced efficiency and government revenue raising are considered in the context of privatization. Whereas the first goal contemplates pricing at marginal cost through open and vigorous industry competition that is designed to increase the level of realized consumer surplus, the second goal often encourages governments to create credible and durable assurances of non-competitive conditions for outside investors, in order to raise the financial returns from asset sales. In the context of the Ontario electric power industry, the danger is that pressing fiscal concerns will cause policy makers to contemplate the adoption of reform measures that seek to enlist public support for industry restructuring by maximizing the level of revenue that can be realized from private investors by reducing the long-term competitiveness of the industry. In our view, the best way for the province to promote its long-term capacity to generate a level of surplus sufficient to support the needs of its citizenry is on the foundation of a competitive power market which requires that paramountcy be accorded to efficiency goals. A related point is that the province should eschew privatization of state-owned assets merely to realize immediate cash revenue where no change in the management or operation of those assets is likely to result. Such policies are merely a form of fiscal illusion where the state is exchanging one set of claims (expected future cash flows) for immediate cash, but often at a discount reflecting the private sector's relatively higher discount rate.

THE SCOPE FOR INEFFICIENT BYPASS AND REGULATORY ARBITRAGE SHOULD BE DISCOURAGED

As we discuss below, the issue of compensation (if any) for stranded costs has been a central, often paralyzing, feature of debates in the United States over the restructuring of the electric power industry. The concern is that if certain classes of customers (large industrials, municipalities) are able to exit the existing system through self-generation or other options, then the utility's capacity to recoup its sunk costs for approved capital investments will be compromised, perhaps completely eliminated. From an economic perspective, it is clear that if existing electric power generation assets can produce sufficient levels of electricity to supply industry demand at short-run marginal costs that are less than the average total costs of new entrants (because sunk costs are sunk, they should be ignored), then it is socially perverse for these new investments to be made.[41] In this setting, a decision by a consumer to exit the system to avoid the embedded costs of existing generating assets in favour of electricity supplied by higher cost entrants (relative to the short-run marginal costs of existing generation) should be deterred either by allowing the incumbent supplier to price down from prices that reflect embedded costs to industry short-run marginal costs or by constraining the consumer's exit mobility. In reality, the latter option is undermined by inter-jurisdictional mobility of large electricity users; if the price of electricity is not competitive for this class of consumers, they will simply relocate their plants to other, more competitive jurisdictions. In any event, industry restructuring should limit, to the extent possible, the stranding of existing (though in hindsight undesirable) investments in socially useful plant and equipment.[42]

Specific Evaluative Criteria

In light of these general considerations, the more specific evaluative criteria which we employ in evaluating the various reform options are the conventional ones employed in industrial organization analysis,[43] i.e., (a) allocative efficiency, which requires that prices be set at marginal or incremental costs, or to the extent that such prices will fail to yield a requisite return on efficiently invested capital, at efficient markups above marginal costs to the extent necessary to recover those capital costs; (b) productive efficiency, which occurs where no more resources than necessary are used to produce a good or service; (c) dynamic efficiency, which requires that firms face incentives to reduce costs over time, and to engage in appropriate levels of innovation and system enhancement; (d) transaction cost and administrative efficiency, which requires that both in terms of private transaction costs and public administrative costs, arrangements be adopted that minimize these costs, consistently with attainment of other policy objectives; and (e) legitimate distributional considerations, which take into account wealth or welfare effects on different constituencies of different structural and accountability regimes. In the light of the general considerations bearing on policy reforms noted above, and within the framework of these five more specific evaluative criteria, we now proceed to a consideration of the three basic models, identified above, of a less vertically integrated and more competitive electricity industry in Ontario.

III. AN EVALUATION OF THE THREE BASIC REFORM MODELS

Model 1: Competition in Generation for Access to the Grid

A minimum institutional requirement to operationalize this model is the creation of a power pool or exchange into which competing generators can bid their power, with bids being chosen on a least-cost, merit-order basis, with downstream distribution companies and customers simply buying their power from the grid. In the absence of any structural changes, the immediate problems raised by this model pertain to serious potential conflicts of interest between Ontario Hydro as grid owner and Ontario Hydro as owner of 90% of the generating capacity in the province. As grid owner, it will face strong incentives to engage in self-dealing in favouring its own generators.[44] A number of solutions short of de-integration of existing vertically integrated utilities have been proposed to deal with the self-dealing temptation. For instance, by segregating the generation and transmission activities of the existing utility into two separate functional divisions or perhaps even subsidiaries with different boards of directors, it is alleged that the potential for self-dealing could be mitigated. However, as in the case of other industries where these governance devices have been utilized (e.g., the separation of investment banking from trading activities in investment houses), the results are equivocal. Not only is it extremely difficult for governance mechanisms alone in the face of common ownership to constrain potentially lucrative self-dealing activity, but even if it were, the functional separation of these entities would subvert the scope for cross-functional complementarities, which is presumably the rationale for common ownership of generation and transmission assets in the first place. In this vein, Pierce has argued that it is highly doubtful that any external agency can devise regulatory counter-incentives powerful enough to overcome this natural incentive, and that any method of creating a competitive market for electricity that leaves intact the present vertically integrated industry structure will be plagued by disputes raised by allegations of self-dealing.[45]

The basic problem is that the complexities of transmission pricing and access, particularly system requirements for out-of-merit-order access that are responsive to congestion externalities, vest considerable discretion in the transmission grid that is difficult to observe and to evaluate from afar.[46]

A radical privatization option in the Ontario setting that would be responsive to the self-dealing temptation would entail Hydro selling off all its generating facilities in perhaps four or five bundles of assets, with appropriate mixes of base load and peak load power plants, which, along with independent power producers and inter-provincial and U.S. utilities, would compete to supply the grid.[47] However, in Ontario, there are serious questions as to whether this option is feasible in light of the large proportion of Ontario Hydro's generating capacity accounted for by nuclear generation. As the British experience reveals,[48] privatizing nuclear generation facilities either on their own or as part of a bundle with other generating facilities is likely to prove extremely difficult, at least without detailed commitments by the state that nuclear power will have preferential access to the grid in order to cover its high average total costs, without further commitments to bear or share the large but extremely speculative decommissioning and fuel reprocessing costs likely to be entailed when the useful life of these plants is exhausted, and without tight restrictions on liabilities to third parties for tortious acts resulting from nuclear mishaps.[49] Moreover, there may well be substantial public concerns over privatizing nuclear generating assets focused on whether private owners will have the same incentives as state actors in adhering to appropriate safety standards in the operation and maintenance of these facilities – especially in a setting of government imposed limitations on liability for tortious conduct. For instance, under private ownership, there may be stronger incentives for managers to chisel on needed investments in asset maintenance, particularly as the remaining useful life of the assets declines. However, to the extent that the propensity to shirk on investments in asset maintenance can (and will need to be) curtailed through detailed and obtrusive levels of regulatory oversight of plant operations, the efficiency gains associated with private ownership will be diminished through such micromanagement. Moreover, as demonstrated by the recent performance of the British Government's state-owned nuclear generating company, Nuclear Electric, it is entirely conceivable that marked improvements in plant operations could be realized without transferring ownership to the private sector.[50] Finally, in a setting of growing public apprehension with reliance on nuclear assets for electricity generation (given concerns over plant safety and environmental risk), it is arguable that privatization will increase political barriers to regulatory policy modifications that will debase the economic value of these assets (for instance, a requirement of earlier than anticipated retirement of assets) because of the legitimate expectations of private investors to realizing a return on these assets commensurate with the level of up-front investment. In other words, were the state to retain ownership, it could change the policy environment respecting nuclear assets to respond to public sentiment but without appearing to act opportunistically against private owners.

If we assume that the nuclear facilities cannot be privatized but only the hydro and coal-fired generators, the British experience again suggests potentially serious problems in engendering a sufficient degree of competition amongst generators in bidding to supply the grid to ensure that prices do not diverge sharply from marginal cost. In order to maximize the prospects of effective competition, it would seem necessary still to be able to privatize the non-nuclear generating facility into at least four firms if this would not forfeit significant economies of scale or scope in their operation. Here, there may be some political objection to sale of Hydro's hydraulic generators on the basis that the net effect of these sales to the private sector would be to increase the cost of electricity to Ontario consumers, who will be forced to pay twice for the capital costs of the plant and equipment of these generators (variable costs are relatively trivial) – once to the state when the plant was originally constructed and a second time to the private operator, who will have to recover its up-front capital investment in these assets (reflecting the expected income stream generated in a competitive marketplace). However, this objection fails to account for the likely use of these funds to retire some of Hydro's current outstanding debt associated with nuclear assets (the market value of which is far below book value), which will result in reduced revenue requirements for servicing the debt on these assets. Another political objection to the sale of the hydro generators relates to the symbolism inherent in selling the province's perceived birthright in water power to private interests. Yet, it is questionable whether these preferences should command the sympathies of policy makers given the wide range of other privately owned natural resource industries (forestry, gas and oil, mining) that implicate similar (perhaps even more intense) concerns over resource exploitation, given their non-renewable status. Furthermore, in comparison to both fossil and nuclear generating systems, the hydraulic generation assets have established and stable technological properties that attenuate the contracting problems surrounding the sale and subsequent operation of these assets. Nonetheless, given that the variable costs of generation for these plants are tied to water rental charges (now set at artificially low prices), the province would be required to enter into long-term contracts with private investors at the time of privatization so as to permit accurate assessment of net future revenues (hence the value of these plants) and to guard against post-purchase opportunism through rental increases.

Assuming that the non-nuclear assets were sold to the private sector, it would also seem crucial that the grid be subject to regulatory obligations to proceed quickly with negotiations over grid enhancement with inter-provincial utilities in Quebec and Manitoba, and U.S. utilities in the northeastern states of the U.S. so that a significant portion of the Ontario electricity market is rendered contestable by extra-provincial generators. On the assumption that these conditions can be met, it would then seem to follow that Ontario Hydro's existing nuclear generating facilities and its transmission facilities should be separated into two district Crown Corporations: Hydro 1 (Genco) and Hydro 2 (Gridco), with independent governance structures, i.e., one should not be a subsidiary of, or in any way formally related to, the other. Then, following the Alberta model, we think it is crucial that a fully independent pool or exchange be set up with representatives of Genco, the MEUs, independent power producers in the province, and large industrial consumers, with Genco having a minority position on the governing body of the exchange. The independent exchange would be set up under legislation that mandated it to purchase power on a least-cost basis from any of these sources and to perform the dispatch function. Further buttressing the independence of the exchange, the separation of Hydro's nuclear generation and transmission facilities into two separate, publicly owned enterprises with independent governance structures would reduce incentives for the grid to distort investment decisions and grid enhancement in favour of its own generating facilities and to the disadvantage of non-Hydro generating sources (including, importantly, extra-provincial sources).

We should note here that we are somewhat agnostic on the efficiency gains that could be realized from privatization of the transmission grid. It is and will remain for the foreseeable future a natural monopoly and thus will require ongoing regulation. As alluded to earlier, the empirical evidence suggests that the economic performance of publicly owned monopolies and heavily regulated private monopolies is typically not sharply different.[51] Moreover, as the "spine" of the system, there may be significant public and political sentiment favouring its retention in public ownership. However, if the political climate changes in a way that makes the sale of these assets attractive (by, for instance, allowing the government to demonstrate tangible and immediate gains from privatization by selling the future cash flows from this asset), we do not believe that efficiency concerns would operate strongly against such a transaction.

With these broad structural features in place, it is then important to identify the indispensable regulatory elements implied by this structure. At the generating level, there will clearly be concerns in the early stages that Ontario Hydro-owned nuclear generating facilities will dominate the market, particularly the base load market, and will require some form of price regulation. We would seek to minimize the need for such regulation in two ways. First, Hydro 1 (Genco) would be constrained by legislation from investing in any new generating facilities, so that over a period of from 20 to 40 years, the share of the generating market held by Genco would decline dramatically as nuclear facilities exhausted their useful life; and correspondingly the share of the generating market held by non-Ontario Hydro generators would increase dramatically.[52] The obvious intention of this structure is to ensure a staged but certain withdrawal of public sector decision making respecting incremental generating capacity in the electricity domain. Critical to the willingness of private sector investors to put their risk capital into the electric industry in Ontario will be the provision of credible commitments by the state that it will exit the industry and will refrain from undertaking heroic measures aimed at prolonging the remaining life of the nuclear assets.

Second, to the extent that Hydro 1 (Genco), as a dominant firm in the generating segment of the industry in the early stages, may raise dangers of monopoly pricing, or of explicit or tacit collusion by the competitive fringe simply pricing up to Genco's monopoly pricing umbrella, we need to acknowledge that this problem is likely to be no more acute, and probably less so, than is the case today, where the Ontario Energy Board exercises a purely recommendatory oversight function with respect to rates, but Ontario Hydro is the final arbiter of its own rates, subject only to Ministerial override. To the extent that Genco, as a publicly owned corporation, sought to abuse its market power in the generation segment, we would expect that political voice of one kind or another, much as is the case today, would operate as a significant constraint on this possibility, so that we are not convinced that there is a case for detailed price regulation of the generation segment under our conception of Model 1, even in the short run. Nevertheless we are open to the argument that an external regulator (probably the Ontario Energy Board) should, as a default option, be empowered to impose price ceilings or caps on generation prices if bids into the power pool reflect price increases over time that are significantly in excess of increases in the CPI. Collusion among generators may also properly engage the attention of enforcement authorities under the Canadian Competition Act, as would predatory behaviour (meaning prices below long-run marginal costs) by Genco designed to foreclose entry by more efficient generators through temporary low pricing policies. Nevertheless, we tend to believe that, in contrast to the problems manifested in the U.K., the effect of greater access to the grid by out-of-province suppliers and in-province NUGs, combined with the privatization and fragmentation of the existing non-nuclear generating assets into at least four different entities, would substantially lessen the prospects for collusive behaviour, thereby enhancing the prospects for competition to supplant the need for regulation.

The transmission grid, would, of course, remain a natural monopoly, but, as argued above, most likely publicly owned. The case for detailed external price regulation of transmission charges under our conception of Model 1 is no greater than it is today, where we have noted that the OEB exercises a recommendatory, but not dispositive, oversight role, and Hydro is not bound to follow its recommendations. In short, again, voice exercised through the political process, given public ownership, is likely to be at least a partially effective constraint on excessive pricing. If political voice is thought to be an insufficient assurance against monopoly pricing, then again, as under the Alberta model, some form of external price regulation will be necessary.[53] Specifically, the problem is that with one monopoly supplier of transmission services, the provider will seek to charge prices above its total costs of production so as to maximize total revenue. Traditionally, this monopoly pricing problem has been dealt with via administered regulation, which permits a return on equity investment calibrated to the actual costs of providing service (cost of service or rate of return regulation). This regulatory scheme requires the regulator to engage in an obtrusive degree of regulatory review aimed at determining an appropriate rate of return, appropriate investments in plant and equipment, and appropriate operating strategies. Since external review is beset by serious and persistent information asymmetries (typically the regulated firm has much greater information than the regulator), considerable resources need to be spent in extensive regulatory proceedings designed to elicit and then to certify the accuracy of information produced by the regulated firm. Compounding these problems are the relatively weak incentives that cost-of-service regulation provides for firm managers (and their principals) to introduce cost saving innovations or to keep production costs down generally. This is because improvements (deteriorations) in productive efficiency redound to the benefit (loss) of customers, while saving shareholders harmless. As well, cost-of-service regulation introduces incentives for over-capitalization (the well known Averch-Johnson effect). And although some have argued that despite its frailties, cost-of-service regulation at least enjoys a relatively clear set of ground rules that promote rational, long-term industry investment, the fact remains that regulatory uncertainties respecting approval or rejection of existing capital expenditures greatly attenuate the certainty of the regulatory environment, thereby subverting long-term investment.

To mitigate the problems inherent in cost-of-service regulation, regulatory lags that permit shareholders to reap temporary excess returns from operating efficiencies can be adopted. However, because the nature of regulatory decisions respecting the duration (hence magnitude) of these returns is unpredictable, private incentives to engage in productive improvements are attenuated. An alternative regime for mitigating the under-investment problem is to use price-cap regulation. Under the RPI-X formula used to regulate utility prices in the U.K., regulated firms are assured of receiving price increases equivalent to the rate of inflation during the period minus anticipated productivity increases. In this manner, customers receive the benefit of price reductions from productivity, while shareholders (and their managerial agents) have incentives to realize productivity savings that exceed the projected levels because that excess is retained by them. However, experience in Britain with price cap regulation suggests that it differs only in degree, and not in kind, from cost-of-service regulation in terms of the information requirements for effective external regulation or in its susceptibility to manipulation by politicians.[54]

Assuming continued state ownership of transmission assets in Ontario, the adoption of either form of incentive based regulation (lagged cost-of-service or price cap) would face certain difficulties. First, neither form of price regulation would be feasible without a prior valuation of the transmission grid assets and an estimation of operating expenses, in order to set an appropriate cost-of-service or price cap.

Second, given agency costs associated with public sector delivery, it is not clear how responsive transmission managers will be to incentive-based regulation. While these problems could be constrained by the adoption of nuanced performance-based employee compensation arrangements, political concerns over excessive compensation may limit their adoption.

As under the Alberta proposals, consideration needs to be given adopting regulatory rules that provide that local distributors and large customers would pay the same rate for transmission out of the pool, regardless of how far they are from sources of generation, while to supply power into the pool, generators and importers would pay charges or receive credits based on location of supply in order to create incentives for suppliers to locate new facilities so as to maximize the efficiency of the grid system. New generators, the exchange, large industrial customers, and local distribution companies may also need to be provided with a legal right to challenge, before a third party regulator, refusals by the grid to extend or enhance the grid, in order to minimize further the risks of Gridco favouring existing state-owned generating facilities, and discriminating against non-Genco generators, including extra-provincial generators.

A final regulatory issue that arises with Model 1 is whether retail rates charged by local distribution companies (MEUs) should be regulated. At present, as noted above, Ontario Hydro sets the retail rates for the MEUs. In our view, as reflected in the recent Alberta government's proposals, as long as the local distribution companies remain owned by municipalities, we see only limited value in price regulation, viewing local electricity prices as analogous to local property taxes, where local forms of political voice and accountability bolstered by threats of exit should be adequate in ensuring that the prices charged are not excessive. However, as we discuss more fully below in relation to Model 2, should the local distribution companies cease to be owned by municipalities and become investor-owned in the future, price regulation would then need to be contemplated.

There are two remaining potential problems respecting Model 1 that need to be addressed. First, the stranded asset problem has fixated utilities, regulators, and commentators in the U.S. in contemplating the transition to a less vertically integrated and more competitive electricity industry in the U.S.,[55] with estimates of total stranded asset costs ranging widely from $20 billion to $300 billion, and has led to elaborate proposals[56] for the imposition of non-avoidable demand charges at either the transmission or distribution levels in the industry in order to allow investor-owned utilities to recover all or a portion of capital investments that have typically been made with the approval or at the direction of Public Utilities Commissions in the past. We do not view stranded asset costs as a problem of central importance under our conception of Model 1 as applied in an Ontario context. A recent internal study conducted by Ontario Hydro concluded that the overall asset value of the Corporation is at least equal to book value and that the net book value of the nuclear assets should be written down by about $11 million whereas the hydroelectric and fossil generating assets should be written up by about $10 billion and $1 billion respectively.[57] The mark-up on the hydro and fossil generating assets would presumably be realized immediately on privatization, whereas the mark-down on nuclear assets would not be sustained immediately (as it would be on a privatization) but would simply be reflected in an expected lower revenue stream over the useful life of these assets in the face of competitive entry by lower cost generators. Of course, the magnitude of revenue losses from the nuclear assets depends on the degree of pressure exerted on Genco from, on the one hand, low-cost entry by non-utility and extra-provincial generators and, on the other, from downstream MEUs and large industrial users.[58] While under Model 1, downstream users are precluded from contracting directly with upstream suppliers (their purchases are mediated through the pool on a merit order basis), it can be predicted that distcos and large industrial users freed from the constraints of the current regulatory system would encourage low cost suppliers to enter the market in order to reduce the costs of electricity charged by Genco. In response, and assuming pricing flexibility, Genco would be forced to price its electricity down to the average total costs of entrants (but not less than its short-run marginal costs) to deter entry.[59] Although in other industry settings, the adoption of low price levels as a barrier to entry could be ephemeral (i.e., just long enough to deter entry), the widespread availability of low-cost CCGT technology, short construction lead times, and the not inconsiderable buying power of downstream consumers should force Genco to maintain price reductions in order to deter entry.[60] Nevertheless, the financial impact of these price reductions on Genco will be muted by its status as a public enterprise, which confers advantages in the form of exemption from corporate income taxes and lower costs of capital by virtue of the provincial guarantee of its debts. Finally, if it is widely believed that Genco as the dominant supplier in the electricity market will be forced to offer price reductions so as to maintain its market share, then the price that investors will be willing to pay for Hydro's non-nuclear assets will be appropriately discounted to reflect the reduced stream of future earnings. In this respect, we emphasize it is essential that the regulatory rules of the game be clearly established at the outset of the restructuring process so that private investors can condition their bids for Hydro's assets on a relatively accurate understanding of the future policy environment.

Although Hydro's internal analysis suggests that no losses will be experienced by Hydro were all of its generating and transmission assets marked to market, prudence dictates canvassing the possibility that losses sustained from market valuation of the nuclear assets will swamp the gains from market valuation of the non-nuclear generation and the transmission assets, requiring further write-downs on Hydro's publicly guaranteed debt. The question is how should these write-downs be paid for and who should bear them? As mentioned earlier, this issue has engendered considerable dissension between shareholders and ratepayers of investor-owned utilities in the United States, with each group attempting to inflict the brunt of stranding costs on the other – shareholders arguing for the imposition of stranding charges on ratepayers, ratepayers arguing for unencumbered access by low-cost suppliers to the existing grid network. While it would, in principle, be possible to design a stranding charge to compensate for foregone revenues on the nuclear assets that could be levied on all ratepayers utilizing electricity from the grid, the convergence of the identity of Ontario ratepayers and taxpayers (shareholders) obviates the need to devise and implement such arrangements in the province. Instead, the write-down of the assets can be gradually imposed on taxpayers through the tax system which, because of its more progressive features, may well be a distributionally more appropriate vehicle for addressing these costs. Apart from the superior distributional features of the tax system, several other factors militate in favour of dealing with stranded assets in this manner, namely the ability to avoid complex determinations of ratepayer contribution (measured in terms of historic load patterns) for past investment decisions that led to over-investment in nuclear assets, the removal of incentives of existing large-load consumers to exit the system through self-generation to avoid stranding charges (a form of regulatory arbitrage), and resultant gains in allocative efficiency from early adoption of marginal cost pricing. In addition, removing the stranded cost issue from the tasks that external regulators would be required to address would greatly simplify the regulatory regime that would be required. While politicians may worry that assumption of these liabilities will increase public scrutiny of the government's finances because it will increase the size of government expenditures, this is simply a form of fiscal illusion which is unlikely to attract any attention by private debt rating agencies, given that they are already likely to have impounded the effects of over-investment in uncompetitive generating assets into their assessment of the government's debt serviceability. In this respect, we would simply follow the Alberta proposal, and ignore the issue of stranding costs as a regulatory issue.

A second potential source of complexity under our conception of Model 1 relates to the pricing levels that are likely to emerge in a competitive power pool. Some commentators appear to assume that if competition to supply the pool is fully effective, all prices will be bid down to short-run marginal cost.[61] If this were so, this would entail no allocative inefficiencies with respect to incumbent generators, whose fixed costs are sunk, but it would presumably deter entry by new generators, who would not find it rational to make new capital investments unless they were assured that prices would cover average total costs (or long-run rather than short-run marginal costs). Reflecting these concerns, the British electricity regime has until recently provided for payment of "capacity adders" to generators on top of system power prices generated through bidding to supply the pool. However, these capacity adders have apparently been subject to serious manipulation by generators.[62] Moreover, the evidence does not suggest that all prices will necessarily be bid down to short-run marginal cost; in fact, the U.K. evidence suggests (albeit reflecting insufficient competition between what are essentially duopolists in the British industry) prices that have been up to 80% above short-run marginal cost.[63] Thus, as with proposals for the imposition of stranded asset charges, we are inclined to dismiss the case for the imposition of capacity adder charges, again simplifying considerably the range and complexity of the regulatory tasks that our conception of Model 1 would entail.

Model 2: Competition for Wholesale Customers

This model assumes that instead of, or more plausibly in addition to, competition for access to a power pool, generators could compete for direct wholesale business with local distribution companies and large industrial users through bilateral contracts. As in the U.K. and in the Wisconsin proposals, bilateral contracting would operate alongside the power pool and would entail contracts for differences – in effect hedging contracts relative to pool prices. Local distribution companies would then pay transmission charges to the grid, as would large industrial customers utilizing the grid. To the extent that the latter also use the local distribution network they would also pay a local distribution charge. At the same time, any constraints on self-generation by municipalities would be removed, as would constraints on retail wheeling through the local distribution system or grid by self-generators with excess capacity (subject only to paying the regulated transmission and distribution charges). We would nevertheless restrict (as in the case of Model 1) the ability of MEUs to merge with the grid; although vertical integration of the wires network may yield clear coordination economies, we worry that the merger of these distcos into the grid would reduce the ability of captive consumers to compare easily the price and quality dimensions of service they are receiving from their local supplier, if some distcos are integrated with the grid and some are not.

There are at least three advantages to Model 2 as a complement to Model 1: First, the pool is likely to become supplemental to bilateral contracts, and simply provide a source of demand for incremental power, with bilateral contracts being struck of whatever length proves mutually acceptable to the parties, presumably ensuring that the power prices paid under such contracts are likely to cover average total costs (or long-run marginal costs). The effect, therefore, of Model 2 is to vest decisions respecting incremental capacity in numerous distributing companies and industrial consumers rather than in the centralized exchange. In this respect, the electricity market will become less vulnerable to non-trivial, system-wide mistakes made by a centralized actor (albeit one, in the case of Model 1, with a diversity of interests having voice in governance). As a by-product of this arrangement, private parties will be better able to secure long-term purchase agreements that correspond more closely, in the case of downstream consumers, to the quality and cost of required service, and, in the case of upstream suppliers, to the nature of the investment risks associated with different generating assets, a shift which should generate tangible improvements in productive efficiency. Second, to the extent that there are any residual concerns over the even-handedness of pool operation, given the continued existence of two publicly owned enterprises – Genco, which would own Ontario Hydro's present nuclear generating assets, and Gridco, which would own the grid – the availability of the bilateral contracting option is likely to keep the exchange honest, thereby reducing the need for costly regulatory interventions. The exchange would, of course, continue to perform a dispatch function, even with respect to generation under contract. Third, by permitting competition amongst generators for wholesale buyers, this will create an immediate incentive for the MEUs to rationalize themselves into units that make bilateral contracting feasible.[64] In order to facilitate such rationalization, which would occur spontaneously rather than through the imposition of some master plan, it is important that property rights issues relating to ownership of the MEUs be clarified, probably most simply by legislation that declares that the MEUs are owned by local municipal councils. Rationalization may occur through mergers of smaller MEUs; formation of cooperatives or buying groups of MEUs; or privatization through sale to private investors who may then choose to aggregate smaller MEUs on a regional basis to realize economies of scale and scope.

In terms of the regulatory requirements implied by Model 2, relative to Model 1, nothing changes at the generation level, although the need for the exercise of the default regulatory powers envisaged in Model 1 for generation may be attenuated. At the transmission level, there would be the same case for regulation of transmission charges as under Model 1. At the distribution or retail level, there may be a more complex potential interplay between structural and regulatory reform. As noted above, to the extent one can view MEUs as simply emanations of local municipal councils, one can regard the prices that MEUs might charge for electricity as analogous to property tax determinations by municipal councils, which of course are not subject to external regulatory oversight, but are disciplined by local political voice and inter-jurisdictional competition, such as is evident in promotional efforts of municipalities today in touting lower property taxes and superior public services as locational advantages. However, the conundrum that arises is that because almost certainly the current MEU structure is inefficient – there are far too many of them and many are too small to function effectively, particularly under a bilateral contracting model and even more so under a retail competition model (Model 3) – under some rationalization scenarios, e.g., where MEUs are sold off to private investors, they would cease to be accountable to local political constituencies, thus strengthening the case for some form of price regulation – ideally some form of incentive regulation, such as price caps. Interestingly, the Alberta Government's proposal contemplates (in our view, astutely) that retail price regulation would only be necessary with respect to local distribution companies that are investor-owned and not for local distribution companies that are municipally owned. We endorse this proposal with respect to Model 2, again in part because of the reduced demands it makes on the regulatory process.

However, it may be thought that Model 2 would raise distributional concerns of various kinds, for example, large industrial users being able to negotiate bulk discounts that lower their per unit power charges substantially below those payable by "captive" residential consumers. We are inclined largely to discount this concern, as municipally owned utilities will also have strong incentives to negotiate the best deals possible for their local customers and citizens. If local distribution companies are privatized, they will similarly have incentives to minimize input prices (while output prices would be regulated). Cross-subsidies from urban consumers to rural residential consumers may be endangered by this proposal. To the extent that local distribution and generation charges are higher for rural as opposed to urban customers, the changes proposed can be expected to undermine cross-subsidies. However, if the Alberta proposals are followed and a standard transmission charge is imposed on local distribution companies, whatever distance they may be from sources of generation, a portion of the current cross-subsidy will be preserved, mitigating somewhat the impact of cost differences.[65] Nevertheless, if substantial reduction in the magnitude of these subsidies generates legitimate distributional concern, it is open to the government to provide supplementary assistance, arguably conditioned on income, for rural customers.

Model 3: Competition for Retail Customers

Under this Model, as reflected in the British regime, Wisconsin Electric's proposals, and the California Public Utilities' Commission's earlier Blue Book proposals, generators would be free to compete directly for retail customers (industrial, commercial, and residential), facilitated by various kinds of energy intermediaries. The retail monopoly presently held by the MEUs would be terminated, creating additional incentives beyond those implied by Model 2, for spontaneous rationalization of this sector. Local distribution companies would, of course, retain a natural monopoly over the local wire network, but not over the retailing of electricity or other energy-related services, such as time-of-day service, metering, etc. The adoption of Model 3, therefore, can be viewed as a way of promoting much greater competition in the demand side management segment of the industry. It can also be viewed as a market corrective for the lack of citizen mobility in response to electricity rates that has been alleged to impair voice at the municipal level.[66] As in Model 2, if local distribution companies become investor-owned they will have incentives to charge monopoly prices for distribution services both to their own customers and to other customers simply seeking to use their local distribution wire network. In addition, there is a further risk that with retail competition local distribution companies may charge independent customers or energy retailers discriminatory rates for use of the local distribution wire system relative to rates that they charge their own customers. Thus, in order to realize the efficiency gains from structural reforms at the retail level, there may be an enhanced need for regulation at this level. This is a case where structural change may not be an effective substitute for regulation, but may indeed intensify the need for new forms of regulation. Ideally, distribution charges would be levied on an unbundled basis (i.e., separate from charges for electricity) and would be subject to some form of price cap or incentive-based regulation to mitigate the dual problems of potential monopoly pricing and potential discrimination between local distribution companies' own customers and other customers.

IV. CONCLUSION

In our view, the Ontario electricity industry should aspire, over the medium term, to the progressive adoption of Model 3, in the short term adopting Model 2, but accompanied by an announced phase-in of retail competition. We believe that this policy objective is most consistent with the four dimensions of economic efficiency that constitute our evaluative criteria (allocative, productive, dynamic, and administrative efficiency) and would not place at serious risk existing legitimate distributional concerns. We also believe that our proposals respect the six basic premises that orient our analysis: (1) by not unnecessarily endangering economies in vertical coordination, through promotion of competition in generation and electricity retailing and thus avoiding the problems of reciprocal opportunism endemic to bilateral monopolies; (2) by avoiding assigning a significant systemic industrial policy role to the electricity industry; (3) by proposing much simplified and more sharply focused accountability mechanisms; (4) by minimizing the need for, and scope of, constraining government regulation by greater reliance on competitive industry structures; (5) by not permitting government revenue-raising goals to thwart efficiency objectives; and (6) by limiting the scope for stranding of socially valuable assets through uneconomic bypass.

It is important to emphasize that once Model 3 is accepted as a logical endpoint of reform, Model 2 is unlikely to persist as a stable equilibrium. If industrial and residential consumers are dissatisfied with the level of service being offered by monopoly distributing companies, they will, no doubt, press for the entry of alternative suppliers or perhaps even attempt to establish local buying groups that will vie for the local franchise. In this respect, the situation is closely analogous to the relatively fragile boundaries established in the U.S. system, which have been only partially effective in denying grid access to large industrial users who have sought access on terms comparable to those enjoyed by municipalities. We also note that, unlike the United States, the issue of retail versus wholesale access is less vexing because it does not implicate issues of federal distribution of power –the provinces' power over both wholesale and retail distribution appears to be well established.[67]

Despite the benefits we have asserted for the proposed model of partial industry privatization and de-integration, it can be predicted that certain stakeholder groups will be opposed to the scheme. For instance, the heavily unionized Ontario Hydro workforce is likely to strenuously resist the privatization of non-nuclear generating facilities, given evidence both of overmanning and overcompensation relative to private sector reference points.[68] Indeed, it is likely that they will strongly inveigh against any program to reduce Hydro's franchise (e.g., open access to the grid) even if it is not accompanied by privatization of existing assets.[69] To some extent, continued public ownership of the nuclear assets should mitigate some of the concerns of the unions with job security, although, to be fair, the rate of layoffs in the notoriously inefficient British nuclear sector post-reform indicates that labour may suffer increased losses even from a state-owned Genco. In this respect, we can only endorse the continuing role for humane and nuanced labour adjustment policies designed to reintegrate into the labour market those employees whose expected long-term careers at Hydro are interrupted by the gale of market forces.[70]

It can also be expected that retail competition will likely engender vigorous resistance from the MEUs. In a recent report by the Ontario Municipal Electric Association,[71] the MEUs oppose any form of retail competition and are extremely cautious about supporting any form of privatization of generating facilities, proposing instead a "managed competition" model under which Ontario Hydro would be insulated from competition in the generation segment until the year 2000, after which date the transmission company would be free to purchase electricity from least cost suppliers, although new generators would face quite elaborate licensing and regulatory hurdles. The Municipal Electric Association's proposals contemplate the splitting of Ontario Hydro's transmission assets from its generating assets and contemplate further division of its generating assets into separate publicly owned divisions that, as state-owned enterprises, would compete against each other. This strikes us as a totally unrealistic competitive scenario. We know of no precedent of vigorous competition among state-owned enterprises, particularly when the putative competitors are operating divisions of the same company. It should be recalled that the principal benefits of industry restructuring are improved operating efficiencies and long-term improvements in investment related to reduced centralization of decision making and increased market scrutiny. We expect that neither benefit will be realized in this scenario. Indeed, we are inclined to see in these proposals a suggestion of a Faustian bargain between the MEUs and Hydro, where the MEUs act as political advocates for insulating Hydro from substantial upstream competition at the generation level, in return for Hydro agreeing to support policies that will insulate the MEUs from significant competition at the retail level. The early history of the electricity industry in Ontario continues to cast a long shadow over current debates about the future of the industry. We believe, in contrast, that structural changes are required at the distribution or retail level that will not only render this segment more efficient in its own right, but also turn it into a strong proponent, on behalf of final consumers (as it should be), of greater competition, more efficiency, and lower costs in the upstream (generation) segment of the industry. A final potential set of detractors from this proposal are various environmental groups in the province. The concern here would be with the loss of levers over resource planning that Hydro would ultimately suffer in a competitive industry. The capacity, for instance, of Hydro to engage in highly centralized integrated resource planning designed to ensure socially optimal levels of electricity consumption would be attenuated. Under either Model 2 or Model 3,[72] downstream consumers (either directly or through their local distributing company) will simply bypass the grid if they are forced to bear electricity costs that reflect payments to other consumers for encouraging more environmentally responsible consumption or for services which are provided directly to them but which are not highly valued.[73] In either case, a competitive market in electricity enables consumers to exit the grid by consuming electricity provided by low-cost producers that is not tethered to environmental adders or negawatt programs. Nevertheless, there is a growing body of literature in the United States that demonstrates the folly of relying on centralized resource planning as a way of ensuring that socially optimal levels of electricity utilization occur; there are advantages to market-based regulatory programs which harness private sector initiative and decentralized local information to achieve appropriate conservation measures.[74] At the core, these programs are based on common regulatory standards that allow private actors to calibrate the social costs of the environmental damage occasioned by their consumption to the private benefits thereby generated, often through reliance on emission fees (imposed on all sources) and/or tradeable emission allowances. We note the existence of support among some segments of the environmental movement, namely Energy Probe, for initiatives of this character.

Despite the various stakeholder groups whose interests may be threatened by our proposal and who might therefore agitate against its adoption, there are several stakeholder groups whose interests would appear to be advanced by it. For instance, although the Independent Power Producers' Society of Ontario (IPPSO) has championed more modest proposals aimed at enhancing the independence of grid access decisions (calling for de-integration of generation and transmission, endorsing greater reliance on centralized integrated resource planning), the interests of at least some of its members would be served by privatization of Hydro's non-nuclear assets (to reduce market power concerns) and by rationalizing the downstream MEU sector (ensuring that the MEUs are an agent for industry change). The Association of Major Power Consumers of Ontario (AMPCO), which represents the province's largest retail electricity users (association members purchase close to 60% of the electricity sold by Hydro) has more obvious interests in achieving rate reductions and in stimulating a rationalization of the electricity sector. Significantly, while AMPCO has called for enhanced industry competition, the association does seem reconciled to some level of public ownership in the foreseeable future, implying the acceptance of the role we have proposed for state ownership of nuclear assets. We suspect that the position of the province's other industrial and commercial interests, represented perhaps by the Ontario Chamber of Commerce, would not depart materially from that of AMPCO. Our proposal can be seen to be favourable to the interests of urban residential consumers, particularly in light of its emphasis on increased accountability of downstream distcos to consumers and because of our expectation that rate differentials between residential and industrial users unrelated to actual cost differences in supplying electricity will narrow in the proposed regime. Nevertheless, as mentioned above, we expect that the adoption of a more competitive industry structure will undermine though not eliminate cross-subsidies from urban to rural customers.

In any event, we have, under our proposals, attempted to keep regulatory involvement in the industry to a minimum and to utilize incentive-based regulation where regulation is necessary. Where regulation is required, we strongly favour reposing all industry-specific regulatory responsibilities in one external agency (probably a refurbished OEB)[75] in order to focus more sharply the diffused, disjointed, and discontinuous oversight process that has prevailed in Ontario in the past, and to enhance institutional stability, continuity, memory, and an integrated industry perspective that is forward-looking, and not primarily crisis-driven. At the same time, the OEB, with an appropriately limited and specific mandate and appropriate staff competencies and resources, would provide a neutral external public forum that can act as an agent for change in the industry, given the various existing entrenched interests that have to be moved and monitored in the process of change. This role for the OEB requires a detailed review of its current powers, competencies, and resources.

A key underpinning of the recommendations that we have made is that they contemplate the formation of mediating market and institutional arrangements that attenuate the reciprocal opportunism problems that beset the industry in an earlier part of the century, and which propelled the industry toward integration. The magnitude of these problems is explained by the properties that have historically characterized the electricity industry: high levels of minimum efficient scale, correspondingly high fixed investment costs, lengthy lead times, and monopsony downstream buyers. Yet, by creating robust competitive generation markets that can exploit the efficiencies of small-scale CCGT technology, by ensuring open access to the transmission grid and efficient pricing of its services, and by increasing the range of downstream buyers of electricity services, both the coordination and opportunistic lock-in problems manifested earlier should be adequately addressed. In this respect, we would not expect the formation of a robust electricity market ultimately to dissolve into a vertically integrated industry, the legacy of which is nothing more than high transition costs that have to be borne by provincial citizens, either as consumers or as taxpayers.

Thus, while our proposals may seem radical in an Ontario context, given the history of the electric power industry in the province, in order to place them in perspective it is crucial to recall the recent evolution of the natural gas industry in Canada and the U.S. since deregulation of natural gas prices in the mid-1980s, of which Ontario was a vigorous proponent. In fact, the course of evolution in the natural gas industry has almost precisely followed the path we have outlined above for the electricity industry. Natural gas producers now compete among themselves to supply under contract both local distribution companies and retail customers or intermediaries acting on the latters' behalf. The inter-provincial pipelines operate purely as common carriers, subject to rate regulation, and the local distribution companies (local investor-owned gas utilities) compete with other natural gas retailers for final customers' business, with distribution charges for use of the local network being subject to regulation to preclude the imposition of either monopolistic or discriminatory charges. We believe that the electricity industry in Ontario will, and should, evolve in a similar direction, and over a time frame not radically different from that which has obtained in the natural gas industry. It bears recalling that a significant part of the problems now faced by the electricity industry in Ontario and elsewhere relates to the low prices of natural gas, both as an alternative fuel in end uses, and as a source of power for gas turbine electricity generation. Vertical de-integration of the natural gas industry, and a much enhanced role for competition, can take a large part of the credit for the enhanced performance of the natural gas industry, which has now become the major competitive threat to the electricity industry. With increasing technological convergence between these two industries, it would be surprising if some broad measure of organizational convergence did not also occur. We have attempted here to map the path that this convergence might follow.

NOTES

1 For histories of Ontario Hydro, see Merrill Dennison, *The People's Power: The History of Ontario Hydro* (Toronto: McClelland & Stewart, 1960); William R. Plewman, *Adam Beck and the Ontario Hydro* (Toronto: Ryerson Press, 1947); H.V. Nelles, *The Politics of Development* (Toronto: Macmillan Co., 1974), chaps. 6, 7, and 10; Neil Freeman, *The Politics of Power: Ontario Hydro and Its Government 1906-1995* (University of Toronto Press, 1996).

2 For a comprehensive review of the industry, see "Ontario Hydro: Moving to a Higher Voltage," RBC Dominion Securities Report 27. August 24, 1994;

Leonard Waverman and Adonis Yatchew, "Regulation of Electric Power in Canada," 53. Working Paper, University of Toronto, 15 April 1993.

3 Data provided by Ontario Hydro to authors. The figures for nuclear generation are lower and for fossil higher when expressed as a percentage of in-service dependable peak capacity: hydro electric 22%; nuclear 47%; and fossil 28%.

4 RBC Dominion Securities, 30.

5 Waverman and Yatchew, 47.

6 Source: RBC Dominion Securities, 28.

7 Waverman and Yatchew.

8 H.V. Nelles, 404-26.

9 RBC Dominion Securities.

10 Ibid., 29.

11 Ibid., 16. It is expected that Hydro's low equity capitalization will improve as a result of significant curtailments in its planned capital expenditure program.

12 For instance, whereas more conventional generation technologies required minimum generating plant capacity of 1000 MW, CCGT plants can be economic at capacities as low as 200 MW.

13 Christopher Flaven and Nicholas Lenssen, *Reshaping the Power Industry* (Washington, D.C.: World Watch Institute, 28 October, 1993).

14 "Ontario Hydro and the Electric Power Industry; Challenges and Choices," Report of the Financial Restructuring Group, Ontario Hydro, 27 June, 1994.

15 RBC Dominion Securities, 34.

16 Scott Hempling, Kenneth Rose, and Robert Burns, "The Regulatory Treatment of Embedded Costs Exceeding Market Prices: Transition to the Competitive Electric Generation Market," 57. Report to the National Association of Regulatory Utility Commissioners, Washington, D.C., 7 November, 1994.

17 Report of Financial Restructuring Group.

18 Ibid., 16.

19 The potential of these new information systems is dramatic. Among the various uses contemplated are: detailed itemized billing including consumption by major appliance or machine, bill projections, instantaneous meter rereads, power off indications, remote turn on/off, and real-time pricing. See discussion in Michael R. Niggli and Walter W. Nixon, III, "A Serendipitous Synergy: Why Electric Utilities Should Install the Information Superhighway," *Electricity Journal* 7, 25 (1994).

20 The role of command and control demand side management systems is lucidly critiqued in Bernard Black and Richard Pierce, "The Choice Between Markets and Central Planning in Regulating the U.S. Electricity Industry," *Colum. L.R.* 93 (1993):1339.

21 Douglas Houston, "Can Energy Markets Drive DSM?" *Electricity Journal* 7, 46 (1994).

22 Ontario Hydro Financial Restructuring Group, 14 ff.

23 Alberta Department of Energy, "Enhancing the Alberta Advantage: A Comprehensive Approach to the Electric Industry," (October 1994.) Under the current system, there are three large vertically integrated utilities: Alberta Power Limited, TransAlta Utilities, and Edmonton Power. The first two are investor-owned utilities, while Edmonton Power is municipally owned. In addition, there are a number of municipal distribution utilities that service local markets (e.g., City of Calgary, City of Lethbridge).
24 Ibid., 14.
25 In the period April 1993 to January 1994, National Power generated 34.7% of total output, PowerGen 25.7%, Nuclear Electric 24.1%, Scottish Power, Scottish Hydro Electric and EdF 8.4%, new entrants 6% and residual 1.1% (source: OFFER).
26 Mark Armstrong, Simon Cowan, and John Vickers, *Regulatory Reform: Economic Analysis and British Experience*, (Cambridge, MA: MIT Press, 1994) figure 9.2.
27 Ibid.
28 See discussion in Armstrong, Cowan, and Vickers.
29 See discussion note 49.
30 Finance, Regulation and Power Supply Policy, Edison Electric Institute, "The British Model: An Assessment," Power Supply Monograph, Issue 2, June 1994.
31 William Baumol, Paul Joskow, and Alfred Kahn, "The Challenge for Federal and State Regulators: Transition from Regulation to Efficient Competition in Electric Power," 9 December, 1994.
32 Jere Jacobi, "Wisconsin Electric's View of a More Competitive Industry," Working Paper of the University of Toronto Electric Power Project, 1 November, 1994.
33 This conclusion appears to have been recently endorsed by the California Public Utilities Commission in its recommendations on reforms for the electric power industry that resulted from extensive deliberations on the Blue Book proposals. In a 3-1 split, CPUC opted for a more modest set of reforms to the California electrical industry that would not require extensive sell-offs of generation assets. Instead, the focus was on the creation of an independent pool operator (under FERC jurisdiction) and direct wholesale access in 1997. However, depending on the resolution of jurisdictional and market issues, retail access could be permitted by 1999. Real-time metering of customer usage is also contemplated. The majority opinion was rendered by PUC President Daniel W. Fessler, P. Gregory Conlon, and Henry M. Duque. The minority opinion of Commissioner Jessie J. Knight Jr. would have allowed direct wholesale access without the interposition of a pool and would have required divestiture of existing utility generating assets where transmission interests were owned. He would have also accelerated the adoption of retail access. See Michael Parrish. "PUC Proposes Creating a Wholesale Market for Selling Electrical Power." Los Angeles Times, May 25, 1995, D1; "CPUC Offers Electric Restructuring Proposals for Comment," *Business Wire, Inc.*, May 24, 1995.

34 Report of the Financial Restructuring Group, 29.
35 Report of Financial Restructuring Group, 30-1.
36 Christopher Armstrong and H.V. Nelles, *Monopoly's Moment: The Organization and Regulation of Canadian Utilities 1830-1930* (Toronto: University of Toronto Press, 1988).
37 John Baldwin, *Regulatory Failure and RenewaL: The Evolution of the Natural Monopoly Contract* (Ottawa: Economic Council of Canada, 1989).
38 George Priest, "The Origins of Utility Regulation and the Theories of Regulation Debate," *Journal of Law and Economics* 36 (1993):289.
39 On the economies of vertical integration in public utilities generally and electricity specifically, see Keith Crocker and Scott Masten, "Regulation and Administered Contracts Revisited: Lessons from Transaction-Cost Economics for Public Utility Regulation," *Journal of Regulatory Economics* (forthcoming); William Baumol, Paul Joskow, and Alfred Kahn, "The Challenge for Federal and State Regulators: Transition from Regulation to Efficient Competition in Electric Power," December 9, 1994, 7-8; Paul Joskow and Richard Schmalansee, *Markets for Power* (Cambridge, MA: MIT Press, 1983).
40 For a discussion of these issues in the context of the newly liberalizing countries of Eastern Europe, see: Ronald Daniels and Robert Howse, "Reforming the Reform Process: Privatization in Central and Eastern Europe," *New York University Journal of International Law and Politics* 25 (1992):27.
41 The incumbent supplier should continue to supply the needs of the consumer until the fixed plant investment is no longer useful. Were exit to occur in this case, new plants would be servicing industry demand even though incumbents were the lowest cost producers.
42 The issue of stranding is analyzed at length in Baumol, Joskow, and Kahn; Robert J. Michaels, "Unused and Useless: The Strange Economics of Stranded Investment," *Electricity Journal* 7 (1994):12; Bernard Black, "A Proposal for Implementing Retail Competition in the Electricity Industry," *Electricity Journal* 7 (1994):58; and in Hempling, Rose, and Burn.
43 Baumol, Joskow, and Kahn; Hempling, Rose, and Burn.
44 Richard Pierce, "The State of the Transition to Competitive Markets in Natural Gas and Electricity," *Energy Law Journal* 15 (1994):323-344; Pierce, "The Advantages of De-Integrating the Electricity Industry," *Electricity Journal* 7 (1994):16.
45 Pierce, "The Advantages of De-Integrating the Electricity Industry."
46 The complexities of efficient transmission pricing using spot market nodal prices are discussed in F.C. Schweppe, M.C. Caramanis, R.D. Tabors, and R.E. Bohn, *Spot Market Pricing of Electricity* (Kluwer Academic Publishers, 1988). See also: William W. Hogan, "Electric Transmission: A New Model for Old Principles," *Electricity Journal* 6 (1993):18.
47 The importance of disaggregating Hydro's existing generating assets into more than two or three generators (as in the U.K.) is underscored by the market

power problems experienced post-privatization in the U.K. See Richard J. Green and David M. Newbery, "Competition in the British Electricity Spot Market," *Journal of Political Economy* 100 (1992):929.

48 Armstrong, Cowan, and Vickers; J. Hewlett, "Lessons from the Attempted Privatization of Nuclear Power in the United Kingdom," *Energy Sources* 16 (1994):17.

49 Recently, the British Government has announced plans to merge Nuclear Electric (which holds the nuclear remnants of Central Electricity Generating Board in England) with Scottish Nuclear, and for the subsequent privatization of the merged entity (amounting to 22% of U.K. electricity market). However, some of the oldest reactors will remain in public ownership (8% of U.K. electricity market). (*INS*, 14 May 1995, "All you need to know about the nuclear sell-off: What will be sold? Why now? Will anyone want to buy it? Will it be more dangerous?") The argument in favour of privatization now is that the issue of future liabilities for plant decommissioning and fuel reprocessing costs has been resolved and the operating efficiency of the plants, particularly those under the ownership of Nuclear Electric, has been dramatically improved over the past five years.

50 Since privatization of the electric power industry in 1990, output of Nuclear Electric is up by 65% and operating costs have been cut by 45%. Further, workforce numbers have been reduced from 14,400 to 8,900, and productivity has more than doubled. (STM 14 May 1995, Andrew Lorenz, "Nuclear: The Ultimate Privatization.") Arguably, much of the benefit from improved operational efficiency can be traced to competition with privatized firms. See M.J. Trebilcock, *The Prospects for Reinventing Government* (Toronto: C.D. Howe, 1994).

51 See John Vickers and George Yarrow, "Economic Perspectives on Privatization," *Journal of Economic Perspectives* 5 (1991):111; Ron Hirshhorn, "The Ownership and Organization of Transportation Infrastructure," 13, 21. Background Study for the Royal Commission on National Passenger Transportation: Directions (Ottawa: Supply & Services, 1992); David Sappington and Joseph Stiglitz, "Privatization, Information, Incentives," *Journal of Policy Analysis and Management* 6 (1987):567.

52 Data provided by Hydro indicates that the average remaining life of existing nuclear generators is 29 years, ranging from 18 years in the case of Pickering A to 38 years in the case of Darlington. RBC Dominion Securities, 26.

53 The purpose of such price regulation is to redress the opportunism problems deriving from the natural monopoly properties of the transmission grid; significant economies of scale (it seldom makes sense to run two or more distinct sets of wires from generators); and low short-run marginal costs (given high fixed capital costs, the variable costs of production are relatively low, requiring second order pricing for suppliers to recover total costs).

54 Deiter Helm, "British Utility Regulation: Theory, Practice, and Reform," *Oxford Review of Economic Policy* 10:17; Armstrong, Cowan, and Vickers, chap. 6.

55 Baumol, Joskow, and Kahn, op. cit.; Hempling, Rose, and Burns.

56 Report by Scott Hempling, Kenneth Rose, and Robert Burns for the National Association of Regulatory Utility Commissioners, "The Regulatory Treatment of Embedded Costs Exceeding Market-Prices: Transition to a Competitive Electric Generation Market," November 7, 1994.

57 Report of Financial Restructuring Group, 22.

58 Indeed, we note that the recent report of the Alberta Department of Energy takes the view that new generating facilities are likely to entail higher power costs than existing facilities, although it is crucial to note that Alberta does not have any nuclear facilities, which entail much higher average total costs than hydro-electric power, although marginally lower than fossil fuel generation (RBC Dominion Securities Reports, 28.), and accordingly proposes no stranding charges in its proposals for the creation of a competitive power pool.

59 The existence of pricing flexibility combined with recovery of losses through the tax base obviates the need to rely on long-term purchase contracts between the exchange and Genco (at supra-competitive prices) to attend to the province's interests in debt serviceability.

60 In respect of this last factor, it is worth noting that the lack of customer choice in upstream generators may limit the capacity of downstream distcos and industrials to discipline Genco (or indeed other upstream suppliers) in the event of strategic pricing behaviour. That is, it may be in the interests of downstream consumers to assure upstream entrants that they will patronize their services even if they are not the lowest-cost supplier in order to guard against the possibility that entrants will build new plants at average total costs that are less than the existing embedded costs of incumbents but greater than their short-run marginal costs. By making firm commitments to entrants, the ability of the incumbent generator to inflict losses on entrants post entry by strategic price reductions (below average total cost) is constrained.

61 Baumol, Joskow, and Kahn.

62 Stephen Littlechild, "Competition and Regulation in the U.K. Electricity Industry," presentation to the University of Toronto Electric Power Project, April 7, 1995.

63 Armstrong, Cowan, and Vickers, chap. .9:303; see also Helm.

64 On the incentives for municipalization in fully vertically integrated U.S. electricity systems created by the potential for bilateral contracting, see Richard Pierce, "The State of Transition to Competitive Markets in Natural Gas and Electricity," *Energy Law Journal* 15 (1994):323, 345. The nature and level of economies of scale in distribution is discussed in J. Stephen Henderson, "Cost Estimation for Vertically Integrated Firms: The Case of Electricity," in Michael A. Crew, ed. *Analyzing the Impact of Regulatory Change in Public Utilities* (Lexington, MA.: D.C. Heath and Company, 1985), chap. 6.

65 It has been estimated that transmission costs constitute only 15% of the total final cost of electricity, with the remaining 85% divided between generation (70%),

and distribution (15%). If so, the impact of location-specific transmission pricing on final electricity prices is unlikely to be severe even if it were to be permitted.

66 Tiebout's model of inter-jurisdictional competition is beset by bundling or tied sales properties, i.e., citizens may find that their electricity rates are only a small portion of their total costs of carrying on a business or forming a household in a given jurisdiction and may, therefore, have only muted voice in disciplining municipal politicians. See C.M. Tiebout, "A Pure Theory of Public Expenditures," *Journal of Political Economy* 64 (1956):416. Indeed, for most Ontario industrials, electricity rates constitute only 2% of total production costs (unpublished data from Ontario Hydro).

67 In the United States, the Energy Policy Act of 1992 makes it clear that FERC only has power over wholesale, not retail, markets, which is remitted to the states.

68 See Waverman and Yatchew, 29-34, although recent substantial lay-offs have reduced the scale of the overmanning problem.

69 In this vein, the Power Workers' Union, which represents the majority of Hydro workers, has made job security the principal concern. The PWU has argued that privatization or restructuring will adversely affect the price, reliability, and quality of electricity service. Specifically, the PWU has argued for enhanced use of integrated resource planning in reviewing the adoption of new NUG capacity and the decision to close Bruce Unit #2.

70 See discussion of this issue in the trade liberalization context in M.J. Trebilcock, Marsha Chandler, and Robert Howse, *Trade and Transitions* (London: Routledge, 1993).

71 Municipal Electric Association, "Restructuring the Electricity Industry in Ontario"; vol. 1 Recommended Strategies, September 6, 1994.

72 It is possible, however, that the institutional arrangements contemplated in Model 1 (centralized purchases and distribution of electricity by the pool) would support centralized resource planning so long as the exchange was willing to support this function.

73 The tension between traditional command-and-control environmental regulation and increased industry competition is discussed in Black and Pierce, and in Paul Joskow, "Emerging Conflicts Between Competition, Conservation and Environmental Policies in the Electric Power Industry," Keynote Address for Public Utility Research Center's Conference on Competition in Regulated Industries, University of Florida, April 29-30, 1993.

74 Reviewed in Black and Pierce. See also A.L. Kolbe, M.A. Maniatis, J.P Pfeifenberger, and D.M. Weinstein, "It's Time for a Market-Based Approach to DSM," *Electricity Journal* 6 (1993):42; and Paul L. Joskow and Donald B. Marron, "What Does a Negawatt Really Cost? Further Thoughts and Evidence," *Electricity Journal* 6 (1993):14.

75 Environmental assessments of new facilities would presumably be undertaken through the more general environmental assessment process in the province.

Comments by

THOMAS ADAMS
DAVID CAMERON
DAVID S. GOLDSMITH
ARTHUR KROEGER
KARL WAHL

THOMAS ADAMS

The authors endorse a competition-oriented power system for Ontario in order to optimize social costs efficiently. Energy Probe believes that the transition to competition must be designed to ensure that the efficiencies gained flow to customers as rate reductions. Power privatization and competition in the U.K., which have produced both efficiency improvements and rate reductions, show the way forward.[1] As the authors suggest, the Ontario experience in gas, where customers both large and small have seen major rate reductions, is also a model from which to learn.

The keystone of all future reforms should be to ensure that all Ontarians gain full de jure and de facto rights to choose among suppliers for their bulk power. The main purpose of breaking up and privatizing at least Ontario Hydro's non-nuclear assets is to achieve the conditions to kick start competition. As the authors warn, privatizers should not seek to maximize realized sale value by hobbling the system with high rates in future. Policy makers should not forget the future tax gains to government from the newly privatized entities, gains which can offset transition costs.

To make electricity competition a practical reality, policy makers need to apply sound economic principles while accommodating the physical operating requirements of complex electricity systems. A spot market for grid-connected power producers and consumers is required to accommodate the needs of short-run dispatch, real-time price discovery, and the clearing of sale/purchase contracts. As in other commodities, the development of a spot market will enable buyers and sellers to develop further financial instruments like future markets and options markets. Simple, bilateral contracts for physical delivery of power without a pool – so called "retail wheeling" – could sacrifice economic opportunities and risk the security of supply.[2]

Creating a deregulated commodity market for power must be coupled with new forms of open and public regulation. These include rate and service quality regulation of transmission entities, the pool operator, and local distribution companies. I do not share the authors' contentment with the status quo among MEUs, which I believe need much closer examination and regulation. Nuclear generating companies, public or private, must be subject to thorough, ongoing, and public regulation of their finances, rates, and safety performance. All conventional generators must be subject to evenhanded environmental regulation.

For a competitive system to flourish with a mix of public and private participants, publicly owned participants in the power business should have to pay the equivalent of their fair share of taxes and full financing costs, burdens they now escape from in whole or in part. Edmonton Power in Alberta is making this change.

Legal academics and practitioners have much to contribute to accelerating the restructuring of Ontario's power system and to lending relief to overcharged power users. As much as one-third of Ontario Hydro's hydraulic assets may be ill-gotten spoils, the underlying water and land rights may rightfully belong to indigenous peoples, not to the Crown and Ontario Hydro. The prevailing view of Ontario Hydro's hydraulic portfolio as being rock solid may be a mirage. Legal scholars might usefully address themselves to the strength of aboriginal property rights claims and what to do about disputes. The recent German experience with solving property rights disputes created by the former East German regime's confiscation of property offers a useful comparative model.

Another useful task for legal scholars would be to explore methods of privatizing municipal utilities and developing appropriate models of regulation. Privatization schemes for utilities must respect any property rights municipalities may have. Reliance on public ownership and the political voice of constituents for regulation has proven to be ineffective. In their current state, Ontario's municipal utilities are generally unfit to manage complex power contracts or converge with water and gas distributors – tasks that may be needed.

Practitioners might undertake to develop viable methods for power customers in Ontario to tap into the huge supply of cheap power available at Ontario's borders. Cornwall Electric, Ontario's best managed and most successful municipal utility, relies on extra-provincial suppliers.

The accomplishment by a single Ontario Hydro customer switching to a supplier from outside the province would do more to accelerate solutions to Ontario's power problems than a thousand consulting studies.

Another useful undertaking for practitioners would be to determine the onus on municipal utilities to secure lowest-cost power. Customers might have class action remedies against municipal utilities who are not duly diligent in securing minimum-cost power for their customers. Providing such a remedy would focus the minds of all managers of MEUs, too many of whom have been complacent in not providing their customers with better alternatives to Ontario Hydro's overpriced supplies.

NOTES

1 Stephen Littlechild, "Competition and Regulation in the U.K. Electricity Industry," presentation to the University of Toronto Electric Power Project, 7 April, 1995.

2 Larry F. Ruff, "Stop Wheeling and Start Dealing," *Electricity Journal* (June 1994).

DAVID CAMERON

Professors Daniels and Trebilcock present a sophisticated analysis. It offers a clear framework for considering a fundamental range of reform options. It is packed with useful information. And it even made me laugh, although not often. At one point the authors speak of the alleged "cognitive infirmities" of consumers. That phrase I have got to remember the next time I want to insult someone without them even knowing it.

I would like to offer two sets of comments which are generated by my reading of the paper. The first has to do with competition and the reform timetable. The second has to do with the role of the public in considering alternative directions for the electricity industry in Ontario.

Competition and the Reform Timetable

I observe that the notion of competition in the present debate is deployed in a number of different contexts, most often to refer to the activity carried on by independent, usually private-sector actors in a market place.

But there are two other contexts within which competition occurs that I would like to allude to briefly: competition between jurisdictions, for example, between Ontario and other provinces and American states; and competition within a single corporate structure, for example, within

Ontario Hydro. I have the impression that competition in the first sense has had a lot to do with the introduction of competition in the second sense.

For all practical purposes, competition between jurisdictions – what Stephen Littlechild has termed, too delicately, I think, "indirect competition by comparison" – is inevitable and unavoidable in today's world, given that large industrial users can plausibly threaten to leave the system and relocate elsewhere if energy costs are too high.

In response to these and other pressures, Ontario Hydro has engaged in a well-known round of cost-cutting measures and has introduced competition into its own corporate structure by creating three business units – nuclear, fossil fuel and hydroelectric – with responsibility for the generation of electricity for the system. These units now engage in competitive bidding via an internal Electricity Exchange to supply Hydro's electricity requirements through the transmission grid.

Does this innovation have any durable validity, or is it to be understood purely as a half-way house, preparatory to the full-scale dismembering of Hydro's generating assets into independent, competing actors? The authors note that the regulation of Ontario Hydro is *sui generis* and does not follow conventional modes of regulation in electricity or in other utilities. Perhaps Ontario is breaking new ground in modes of internal competition as well.

At any rate, Hydro's aggressive attack on its problems, together with the decline in the Canadian dollar, have fended off the forces of interjurisdictional competition, and the forecasts I have seen suggest that this situation is likely to prevail for several more years, possibly until the turn of the century.

This brings me to my second set of remarks.

Public Participation in Reforming the Electricity Industry

If the forecasts are correct, Ontario has won some time to sort out how it wishes to proceed and to introduce the changes it decides upon incrementally.

My point here is to raise the question of what role the "attentive public" should play in shaping the future direction of the electricity industry in Ontario. I do not assume that the public interest is synonymous with the interest of individual consumers.

These conference proceedings, and a host of industry task forces, colloquia, academic papers, stakeholder consultations, etc., make it clear that the policy community is aroused and paying rapt attention to the future evolution of the electricity industry.

Is the public? No, I don't think so. Once the rate increases were brought under control, Hydro receded from the centre of public attention. Will the attentive public wish to have a voice in setting the direction for the potentially radical restructuring of the electricity industry in the Province? I frankly don't know the answer to that question, but it seems to me it is a question the policy community would do well to ponder. The authors of the paper themselves note that "from its early history Ontario Hydro has rarely been far from the political spotlight."

When I consider the central role Ontario Hydro has played in the development of the Province, when I look at the list of special inquiries and legislative committees that have examined aspects of the electricity industry (the authors list 16 since 1960), when I reflect on the magnitude of the interests at stake, and when I think of the public's concern with nuclear power, I conclude that Ontarians might very well insist on having their voice attended to prior to any radical transformation in the electricity system in the Province. We have had lots of experience in this country recently with experts and politicians not getting it right so far as the public is concerned.

Professors Daniels and Trebilcock list the last major public inquiry as occurring in 1990. Given that what's being considered within the policy community is arguably the biggest change in the electricity industry since the creation of the current system in the early years of this century, perhaps it's time to bring the public back in yet again.

DAVID S. GOLDSMITH

I would like to open by expressing a concern: opinions and statistics about AMPCO are quoted without ever actually having spoken to us. While you have not materially misrepresented our position, I would like to point out that the media are a notoriously inaccurate source of information, particularly on a subject as complex as this.

To set the record straight, the Association of Major Power Consumers in Ontario, or AMPCO, represents the interests of electrical energy-intensive and large electricity-consuming industrial customers of Ontario Hydro and the municipal utilities in Ontario. Our members, over sixty of them at this time, come from a cross section of Ontario industry sectors including steel, pulp and paper, mining, chemicals, automotive, and abrasives. About 30% of our members are large users in municipal utilities; the balance are direct customers of Ontario Hydro. We purchase about 15% of Ontario Hydro's sales, which is about half of the industrial load in the province. Contrary to one of the notes in the paper, electricity costs represent 15% – 20% of production costs for most of our members, although the range is from about 8% to in excess of 60%. I would also emphasize that we pay bills, not rates. When your electricity bill is in the millions of dollars per month, as it is for many of our members, it is an important component of cost, regardless of the rate or the percentage of the total cost.

In terms of the imperatives facing the industrial customer in Ontario today, competitive rates have to be number one in electricity. The massive rate increases imposed by Ontario Hydro during the recession of the early nineties took Ontario from being one of the lowest-cost jurisdictions in North America in the mid-1980s to well above the mean. Ford Canada points out that only 25% of its costs vary by plant location, but electricity represents 25% of that number. Falconbridge, Ontario Hydro's largest customer, has found that relocating some of its facilities to Quebec would save it eight million dollars per year in electricity alone. This ability of industry to relocate is much more relevant and important than the authors allow. These statistics are given as background to my comments.

At first, I was afraid I was going to see a replay of Isaac Asimov's *Foundation Trilogy*, where historians of the future relied on reading history books rather than on viewing the real world to determine policy. I was very pleased as I read on to find what I would view as an excellent analysis of the options and rationales for change in Ontario.

Model 1, competing generation feeding a pool, is clearly a non-starter from the industrial perspective. Forcing all buyers to buy through a power pool will not allow industry to operate competitively. Although it is true, as stated in the paper, that large industrial users freed from the constraints of the current legal system would encourage low-cost suppliers to enter the system, this would only result in averaging down the cost of electricity. Much more rapid response to competitive forces is required in today's internationally competitive environment.

Given that Model 1 forms a basis for the other options, one additional point is worth making. The authors correctly point out that four or five competing generators are required to ensure a true competitive market, unlike the situation in the United Kingdom. One refinement to this concept is that there should, where feasible, be competition within a generation sector (hydraulic, fossil) as well. This would force production efficiencies on all generators, rather than, for example, insulating hydraulic generators from the need to become more competitive just because they have an inherent cost advantage over other types of generation.

Model 3, with competition for retail customers, introduces another level of regulation, a situation to be avoided since the government is no good at regulating anything. It is also a more complex situation, and will tend to be driven by a variety of forces, including government, the MEUs, and the people.

Model 2 is clearly the preferred option for the immediate future. As the authors point out, it will allow purchase agreements which correspond more closely to the quality and cost of required service. At the generation end, it will allow contracts which correspond more closely to the nature of the investment risks with different generating assets. This puts decision making in the hands of those spending the money – right where it should be.

We have made a couple of assumptions in reaching the conclusion that Model 2 is a necessary first step from an industrial perspective. First, there must be a change in who serves large users within MEU boundaries. For this system to function adequately, these users must have full and unhindered access to the generator of their choice as would the direct customers. The only difference would be the possible additional cost of local distribution.

Second, where regulation is necessary, we support the use of incentive regulation. As much as it has its problems, it is superior to rate-of-return regulation, provided the incentives are set properly (unlike in the United Kingdom) and provided allowance is made for system expansion when required.

Also, the authors have made some assumptions in defining Model 2 which we are not prepared to accept as givens. First, it is not clear that postage stamp transmission rates are the ideal option for Ontario. This is certainly not the basis upon which Ontario Hydro was founded. In its early days, "power at cost" was defined almost on a customer-by-customer basis. This may no longer be viable or desirable in the days of integrated power systems, but this does not mean that the other extreme is the necessary result. It should be noted that true power at cost was first compromised in the early 1920s with government subsidies for rural expansion, and finally discarded almost entirely in the post-depression days, when it was realized that it was counterproductive in a recession era to raise rates to recover costs from a dwindling customer base, thereby causing the base to dwindle even more rapidly. My, how history repeats itself... Even the OMEA, precursor to the MEA, said in a pamphlet published in 1922 that a uniform rate would be a distinct encroachment on and violation of the rights of the municipalities. The *Globe* stressed that Hydro's finances were so delicately balanced that movement toward a uniform rate would bring public ownership crashing down into bankruptcy. And the visionary chief engineer of Hydro warned that municipal ownership would succumb to provincial ownership, and Hydro would become just another government department.

Second, the authors have assumed continued public ownership of Gridco. I point out that oil and gas pipelines are natural monopolies which are privately owned. There is no reason not to sell the transmission system. It is probably worth much more than its book value, thereby allowing its sale to contribute to the reduction of Hydro's debt.

Third, we see no reason to continue the cross-subsidy for rural customers through the rate base. Just as Ontario Hydro should not be an instrument of industrial policy, it also should not be an instrument of social policy. These subsidies were historically provided by the government. However, if they are continued in rates, they should at least be on a consistent basis. By comparison, large users within municipal utilities are subsidized to allow them to pay the same rates (plus local distribution charges) as direct customers. This subsidy is paid by the direct customers i.e., by other customers taking the same class of service. In the same way, any subsidy for rural residential should be paid exclusively by urban residential customers, not industry.

Finally, we would support the authors' comments on the need for a detailed review of the Ontario Energy Board's current powers, competencies, and resources, with a view to changing its mandate to make it more sharply focused.

In summary, AMPCO is strongly in favour of an electrical industry restructuring along the lines of the Model 2 discussed above as a necessary first step for Ontario from an industrial perspective.

ARTHUR KROEGER

Ron Daniels and Michael Trebilcock have produced a sophisticated, carefully worked out analysis of the situation in Ontario's electricity sector.

Their contribution demonstrates both the need for change and the complexity of bringing it about. Neither is widely understood by the public at present. Unless a greater degree of public comprehension and "buy in" can be brought about, change may prove very difficult.

Among the direct participants in the sector – the industrial users, the municipal distributors, the independent power producers, the unions, and certainly on the part of Ontario Hydro itself – there is now a recognition that competition is here and is going to require a response sooner rather than later. There are differences among these groups about *what* response is called for, but there is general agreement among them that the status quo has a fairly short future.

AMPCO and IPPSO both want increased competition and, whether formally or informally, are inclined to favour some form of privatization. The municipal distributors also recognize the potential benefits of competition. They dislike privatization, and have proposed instead that Hydro's generating units be split off from the rest of the system and grouped into publicly owned entities that would compete with one another.

The unions are in a more difficult position because they correctly foresee that restructuring of whatever kind would likely have an adverse effect on their members. Nevertheless, the Power Workers in particular have recognized that change is going to come, and have demonstrated a readiness to see future options examined. If I have understood their position correctly, they have come to the view that in the medium to long term their best interests lie with an Ontario Hydro that is fully competitive with a solid financial structure, and they are therefore prepared to consider certain kinds of change that would move Hydro in this direction. I speak here of the leadership, recognizing that many of the members may be inclined to a more conservative approach.

In all of this there is cause for some encouragement. However, the state of knowledge and comprehension on the part of the public are another matter altogether. While electricity got a good deal of unwelcome attention in response to the three annual rate increases earlier in this decade, this has largely abated as a result of the subsequent freeze. We are now back to the traditional situation in which the public think of electricity only when they throw a switch and the lights fail to go on – something that is commendably rare in Ontario.

While it is true, as has been said, that the surgery on Hydro has been completed and the patient is now in the recovery room, this is only an interim situation. For example, industrial users are of the view that Hydro's rates are too high and must be reduced, not merely held at their present level. Given the importance of electricity for Ontario's competitiveness, this is a concern that has to be taken seriously – the more so because industry has the option of moving to other, more favourable jurisdictions.

Among the public, however, there is little awareness of the new technology, of the pressures for open access, of the provision in NAFTA that would give U.S. utilities the right to compete with Hydro and its successors if access were eventually permitted, or of the risk that some of Hydro's assets could be stranded. Yet all of these are current realities that cannot be put to one side for very long.

In public policy it is a good rule that if you want to get customers for your proposed solution, you first have to demonstrate that there is a problem. This is certainly true of any moves to restructure or privatize Ontario Hydro. Public ownership, "the people's power," and the like are still in the nature of icons for much of the population, and any moves to tamper with them are bound to generate controversy. Any provincial government, whatever their views, will need to look for ways of increasing public recognition of the problems faced by Hydro and of developing public support for such change as may be settled upon.

There are some elements in the picture that could make the government's job somewhat easier than it might otherwise be:

1. A survey by the Canadian Electrical Association found a significant degree of public willingness to switch to lower-cost suppliers if presented with the opportunity, together with an acknowledgement that private ownership could result in lower prices and improved service.
2. There is a good deal of public disillusion about government promotion of industrial development; too often, the outcome in one part of the country after another has been high costs and disappointing results. Consequently, there may be less resistance than in the past to measures that would reduce the ability of governments to use Hydro to pursue particular industrial objectives.
3. How much pressure the government comes under to undertake drastic changes will depend in some degree on the general business climate. In assessing Ontario's attractiveness as a place to be in business, industries take into account a number of factors in addition to electricity rates: the tax regime, the state of labour relations, environmental regulation, and transportation infrastructure, to name a few. To the extent that the Ontario government takes steps in these areas that could be regarded as favourable by industry, it could buy itself some additional time in which Hydro's debt-equity ratio could be improved, and options for functioning in a competitive world could be carefully assessed.

The analysis in this chapter represents an important contribution to any such assessment. What will also be required, however, is some process to increase public comprehension of the need for significant change in the electricity sector. How well this is done will have an important bearing on the political price that will have to be paid for measures to position Hydro to hold its own in a competitive world.

KARL WAHL

As a Municipal Electric Utility, some might expect us to be opposed to the recommendations in this paper and might even expect us to be fearful of the new impending world of competition and the threat that it brings to our very survival.

Those who are familiar with Hydro Mississauga will not be surprised to hear that we are in agreement with most of the views and opinions expressed. We not only accept impending competition but welcome it and will embrace it when it comes.

Present Situation

As a utility with high reliability, no debt, and the lowest rates and margins in the GTA, Hydro Mississauga customers are well served. We got in this position by making tough decisions, and this situation is not good enough for us. We certainly will not become complacent or support the MEA position that the status quo is satisfactory.

We are completely customer focused. Until our customers have the very best available in terms of choice, price, flexibility, and value for their electricity dollar, we will support any effort to revise the regulatory environment in which we operate.

The industry structure that served Ontario well for many years no longer works. Our one time price advantage over the U.S. and other provinces is gone. We see the necessity for a major effort to work with our customers and our supplier (Ontario Hydro) to obtain more cost effective rates, an effort that must overcome numerous barriers in today's environment. There are regulatory barriers that only competition can overcome.

Customer Demands/Service

Change in regulated industries begins before regulatory decree. It begins when customers perceive a need for new choices or control and is usually manifested as interest in a new service or product. The threshold question is: Who will meet the market demand, the traditional supplier or someone else?

Customers want choices – product choices, service choices, and price options, and companies must offer choices or risk losing market share. Unless customers receive the choices they demand, the utility's market share can only go down.

Large commercial and industrial customers will likely be the single largest battlefield in a retail access environment. Price pressure to these most desirable customers will be fierce, and electricity will be priced at levels approaching the utility's variable cost structure. Programs and options demanded by the largest customers will be echoed by other customers. Successful utilities of the future will organize around their customer base, with the entire organization focused on marketing and emerging as the focal point of the company.

Competition

Competition drives prices toward marginal cost. Competition will establish electric power generation as a separate industry with new low-cost entrants seeking to earn fair returns. Competition will have the effect of forcing electric rates down towards the cost of the most efficient producer in a particular geographic region. Rates will gravitate towards the cost at which electricity could theoretically be supplied by a new competitor. Utilities will need to cut costs to a level where they can price to beat the competition and stay within acceptable levels for their customers. At the end of the day, every cost is variable.

Pricing

In the future, more and more of our business will be conducted on a contractual basis. Term structures of contracts will become a principal area of focus in matching supply with the requirements of the market. Flexibility is a key to a utility's ability to compete. The time necessary to get the "competition" rates approval by a regulator must be eliminated or at least greatly reduced. In the future, rates will evolve from being totally cost based to being primarily market driven.

Access to Better Pricing

The needs of customers and their expectations are rising. Customers will increasingly respond to competitive market pricing. At Hydro Mississauga, we operate with an 8% gross margin as compared to an industry average of about 17%. We consistently have the lowest rates in the GTA, but there is no level playing field. A well-managed utility can offer its customers a significant price advantage even now when we all pay the same cost for power. But at this stage we can only offer our customers an advantage based on our own local distribution efficiency. This is not good enough.

We want to keep working with our supplier to reduce the cost of raw material. We need to serve our customers. Raw material represents 92 cents of every dollar that our customers pay. We must reduce this for the benefit of our customers. And why should we not be able to do something?

Our cost of power is $75 million and we represent approximately 5% of Ontario Hydro's load. We are one of the top three customers of Ontario Hydro and would like to be viewed as the major customer that we are. We are Ontario Hydro's customer and Mississauga customers are Hydro Mississauga's customers. Hydro Mississauga needs pricing options to retain our large users.

Should we not obtain concessions on pricing, we will have to ask for open access to supply from other providers that will offer better value to our customers.

The final alternative, which we really don't want to do unless pushed, is to invest new capital for generation of electricity right here in Mississauga and not be dependent on the Ontario Hydro supply system.

Open Access/Wheeling

Open access to transmission is not new. It is simply a restatement of customer demands for improved service at lower cost at a time when the technology exists to satisfy the demand. The question is whether the customer's traditional energy supplier will meet the service and price demand or whether a competitor will.

Technology permits open access, customers demand it, and the economics encourage it. Regulatory directives simply increase or decrease the speed of the inevitable; open access defines what marginal costs should be. Ontario Hydro has excess capacity but if they don't price it right, the market will.

Ontario Hydro

Under the provisions of the Power Corporation Act, under which Ontario Hydro is constituted, Hydro is required to provide "power at cost." We say that power should no longer be at cost but at competitive cost, recognizing the economic realities of the market place:

Ontario Hydro sets wholesale rates for the supply of electricity to the MEUs and regulates retail rates that may be charged by the MEUs.

We want to set our own rates:

Ontario Hydro possesses substantial excess capacity. We want to use that excess capacity and work with our supplier to arrive at the most effective rates for each of our customers.

Incremental Pricing will make this happen:

Let us have access to other power sources that will price incrementally. If Ontario Hydro won't price incrementally or let us have this access, we will put in our own generation to get better prices for our customers. We see this as a terrible waste of capital and would only do this as a last resort.

Municipal Electric Association

"The MEA opposes any form of retail competition and will not support any form of privatization of generating facilities, proposing instead a 'managed competition' model." We say – No way.

"MEA contemplates state-owned enterprises competing against each other. This strikes us as a totally unrealistic competitive scenario." We agree.

"There is no precedent of vigorous competition among state-owned enterprises."

We already see a bargain between the MEUs and Ontario Hydro, where the MEUs act as political advocates for insulating Hydro from substantial upstream competition at the generation level, in return for Hydro agreeing to support policies that will insulate the MEUs from significant competition at the retail level.

Privatization

"The evidence suggests that the economic performance of publicly owned monopolies and heavily regulated private monopolies is typically not sharply different." We say – Right on. Competitive pressure will keep prices in line.

Hydro Mississauga favours Privatization. We have in fact already studied proposals for the privatization of Hydro Mississauga. As you can appreciate, many laws need to be rewritten before such an undertaking can be accomplished by a municipal utility.

Critical Mass

An approach to competitive viability is to achieve "critical mass." This is the process by which companies adjust their size to attain a new, larger "mass." Critical mass does not mean that size alone will permit a company to survive.

The key aspect of critical mass is that it provides an opportunity to reduce unit cost of service through operating and financial synergies, thereby providing a company with the flexibility to compete by offering better value to customers. Being a low-cost operator will be a key to survival.

The pace utilities are moving toward critical mass through mergers is quickening in the U.S. and should not be avoided in Ontario. The prospect of consolidations of distribution in Ontario is not radical but reflects the results of continually evolving economic, technological, and customer-based realities. Under this critical mass scenario, small companies become medium-sized and medium-sized companies become large. The largest companies become the strategic providers in the industry with the smaller ones becoming niche providers. The smaller companies also may be relegated to serving the less accessible or smaller market share areas.

Critical mass will largely determine whether each company is the acquirer or the acquiree.

"Almost certainly the MEU structure is inefficient – there are far too many of them and many are too small to function effectively, particularly under a bilateral contracting model and even more so under a retail competition model."

We are in total agreement. There are far too many MEUs to fully realize economies of scale and scope. Let us rationalize in sizes. Hydro Mississauga has recommended that Ontario could be well served with eight utilities with no central control.

The GTA study that recommended one utility for the entire GTA in order to reduce operating costs by 25% and save $95 million per year was authored by Hydro Mississauga. Most of the bureaucracies of the 312 Municipal Utilities throughout the province are an endangered species and we may as well face that fact now. Set tough targets, find the structure that can achieve them, and shed the excess cost from the distribution system.

Regulation

The current regulatory approach has fostered a situation in which utilities do not behave competitively. Power at a cost-plus mentality has the effect of greatly diminishing the incentive to be efficient. First, the regulators should ensure that any changes are fair: fairness may include reducing rates for all customers, not only those with bypass or bilateral options. Without rate reductions, small consumers will not perceive the industry changes as being fair. Second, if regulators are committed to promoting "choice" for customers, then such choice must be meaningful choice for small customers. Third, it is essential that regulators place a high priority upon protecting captive customers from cross-subsidies and stranded costs, including those costs attributable to the introduction of competition into other customer classes and industry restructuring as a whole. Fourth, regulators should direct their efforts to use structural changes to encourage greater efficiencies in utilities' non-competitive utility activities.

We were told by Ontario Hydro that some Ontario Hydro pricing options were not available to Hydro Mississauga due to the fact that we are an MEU. Why does Ontario Hydro favour their direct customers over Hydro Mississauga? Why does Ontario Hydro continue to offer pricing options to Ontario Hydro directs without allowing Hydro Mississauga the same access to these options Why can't we have access to Ontario Hydro's direct customers' pricing options so that we can distribute and price to our customers according to the economics of the marketplace?

Conclusion

We agree with the conclusion that the Ontario Electricity industry should aspire, over the medium term, to the progressive adoption of Model 3, but in the short term adopting model 2, accompanied by an announced phase-in of retail competition.

Certain stakeholder groups will be opposed to the scheme. Hydro Mississauga is not one of them. We agree that structural changes are required at the distribution or retail level that will turn it into a strong proponent on behalf of final consumers of greater competition, more efficiency, and lower generation costs. The MEU structure in Ontario must be rationalized. Competition is the foundation of our economic system, and the sooner a utility recognizes that the natural monopoly marketplace is dying, the sooner that utility will be positioned to be successful.

Resisting inevitable market forces only results in faster loss of market share and ultimately leads to the failure of the business. Competition for customers means that it is no longer feasible for some classes of customers to subsidize other classes of customers if the traditional utility is to compete with other suppliers entering the market. As the natural gas industry has done, so too the electric utility will unbundle rates and competitively market generation, transmission, and distribution and create value-added customer services. Accept the future – it is competition.

For those who see this as an opportunity, success is more likely than for those who bank on the status quo. To survive, the twenty-first century utility must take control over its own destiny. Being a commodity service provider always has a limited life when customers and competition demand change. Our customers demand competitive pricing. We want open access to competitively priced power on behalf of our customers and to avoid having to install additional capacity.

If there is to be any retention or creation of jobs in Ontario in the emerging trade balances of the global economy, we must pursue a strategy that will develop the lowest, most competitive cost in the entire electricity system that serves our economy. Not only is the future of our industry at stake, but also the quality of life and standard of living in Ontario.

Regulation of Transmission and Distribution Activities of Ontario Hydro

LAURENCE BOOTH
PAUL HALPERN

I. INTRODUCTION

Ontario Hydro has been under increasing pressure over the last four years as its plan for aggressive facilities expansion has coincided with both a stabilisation in the demand for electricity and increasing concerns over the impact of Hydro's borrowing on the province's debt capacity, as Ontario's DBRS bond rating has declined from AA in 1990 to A(High) in 1993.[1] Until very recently Hydro worked on a "bundled" service basis without explicit prices for the different services provided by its constituent parts. However, without the signals provided by those prices, management behaviour was not necessarily consistent with either efficient cost management or the efficient provision of electricity services. In part, this resulted in the rapid escalation of electricity prices in the late '80s and early '90s that were the trigger for the Hydro "crisis."

The crisis has had several effects. Internally it lead to a dramatic restructuring in 1993-4 within Hydro, both with the establishment of the "Electricity Exchange" as an internal entity to act as a surrogate for competition, and the increased separation of generating and transmission functions into separate business units. The explicit goal of this restructuring was to provide the right internal signals to generate business incentives and "to prepare Ontario Hydro for a more competitive market in an evolving industry."[2] Externally, it has led to increasing concern as to whether or not the legal framework that governs the electricity market in Ontario remains optimal. For example, the Municipal Electric Association has recently completed its own appraisal of the options open for restructuring the electricity industry in Ontario.[3]

These modifications and suggested alterations in structure are being considered within the context of major changes occurring in the electricity market. It has become clear that the minimum efficient scale for generating electricity is no longer the compelling argument for the ever-increasing size of generating plants, as in the past. Many countries have introduced bold new experiments in restructuring electricity markets, privatizing some or all of the restructured entities, and introducing competition in generation. Regulation, which at one time was focused on protection from monopolistic power, has instead begun to focus on the promotion of competition.

Associated with the above changes, and really a result of them, has been the evolution of markets in which electricity can be priced separately from distribution charges. Underlying these changes has been the dual attempt to obtain both the benefits of competition on costs and efficiency, and the use of the correct signals to generate the optimal short- and long-run decisions concerning electricity generation, consumption, and locational choice. These changes have not been without problems, which are currently being addressed by regulators. However, the structures, problems, and solutions found in other countries are useful information for the Ontario debate.

These broad questions of the optimal structure for the electricity market in Ontario is the focus of this chapter. In both Hydro's reforms and the MEA appraisal of the options there is the evident constraint imposed by history and the existing legal and regulatory framework. In our opinion neither of these factors should constrain the initial establishment of options. To do so is to close off potential solutions prior to their considered evaluation. The present objective is therefore:

- to consider the overall form of regulation
- to consider the correct introduction of incentives
- to determine the basis for prices
- to consider connection and exit charges.

We also assume a specific structure for the electricity industry. Given the changes in structure and ownership observed in many other countries, our assumed structure is a minor variation on other systems. There will be three groups in the system: generating companies, the transmission grid, and the distribution companies referred to here as the Local Distribution Companies (LDCs). We assume that there is no cross-ownership of companies within the three groups. This assumption minimizes the potential for wealth transfers through transfer pricing strategies and limits the purview of any regulatory authority. Generation and transmission will be privatized.

Implicit in the assumption of an independent generating sector is that all generation will be privatized with all generating modes, including nuclear, owned by investors. The sale of the nuclear generators to private investors is counter to the experience in the United Kingdom. While it is expected that the nuclear plants could be sold off separately in a privatization, the price received, given the environmental risks and expected costs of maintenance, may be very low without some residual "insurance" provided by the province. The assumption of the privatization of nuclear is not crucial to the analysis. In addition, the structure of the resulting privatized generating industry is not specified. Clearly it must be competitive. There has been some criticism of the structure in Britain with its two major non-nuclear generating companies and the resulting cooperative behaviour and abuse of the process. Our preference is to have a large number of companies and ease of entry to establish potential competition.

Finally we do not address the structure of the retail end of the market or its regulation; suffice it to say that distribution is likely to have economies of scale and the monopoly elements require some form of regulation. Since we are not addressing the distribution issue at the local level, we do not impose privatization on the LDCs. However, this may be the preferred solution given that this part of the industry will be regulated either through traditional base rate of return or price cap regulation. Under the existing structure in Ontario, the LDCs are regulated by Ontario Hydro, but the form of regulation is not well specified and neither is the objective of the regulation. Clearly, under a restructured system, regulation will have to be undertaken by an independent, economically sophisticated regulatory authority.

In section II we consider the evolution of the natural gas market in Canada, since not only is natural gas a major competitor fuel, but it has also gone through many of the major changes now faced by Hydro and thus provides a good basis for the subsequent analysis. Section III then considers the major technological differences between the electricity and natural gas markets and the types of structural changes that these introduce. In section IV we compare our proposed structure to that presented by W. Hogan in a number of papers and presentations. Section V gives our conclusions as to an appropriate industry structure.

II. EVOLUTION OF THE NATURAL GAS MARKET

The natural gas market is dominated by two critical factors: natural gas production occurs in defined distant supply basins, while demand is geographically less concentrated, and that transmission and distribution of natural gas through pipelines is a natural monopoly. By natural monopoly we use the standard definition that at a particular point in time the average cost curve is decreasing throughout the relevant range. As a result, duplication of service produces higher average costs than does an efficiently operating single pipeline. For this reason, the National Energy Board has licensed individual Canadian pipelines to connect the Western Canadian Sedimentary Basin (WCSB) to different markets, even when "competing" pipelines have initially requested permission to build. The local distribution companies (LDCs) then distribute the natural gas from the mainline transmission pipelines to individual customers.[4]

One of the features of the natural gas industry is that natural gas production is largely in private hands and unregulated as regards profitability and supply. Government intervention has been largely based on macroeconomic "national interest" grounds, rather than the typical microeconomic "public interest" grounds for intervening in natural monopolies. The reason why the production of natural gas is in private hands is that there are no significant economies of scale, either in the exploration or development of natural gas, beyond a point that allows for the operation of many individual companies. Although PetroCanada was created as a government "window" on the industry, no information has subsequently become available indicating the existence of a natural monopoly.[5]

The production of natural gas is initially distributed on "sour gas" lines to gas processing plants and then as system gas to feeder pipelines to hook up to the main transmission line. In Alberta, the sour gas lines and processor plants are in private hands, while most of the feeder lines are owned by Nova Gas Transmission, which also handles the transmission of all Alberta gas out of the province. In British Columbia, Westcoast owns the sour gas lines, the gas processing plants, the feeder lines and the main transmission line. All of Westcoast's system is regulated by the NEB.

The distribution of system gas from the WCSB is initially through the mainline transmission pipelines. TransCanada Pipelines (TCPL) takes the gas east to Eastern Canada, with Trans Quebec and Maritime Pipelines (TQM) on to Quebec, Westcoast Pipelines takes the gas to B.C. and the Pacific Northwest and Foothills and Alberta Natural Gas takes the gas to the U.S. border for delivery to California. The final leg of the distribution of natural gas is via the local distribution companies. In Ontario these companies are Centra Gas Ontario, Union Gas and Consumers Gas. These LDCs interconnect with TCPL and some U.S. pipelines with Union Gas also operating a critical part of both the transmission system and the connection to the main storage facilities in southwestern Ontario. The LDCs are regulated by the Ontario Energy Board, since they too are natural monopolies.

Prior to 1985 the services of the pipeline and distribution companies were bundled into a single commodity charge for natural gas. Natural gas in the WCSB was aggregated by long-term contracts with TCPL and sold to the LDCs, who then sold it to individual customers. After the Western Accord, natural gas prices were effectively deregulated as of November 1, 1986, and the functions of both the transmission and distribution companies reorganised, since the commodity price for natural gas became separable from the storage and transmission services. TCPL created Western Gas Marketing Ltd. (WGML) to take over its aggregator business and its regulated mainline became a pure transmission pipeline.

The operation of these pipelines is almost totally via fixed costs. As a result the National Energy Board now determines pipeline tolls in a way to recover fully these costs on a straight fixed/variable tariff (SFV) with the fixed toll covering about 98% of the total costs, and with a very small commodity toll to cover compression and some other incidental costs.[6] The NEB allows significant system facilities expansion only after hearings demonstrate that enough shippers have signed up capacity to justify the expansion. Normally, the transmission pipelines will not advocate expansion unless they have long-term contracts backstopping the expansion, and have depreciation rates tied to the economic recovery of natural gas. This is a sensible strategy for risk sharing, since the pipeline runs the risk of losing a customer and hence the possibility of "stranded assets" if they are unable to raise sufficient revenues to pay their capital costs.

With costs now largely recovered on a fixed toll basis, this unbundling of natural gas prices has exposed the impact of low average loads on the different LDCs. Loads are a critical factor in natural gas markets, since a significant component of demand is the heat sensitive residential and commercial sector. Given the seasonality of consumption and the maximum capacity of the pipeline, the SFV toll makes it expensive to meet seasonal demands through fluctuating throughput. This is because the capacity demand charges have to be met even in the summer period, when poor loads mean that little gas is being shipped.

The unbundling of natural gas prices has therefore meant the development of separate charges for natural gas itself as well as for the derived transmission and storage services. The result has been a dramatic increase in the development of storage services by Union Gas to sell to other LDCs, as well as the purchase of Tecumseh Gas Storage by Consumers Gas. Seasonal peaks in Ontario demand can therefore be partly met from storage. As a result, the average loads on TCPL and Union's system have increased, as the cost of storage is less than the cost of spare capacity on the TCPL mainline.

In areas where storage in the geographical market centre is very expensive, such as the lower B.C. mainland, peak seasonal demands can not be economically met through storage. As a result, peak seasonal demands are met by fluctuations in production and storage in the WCSB. Therefore, average loads on the transmission pipeline to the lower B.C. mainland, are relatively poor. Simply put, the costs of spare capacity on the mainline, in this case Westcoast, plus the upstream storage and production capacity is less than the cost of downstream storage. For the lower B.C. mainland this makes sense primarily because of its proximity to the main producing areas, a situation that is in stark contrast to Ontario, which is a considerable distance from the WCSB.

The important point, however, is that the unbundling of natural gas prices has allowed for a more efficient overall distribution system. This overall distribution system is made up of an integrated transmission and distribution system plus a mechanism for handling seasonal and peak demands. The tradeoffs between upstream and downstream storage and pipeline capacity is then made on grounds of economic efficiency.

The unbundling of natural gas prices has therefore promoted restructuring on the supply side of the natural gas market. It has also had some impact on the demand side. Initially, unbundling simply meant the availability of direct purchase from WCSB producers. This was implemented either as the purchase of gas and its subsequent sale and repurchase to the gas LDC, or by the purchase of Transmission ("T") service on the mainline and LDC. At certain times, this was an extremely attractive proposition, given differences between spot gas prices and the price available from the LDC. This has also revealed implicit subsidies in some LDC rate structures, as some companies close to the TCPL mainline have threatened to bypass the LDC altogether. This has led to the development of special "bypass" rates for some large industrial users which provide some contribution to the fixed costs of the system.

Some have speculated that in the future the development of real time metering from the wellhead to the residential burnertip will result in day-to-day and even hour-to-hour prices.[7] This would allow the complete unbundling of the natural gas market, so that both supply and demand sides of the market can reflect the seasonal and time value of natural gas.

III. THE ELECTRICITY MARKET IN ONTARIO

The previous comments on the natural gas market are made to point out the similarities between the two industries. The electricity market in Ontario is still in the same "unbundled" state as the natural gas market prior to 1986, in that the price paid for electricity is an average price, based on simple seasonal and peak load classifications, while the consumer has little or no choice of supply or contracting relationships. Like natural gas, electricity is a commodity, since whether generated from nuclear, coal fired or hydro facilities, it is identical as far as the consumer is concerned.

Electricity generation and distribution has been bundled primarily because both parts of the industry were considered to be natural monopolies. However, the apparent natural monopoly of the grid arises from the large minimum efficient scale for a generating plant. If the natural monopoly in generation were to disappear and a fully distributed transmission system were to evolve, the investment in the grid itself could be stranded, since the grid itself does not have any long-run natural monopoly elements. In fact, the main structural change driving the industry has been the decrease in the minimum efficient scale for a generating plant. Compared to just twenty years ago, there no longer seem to be the obvious economies of scale that cause electricity *generation* to be a natural monopoly. Hence, we take as given the observation that there are enough generating stations in Ontario that the generating side of Ontario Hydro can be privatised.[8] In turn, this implies that the economics of electricity generation no longer result in decreasing average costs. This explicitly removes one of the main factors requiring complicated tariff structures.[9]

With similar commodity products and potentially competitive supply, the main differences between natural gas and electricity markets are the transmission and distribution systems. Here, two main features differentiate the two markets; *the absence of storage* and *unique supply pools*.

Storage plays an increasingly important role in natural gas markets, but as noted, some markets, such as the lower B.C. mainland, have very little storage. In this case, demand variability is largely met by load variability and increased deliveries from the supply basin, both from storage and from extra production. The electricity market, with instantaneous equalisation of supply and demand from the grid, therefore differs in degree, but not in substance from the types of results in natural gas markets. In electricity it is *normal* to meet seasonal and peaking demands from additional supply and load variations, whereas in natural gas this is relatively rare with expensive transmission capacity.

The second feature is that Canadian natural gas comes from one main supply pool, the WCSB. Hence, natural gas consumption in Ontario of necessity involves very expensive transmission charges, which motivates the demand for storage. In contrast, electricity in Ontario is generated in several different locations. Some of these locations are natural (hydro), some are the result of other economic decisions, such as cogeneration and proximity to the TCPL mainline, and some are political (perhaps nuclear). Moreover, the location of these generating plants is most likely not optimal, in the sense that they are not where they would be if the system had been designed from scratch with unbundled pricing in mind. For our purposes we take these locations as given and treat them as fixed locations of supply, similar to gas fields. The result is a *partially distributed* supply system, as compared to a fully distributed supply system when generation is market-centred, or a non-distributed system as in natural gas.

Given these locations, Hydro has constructed an integrated grid that distributes electricity generated to all the local distribution companies (LDCs)[10] at an average pooled price. This grid has certain facilities that are "generating rich," since they are designed to take electricity from remote generating facilities, for example Northern Ontario, to market centres; others are "load rich," since they are designed to absorb electricity in market centres. The system is also affected by the interfaces with U.S. transmission services that differentiate the design of the system in Southern Ontario from that in the North. Hence, the grid is designed with a specific supply/demand structure in mind, and in the short run is relatively inflexible.

The technical integration of the grid with generation poses unique problems, not apparent with the deregulation of natural gas markets. With the separation of transmission from natural gas sales, the physical transmission of natural gas remained essentially unchanged. While new markets have prompted system expansion, natural gas still flows, for example, from the WCSB to Ontario. With a similar opening up of the electricity market, the "flows" on the system are potentially much more complicated and additional relocation of demand and supply may result. These changes can occur both in the short and long run as demand and generation shift in response to pricing signals and the 'flows' on the system are altered. This requires technical support for a competitive electricity market not required to the same degree in natural gas markets.

With these comments in mind we will consider the electricity market under two specific sets of assumptions. The first scenario will assume that congestion on the grid is not a factor and that the grid is a passive player in the electricity market, in that supply can reach all potential markets in amounts sufficient to satisfy demand.[11] The second scenario will assume that congestion exists and that the input of the grid or the Electricity Exchange is necessary to ensure that potential transactions are feasible.

No Congestion

By assuming that the grid is a passive player, direct intermediation between generating and market players is possible on a decentralised basis. The obligation to serve is assumed to exist with the LDCs, who are the front line representatives of the electricity market for the small user, who is at most risk in both a monopolistic market and an emerging complex competitive one. The LDCs will then contract with particular generating plants for the provision of electricity. These contracts can consist of bundled contracts or more likely unbundled contracts for specialised services.

A key feature of the emerging competitive nature of electricity supply is the competitiveness of alternative generating facilities. These broadly consist of base load generators, such as nuclear and hydro, with very large fixed costs and very low short-run marginal or variable costs. In contrast, peak load generators, such as coal fired plants, have relatively low fixed costs and high marginal or variable costs. Non-Utility Generators (NUGs) are primarily cogenerators producing electricity and steam by burning natural gas, and they occupy a middle ground. The different short-run average and marginal cost structures of the different plants make them suitable for different services.

With the privatisation of generating plants in this scenario, we see no problem with the complete unbundling of the market for electricity services. Base load contracts can be purchased by LDCs directly from hydro, cogeneration, or nuclear facilities. Peaking supplies can then be purchased from coal fired plants or elsewhere. Each LDC will contract for a portfolio of electricity services which depend upon its demand characteristics and its risk preferences. Residual supplies can then be sold through a spot market, and be sold for either base load or peaking requirements. These residual supplies can be supplies not contracted for as base load or peaking supplies of electricity that is currently not required under those contracts. Currently, Hydro has reorganised its business operations and transfer pricing around the above contracts.[12] The Hydro reorganization has the Exchange signing contracts on behalf of consumers. However, there is no reason why LDCs and direct purchasers can not sign their own contracts on the demand side and independent generators their own contracts on the supply side.

The key issue in the above scenario is dispatch and the operational implementation of the system. As currently envisaged by Hydro, it is the Electricity Exchange that is the "transmission system operator" (TSO)[13] with the following specific responsibilities:

- negotiate and administer contracts on behalf of the grid, generators, and international customers
- ensure operation of an integrated system
- schedule and economic dispatch according to contractual obligations
- maintain a spot market for electricity
- maintain the information system for billing, forecasts, planning, and management of the system.[14]

While most of the "contracting" functions can be devolved onto the LDCs and independent generators, what will still be required are the "market-making" functions similar to those of, for example, a futures exchange. When someone exercises a futures contract to take delivery, it is the futures exchange that ensures delivery; the actual supplier is normally a matter of indifference. Similarly, with a privatised electricity market such as the one envisaged above these same functions will have to be performed. For convenience we will describe such an entity as the Exchange. In principle, the Exchange can perform the market-making function as well as the actual operation of the grid.

The generators and LDCs can come to a contractual understanding for base load and peaking supplies and then provide the Exchange with a standard description of the technical requirements and contract prices. The Exchange will then provide summary data on contracted capacity and prices in statistical schedules available for all parties. In this way price and quantity data will be readily available in an anonymous format, and market participants can observe market prices for spot delivery, base load supplies and peaking supplies for different points in time. The result would be a "term structure" of electricity prices differentiated according to base or peak load characteristics.[15] A complete market for electricity supply can therefore develop to foster the efficient production and use of electricity. In this way, the Exchange will become the centre for the commercial side of the electricity market in Ontario.

With our assumption that the grid is a passive participant, all areas of the province are effectively integrated and there is a single electricity market in Ontario. All purchasers can then be regarded as facing the same commodity price for electricity. Competitive pressures would then lead to transparent market prices, and differences across LDCs would result solely from contracting differences; for example, one LDC's reliance on spot markets versus another's reliance on long-term contracts.

The "twin" to the Exchange's commercial operations is the dispatch and scheduling services of the TSO, which ensure that contract terms are upheld and, if not, supply the technical information required for compensation clauses in contracts to be activated. The dispatch function is required, since LDCs will simply draw more electricity from the grid to meet their peak demands, indicating that these supplies are from the spot market or peaking contracts. The information then has to be given to particular generators to fulfil their contracts. This service is presently being done as part of Hydro's integrated operations; the only difference is that specific generators, as determined by prior contracts, would be required to generate, rather than the TSO's view of the least-cost economic dispatch. That is, commercial contracting decisions would take precedence over the current centralised decision making. It, therefore, requires that more detailed records be kept for individual generators and LDCs, since their commercial profitability will depend on the information.

Note, however, that economic dispatch should still occur in our scenario. First, the LDC, in choosing additional supplies, always has the choice between activating a peaking contract or going into the spot market. If there is more efficient spot supply available, the LDC will simply purchase that instead of relying on the option contract with one of its peaking suppliers. Alternatively, even should the LDC activate a contract with an inefficient peaking supplier, that supplier can always make good on its contract by purchasing in the spot market, rather than producing itself. It would then of course pocket the difference between the spot price and the contract price from the LDC. As a result, efficient dispatch in a decentralised market should be the same as centralised dispatch in a pool market.

Similar to the current situation, the TSO would also be required to adjust for unscheduled system flows and losses by buying and selling electricity in the spot market. The balance of these transactions would be allocated to the grid and included as part of the grid's costs to be recovered by the fixed toll. We will address the correct regulatory treatment of the grid later.

In our judgement, the Exchange and TSO can be either separate entities or a single unit internal or external to the grid. Like the grid, the cost of the exchange will then be largely of a fixed cost nature and recovered by a fixed charge on all *purchasers* of electricity. In this way, the costs would be treated similarly to regulatory costs for rate-of-return regulated utilities.

The key differences between our model and Hydro's current plan are the unbundling of the buy side of the market and the abandonment of average pooled prices. In our judgement there is no compelling economic reason why all consumers in Ontario in the same rate class should pay the same electricity rate. Such a solution represents a denial of the value of explicit market prices in allocating electricity and generating and transmission facilities. Electricity is no longer an innovative commodity that requires government involvement to stimulate markets.

Individual LDCs should contract for electricity and be responsible for the management of their purchases of electricity. Individual ingenuity and good management will then lead to imaginative load management schemes and rate structures that can lower costs in particular jurisdictions and lead to healthy competition to produce lower electricity rates in different parts of the province. This will also open up LDCs to more competitive pressures, since their performance can be benchmarked relative to the performance of other LDCs.

The key problems engendered by our recommendation are twofold: bypass and stranded assets. We have assumed that the grid is a natural monopoly. If this is true, then no one can economically bypass the grid and set up their direct connection to a generator. However, if generation is no longer a natural monopoly, it may be economic to locate new generation facilities closer to markets and avoid the transmission costs entirely. This is already the case for some small cogeneration facilities. This might lead to increased pressures for a fully distributed system with complete local generation and the bypass of the grid. This would strand not only distant generation facilities, but also the investment in the grid itself.

One way of avoiding this problem would be to ban bypass of the grid. However, we are uncomfortable with restricting individual choice in this manner, since it means suppressing market prices and removing their important signalling function. More fundamentally, we are also somewhat sceptical of the danger of bypass in the first place. If a fully operating market in electricity is developed with explicit prices, then the charge for electricity will consist of the fixed charge for the grid and exchange, plus a charge for electricity, which itself may consist of a fixed capacity charge and a variable commodity charge. In a competitive market the generating costs will, by definition, be competitive. The only reason for bypass would therefore be the fixed Grid charge.

Hydro currently operates a five-zone postage stamp toll rate structure. This offers some economic incentive to make the location decision on non-economic grounds. By either locating close to a generator or locating a generator close to a plant, this decision can be overturned. However, in almost all cases access to the grid will still be required for peaking and reserves. In our judgement such cases will be relatively rare and can be treated in an analogous manner to those in natural gas markets by the setting of special "bypass" rates. These special rate classes make it uneconomic for individual users to bypass the grid, and yet still provide incremental revenues to cover the grid's fixed costs. These rates are also consistent with "Ramsey prices" and the inverse elasticities rule, since bypass implies high-price elasticity of demand and a relatively small distortion from marginal cost pricing.

A more normal motivation for bypass is to access lower generation costs. In our framework such a motivation does not exist, since generation is competitively priced. As a result, we do not believe bypass to be a significant problem once the generating business is opened up to competition. However, in our model there could also be significant financial losses in the system. For example, nuclear plants have very low short-run marginal costs, and it is inconceivable that they would not operate at optimum load in an efficiently functioning electricity market in Ontario. This is because they can undercut almost all other base load suppliers, based on their short-run marginal cost function. The problem is that the revenues obtained by the nuclear plants may not be sufficient to cover their fixed financial charges, based on the debt used to finance their construction.

Given the difficulties experienced in the U.K. with the privatising of nuclear generating plants at a time when competitive markets were being introduced, it may be several years before nuclear plants can be privatised in Canada. This is because investors may want to see the behaviour of prices in competitive electricity markets, prior to assessing the debt capacity and market value of the nuclear plants. However, assuming that only a portion of Ontario Hydro's debt can be offloaded to the generating and transmission assets, the question is what to do with the residual debt?

The "Catch 22" is that rolling-in the costs of the debt that cannot be offloaded onto a privatised competitive system raises system costs. This in turn encourages bypass and the development of even cheaper alternative capacity. On the other hand, going to a competitive market avoids bypass and the development of extra capacity, but at the cost of making the excess debt highly visible. What is needed is a solution that covers all of Hydro's debt costs, and yet gives the correct market signals to avoid the repetition of such problems in the future.

For an economist, the classic solution that meets the dual objectives of maintaining economic efficiency while increasing revenues, sufficient to meet the additional interest payments on the debt, is the imposition of a fixed tax. This is, after all, the solution to pricing in the face of decreasing average costs. The easiest and "fairest" way is simply to recognise that these costs were the result of lax political control over Hydro's operation; the responsibility for which lies with the voters of Ontario. In this case, we would recommend privatising all of Hydro's operations and offloading as much debt as possible onto these operations, and then consolidating the residual debt with the provincial debt.

The disadvantage is that a huge Hydro debt bill would then be added to the provincial carrying costs. If this is politically unacceptable, the alternative is a fixed charge on all electricity users in the province. Given that we should strive to avoid altering prices from their competitive solution, this charge could be a flat *generating* tax. A generating tax rather than an electricity tax is needed to avoid discrimination between base and peaking requirements. We would therefore recommend that all generating capacity above some minimum size be assessed a flat kWh tax, regardless of whether that capacity is nuclear, private, or public.

We would recommend that this tax be imposed on all generating capacity, even if privately installed and operated for internal needs. The only requirement would be that the capacity has to be above a certain minimum level. This ensures that there is no "scramble for the door," which might require an exit tax. This also affects those entities that have already moved to self-generation to avoid the effect of higher electricity charges.

We would also prohibit imports of electricity except for short-run load management reasons, otherwise such a tax would artificially encourage imports. Alternatively, any base load or peaking contracts for imports of electricity into the province could also be assessed the generation tax based on peak contract capacity. In this way, all electricity consumed within the province would pay the generation tax and cover the excess hydro debt.

The generation tax would of course find its way into electricity prices. Peaking supplies would have higher fixed part tariffs with the variable costs unchanged. Base load supplies would see higher average prices to cover the higher fixed and variable costs. However, there would be no incentive to leave the grid, since internal generation would face the same fixed cost per unit of capacity. As a result, capacity decisions would be made on economic grounds, and there would be no incentive to increase the existing excess generating capacity within the province.

If the tax is imposed either as a per unit charge or a tax on the grid itself, uneconomic decision making will result. For the per unit charge, self-generation will be encouraged, while an increase in grid charges will encourage bypass and more generating capacity.

Congestion on the Grid

Our second scenario assumes that the grid is not passive. The reason for a non-passive grid is simply that some contracts may not be feasible. Suppose, for example, that an LDC in Southern Ontario signs a contract for base load with a generator in Northern Ontario. In the passive case, the exchange simply logs the quantities and prices and informs the generator when the LDC increases its take. Actual consumption will take place by displacement, since no one actually matches generation and consumption. In this case, an inefficient generator in Southern Ontario could be idled in preference to an efficient generator in Northern Ontario, and the grid effectively creates one unified electricity market in Ontario.

However, in practice the grid is designed for particular flows based on an integrated generation-transmission system. It may, for example, have been designed assuming that the idled generator in Southern Ontario was used. As a result, it may be impossible to transmit power from Northern Ontario to Southern Ontario according to the contract signed between the generator and LDC. Historically, this may have mattered very little since electricity was sold in the province on an average-cost basis with the same rates for the same customer classes pretty much throughout the province. However, unbundling may give market signals to mothball local generators to the advantage of more distant generators. Unfortunately, such signals may indicate transactions that are infeasible because of the design of the grid.

The situation where the design of the grid conflicts with free contracting is one of *congestion*. This gives rise to two basic problems, one economic and the other managerial. The economic problem is that congestion serves to segment the electricity market. The essence of a passive grid is that arbitrage creates one price for equivalent electricity in Ontario. For example, if new hydro for base load is developed, it is available for all of Ontario and will thus alter all market prices in the same way. However, in a congested market new hydro in one area of the province may not increase capacity in another region, simply because transmission of that capacity is limited by the existing structure of the grid. In this case, free contracting will result in lower electricity prices in areas where the new generation can be brought to market and have no impact elsewhere. In economic terms, congestion implies an imperfect market, where prices can diverge across the province due to the absence of commodity arbitrage.

Politically, it is understandable why Hydro has opted for average or pooled prices. However, there is no reason why electricity prices should be the same across the province, anymore than gas or oil prices or the price for milk and bread. In the above example, the segmentation of the electricity market caused by congestion gives rise to powerful economic forces:

- industry in high-cost areas will have an incentive to move to low-cost areas
- generation will shift to higher-cost areas
- consumers in high-cost areas will lobby for increased grid infrastructure to reduce the impact of the bottleneck.

The first two effects will most probably be of lesser importance than the third. LDCs in high-priced and generators in low-priced areas will come to the grid with long-term contracts underpinning a requested facilities expansion.

If Ontario's system has been efficiently designed there may be very few such "transmission bottlenecks" and average prices across the province may be close to uniform. However, if there are bottlenecks the TSO has to deal with them. There are two polar solutions to the problem; the first is to have the grid take a proactive role in transmission charges and the second is to let it take a reactive one.

The first solution has been proposed by Hunt and Shuttleworth. They propose that LDCs purchase transmission rights when they purchase electricity from generators. This is similar to the natural gas market, where purchases are not made until capacity on the mainline has been purchased. In this case, the grid recovers its costs by selling capacity rights, rather than through a fixed toll on all users. The grid then has a responsibility to allow transmission, even if total rights exceed system capacity. If this happens, the TSO is required to make offsetting transactions and balance the load with the losses allocated to the grid. This might happen as a result of system failure or deliberate overselling of capacity.

In the above example, if more electricity is contracted from Northern Ontario than can actually be transmitted to Southern Ontario, the TSO may have to take offsetting transactions. The word "may" is operative, since peak demand from different contracts may diversify the risks from overselling capacity, so that the transmission bottleneck may rarely be met in practice. However, once the constraint is activated, the TSO would purchase more expensive supplies in Southern Ontario and thus meet contract demands in the South, and take delivery of the excess supplies in the North to resell to Northern LDCs in the spot market. The losses from these "swap" contracts would be paid by the grid as a cost for system failure or overselling capacity. By making the grid responsible for these losses, there is an incentive for the grid to manage the sale of rights efficiently and to avoid system failure. Hunt and Shuttleworth even go so far as to insist on electricity traders making a two-way market in electricity. This means quoting bid and ask prices to make load management easier for the TSO. In addition, it will be the TSO which makes decisions on the installation of new capacity to remove congestion.

The problem with this solution is the incentive structure faced by the grid as an assumed natural monopolist. If the grid is unregulated, it will simply increase charges and avoid overselling capacity in the first place. This ensures that it will not be forced into disadvantageous swap transactions. There is also the difficulty of the grid establishing transmission charges when there are no marginal costs of transmission, and there is the built-in incentive to reduce transmission capacity. We are therefore hesitant about recommending the Hunt and Shuttleworth solution.

The second solution, which we prefer, is for the grid to charge a standard fixed capacity charge to all users of the system. LDCs and generators would then have all contracts vetted by the TSO and the grid to ensure deliverability. The TSO can still oversell capacity in the sense of allowing more contract capacity than the grid can actually handle. However, not all contracts will be met, leaving differential prices in different markets as a result of the transmission constraints. The scarcity will be handled by some priority rationing system or pro-rating bids. The TSO will then continue to have the same balancing function when contracts are activated, and will continue to allocate the losses to the grid.

The essential difference between the two solutions is the role of the grid. In our recommended solution the grid does not sell capacity and does not extract the rents from transmission bottlenecks. As a result, it has no incentive to reduce transmission capacity. All the forecast costs, including the costs of balancing swap contracts, would then be forecast and incorporated into the fixed transmission charges. As with many pipelines, this could be accomplished on a cost-of-service basis with month-to-month differentials rolled into the charges for the following month. However, this would also remove the incentive for the grid to expand.

Similar to pipeline expansion, the initiative would lie with the generators denied access to the higher-priced constrained market and purchasers faced with higher costs. Both sets of participants will sign contracts and bring these to the Grid in a facilities' hearing to justify expansion. The existence of market prices in the constrained market for delivery at different points in time will provide the basic information to justify expansion. The basic rule is that the present value of the incremental revenues from the existing restriction has to be greater than the cost of the increased capacity. In this way, the effect of the relocation of capacity is also accounted for. For example, if it were economic to relocate generating facilities closer to the high price market, thus avoiding the transmission constraint, the forward sale of the output would already have depressed the price premium making increased capacity uneconomic. A complete market for electricity across time therefore facilitates the analysis of incremental capacity.

The only remaining consideration is how the costs of the expansion should be shared. Two alternatives are normally available: *incremental* or *rolled-in* tolling. Incremental tolling would take the cost of the transmission expansion and add the annual costs as a surcharge to the users of the new capacity. In this way, the price premium in the constrained area would still to some degree remain, since the transmission charges would be higher. In contrast, rolled-in tolling just adds the increased capacity costs to the embedded grid costs to be recovered from all users.

In our judgement, incremental pricing would be unfair, since the original grid was designed on an integrated basis, not on the basis of unbundled prices. As a result, some areas may be constrained and pay premium prices purely because of the accidents of system design. This would arbitrarily penalise some consumers and reward others. We see no economic grounds for endorsing these wealth transfers. We would therefore recommend that any facilities expansion result in increased costs that should be recovered in a rolled-in, increased fixed charge on all users of the grid.

The final aspect of the proposed system is the role and scope of regulation. The only part of the system that needs to be regulated is the TSO/grid function. The regulation would cover prices charged for transmission, including rolled-in costs expected to be incurred in the load swap under bottlenecks, approval of the rolled-in costs, facilities expansion hearings, setting of bypass rates, and the identification of the rules by which certain transactions are considered infeasible. The actual form of regulation could be cost of service/rate of return or price caps. The latter approach is not free of problems, but it has been designed to ameliorate a number of the problems with cost of service regulation. However, it works best in a privatized company, where share prices perform an important incentive function for management. Hence, the privatization of the exchange function is important. Finally, in order to guarantee equal access to the grid and exchange, the generation facilities should be separate from the transmission and TSO functions.

IV. AN ALTERNATIVE APPROACH: THE HOGAN MODEL[16]

The Hogan model and our preferred model are based on the same premise: competition in generation, and the existence of a company which maintains the grid and identifies prices for electricity. In addition, both models are structured to provide efficient allocation of generation capacity and electricity consumption. While our model permits direct bilateral negotiations between demanders of electricity, including middlemen and LDCs and the generators, the Hogan model in its initial stage constrains the access to the distribution companies. In contrast, the current Hydro model has the Electricity Exchange representing the whole of the demand side of the market.

The interesting part of the structure of the Hogan model is the operation of the pool company or "Poolco," which receives firm bids from distribution companies to purchase defined amounts of energy at specific prices and times and firm offers to supply electricity at various times and prices. The net result is an equilibrium spot market price for each hour of the day. The pool company purchases energy at the clearing price and sells the energy to the distribution companies at the pool price plus a mark-up. The electricity is dispatched in a least-cost manner by the pool company. With no congestion there is one price for spot market transactions over the entire system, similar to our model.

Of course, some participants may be risk averse and desire contracts that have less variability in price. These long-term contracts can be negotiated and are referred to as contracts for differences. These contracts in essence are equivalent to a swap transaction, in which the variability in the spot market price is removed and the supplier and purchaser receive and pay the agreed upon price. Note that the spot price is the important price and the terms of the longer-term contracts need not be market clearing prices. In our model, with no congestion costs and the existence of the TSO, we also obtain economic dispatch but with all contract prices posted, the signals for long-term and spot contracts are available.

With the introduction of congestion, it is possible that prices are not equalised across all points in the system. In this case, the poolco uses least-cost dispatch and buys in the market in which there is an excess supply at low prices and sells at a higher price. The congestion revenue generated is not kept by the poolco, but is provided to those who hold the compensation rights. Thus the pool company does not have any incentive to generate congestion and equally has no interest in alleviating it by increasing capacity. Our model also generates similar results but in a different manner. In the event of congestion, and the existence of transmission rights, the TSO is required to buy at the high price and sell at the low, thereby generating losses. If the TSO is regulated, then the implicit insurance function of eliminating congestion is compensated and the pool company has no incentive to alleviate congestion by introducing new facilities. In both models, the new facilities are introduced at the initiation of the suppliers and users of electricity. Note that in each model there are incentives provided for the economic location of generation and load.

A major difference between the models is the function of the pool company and the price signals generated. In the Hogan model, the pool company purchases and sells in cases of true congestion and obtains revenues that are returned to the generator with capacity rights. In our model, the TSO does not actively purchase and sell for participants, but they contract on their own. The TSO does maintain the grid and post prices for transactions undertaken on all contracts. Thus, unlike the Hogan model where spot market prices are identified, the whole set of prices is available under our model. This provides a richer set of price signals. For example, the influence of a spot market price signal for a long-term contract is not obvious.

V. CONCLUSIONS

The essence of our proposal is to facilitate contracting between users of power (LDCs) and suppliers of power on terms and conditions that are mutually beneficial. The purchasers could be large industrial users, the local distribution company, or even an agent representing a retail group. Any purchaser could approach any supplier to obtain the power needed at the times needed. Thus the purchaser would contract for a portfolio of energy supply based on its needs.

For all eligible contracts, the grid will provide transmission at a regulated tariff. There are some transactions which are not feasible since they cannot be undertaken under the existing configuration of the grid. The rules under which these transactions will be declared ineligible must be specified in advance and will be part of any regulatory process. This will ensure economically equal access to the grid for buyers and sellers of electricity. The transmission function is currently a natural monopoly and thus will have to be regulated. The form of regulation could either be price cap or rate base-rate of return. Under either approach, it is likely that the transmission charge will be primarily a fixed charge plus a small portion for a variable charge. This reflects the economic structure of the grid in which there are primarily fixed assets and few costs that are sensitive to the flow of electricity. Third, the grid company and/or the Transmission System Operator (TSO) will post prices on completed transactions and their terms and conditions. This will result in access by buyers and sellers to information on recent transactions. Notice that the prices will not be based exclusively on spot market transactions, but will be the result of negotiations for delivered power at a particular location. These prices will be used to provide signals for efficient generation, construction of generation, and the location of either load or generation. Finally, the grid company and/or the TSO will be responsible to ensure that the system itself does not face any technical problems; this can be accomplished through contracts for flexible power with suppliers.

NOTES

1 Hydro borrows under a provincial guarantee for which it is charged 50 basis points per dollar of debt.

2 D. Goulding and A. Poray, "Implementation of a Transfer Pricing Scheme to Emulate and Prepare for a Competitive Market," Transmission and Planning Conference, November 1994, Denver, Colorado.

3 "Restructuring the Electricity Industry in Ontario," Vol. 1 & 2, MEA (September 1994).

4 In some cases the gas LDC, for example Union Gas, owns part of the transmission mainline.

5 Halpern P., A. Plourde, and L.Waverman, *PetroCanada: Its Role, Control, and Operations* (Economic Council of Canada, 1987).

6 FERC order 636 (April 1992) has introduced a similar tolling methodology into the U.S.

7 K. Vanderschee, "The Newly Competitive Natural Gas Marketplace," *Canadian National Gas Focus* (July 1994).
8 This may or may not mean that the nuclear plants can also be privatised.
9 Block tariffs and multi-part tariffs have traditionally been used to price discriminate and enable the electricity generating company to recover its total costs.
10 All participants on the buy side will be referred to generically as LDCs, these are taken to mean the LDCs themselves, independent power purchasers, and brokers.
11 Matching supply and demand is meant in a contract sense. Actual electricity flows around the grid are determined by the laws of physics, not by the "laws" of economics or contracts.
12 These are the same types of contracts envisaged by Hydro, "1995 Transfer Pricing" (March 1995):5.
13 This is the term used by S. Hunt and G. Shuttleworth, "Operating a transmission system under open access: the basic requirements," *Electricity Journal* (March 1995).
14 Our paraphrase of pp. 1 and 2 of "The Electricity Exchange Business Unit," *1995 to 1997 Business Plan* (March 1995).
15 Peaking contracts would have to indicate average expected prices and minimum takes, otherwise there would be little comparable information with base load contracts, given the different expected consumption patterns of the purchasers.
16 We refer to the model as the Hogan model but there are a number of individuals who have proposed similar models either separately or in conjunction with Hogan.

BIBLIOGRAPHY

Armstrong, M., S. Cowans, and J. Vickers. *Regulatory Reform: Economic Analysis and British Experience,* chaps. 6 and 9. Cambridge, MA:MIT Press, 1994.

Doane, M., and D. Spulber. "Open Access and the Evolution of the U.S. Spot Market for Natural Gas." *Journal of Law and Economics* (October 1994):477-517.

Garber, D., W. Hogan, and L. Ruff. "An Efficient Electricity Market: Using a Pool to Support Real Competition." *The Electricity Journal* (September 1994):48-60.

Goulding, D., and A. Poray. "Implementation of a Transfer Pricing Scheme to Emulate and Prepare for a Competitive Market." Transmission and Planning Conference, November 1994, Denver, Colorado.

Halpern, P., A. Plourde, and L. Waverman. *PetroCanada: Its Role, Control, and Operations.* Economic Council of Canada, 1987.

Helm, D. "British Utility Regulation: Theory, Practice, and Reform." *Oxford Review of Economic Policy,* 10: 17-39.

Hoffman, M. "The Future of Electricity Provision." *Regulation* (1994):55-62.

Hogan, W. "Electric Transmission: A New Model for Old Principles." *The Electricity Journal* (March 1993):18-29.

__. "Efficient Direct Access: Comments on the California Blue Book Proposals." *The Electricity Journal* (September 1994):30-41.

__. "Reshaping the Electricity Industry." Prepared for the Federal Energy Bar Conference: "Turmoil for the Utilities," November 1994, Washington, D.C.

Hunt, S. and G. Shuttleworth. "Operating a Transmission Company Under Open Access: The Basic Requirements." *The Electricity Journal* (March 1993):40-50.

Municipal Electric Association. "Restructuring the Electricity Industry in Ontario." Vol. I: Recommended Strategy (September 6, 1994).

Ontario Hydro. "Providing the Balance of Power." Demand/Supply Plan Report.

__. "1995 Transfer Pricing." (March 1995).

__. "1995-1997 Business Plan." *The Electricity Exchange* (March 1995).

Shepherd, W. "Reviving Regulation and Antitrust." *The Electricity Journal* (June 1994):16-23.

Vanderschee, K. "The Newly Competitive Natural Gas Marketplace." *Canadian Natural Gas Focus* (July 1994).

Comments by

JAKE BROOKS
DAVID GOULDING
FRANK MATHEWSON

JAKE BROOKS

To put IPPSO's contribution in perspective, it is useful to note that IPPSO is not composed exclusively of private sector companies. Our 700 members include municipal utilities, schools, hospitals, conservation authorities, and a range of non-profit/third sector organizations, in addition to private companies. This is a kind of diversity that may be instructive to keep in mind as we consider demonopolization of the electricity sector.

The electricity business is changing rapidly, and fundamentally, in Ontario. Independent power has demonstrated that it has short lead times, flexibility, diversity, and declining costs. Dispersed generation has dispersed capitalization that prevents the accumulation of massive amounts of publicly guaranteed debt. Non-Utility Generators (NUGs) have now eclipsed whatever economies of scale may once have been enjoyed by the mega-projects that utilities used to build. As many people have said, the future of generation is NUG. To get the kind of future we want, it is important to look at what NUGs need and how they behave.

There is an urgent need for a new kind of overall regulation of the electric power sector, in order to maintain Ontario's competitive position. I would also suggest some principles to use in designing a system of regulation. By regulation I mean an impartial overseer, structured to maximize the public interest, and with binding authority to issue certain kinds of directives to the various companies in the field.

From many people's points of view, this discussion is all about access, access to transmission. Increased access to the transmission system for new power producers is quintessential. It is the key to competition leading to improved economic efficiency – both in the electricity business and in the provincial economy as a whole. It is also the way to introduce innovation and new, cleaner power to the system. But we cannot have that access, and the urgently needed benefits, until there is an effective regulator for electricity in Ontario. Regulation is a prerequisite for access. It is almost a truism that regulation is necessary wherever there are significant economic externalities. Running an electric power system without regulation is like trying to run a stock exchange without a securities commission. You can do it, but you will not get optimal deals. Regulation is necessary to achieve economic efficiency wherever the terms of the transaction are complex, but it needs to be managed fairly and impartially. It will be the regulator that determines and codifies the terms of access to the grid.

The new electricity business needs an effective regulator for the whole industry, not just for Ontario Hydro. Since the industry is already in the process of rapid change it is urgent that government address, immediately, the issue of setting up an effective electricity regulator.

Principles of Regulation

So we need regulation – but what kind? The type of regulation cannot be much like the command and control systems applied in the past. Binding authority may remain a necessary component of regulation, but regulation must also accommodate the free use of new ideas, making use of the power of entrepreneurial ingenuity in producing new ways to meet public objectives, without relying on prescriptive solutions as binding regulation has often tended to do. There are some basic models and underlying principles that are worth consideration.

Professors Booth and Halpern have made some reasonable suggestions about the optimal model. It allows access, separates generation and transmission, and allows the price system to work. It is interesting to note however that their paper focuses primarily on the process for negotiating transmission prices, which is strictly speaking not a question of regulation, but one of restructuring. I say that that is interesting because later I will argue that the right kind of restructuring can effectively reduce the need for regulation. My only concern about their paper is that it suggests that "the form of regulation could be price cap or rate base rate of return." I am concerned that this looks at the field too narrowly. There are a number of other types of regulation that could be possible and desirable, ranging from adapted integrated resource planning to one which focuses on alternative dispute resolution processes between the interested parties. Ontario Hydro itself has suggested that incentive-based regulation would be optimal, and under the right set of circumstances, I tend to agree. The point is to reconcile the firm's interests with a well-defined set of public objectives to the optimal extent. The form of regulation must leave as many options open as possible.

In terms of the models for restructuring the system, some NUGs have said it should be open competition, as in the auto industry. Entry is unrestricted. Price, quality and service reign supreme. Players who cannot meet the competition go broke. You can justify junking the old clunkers whenever something more competitive comes along, and overcapacity is not considered a social problem.

At the other extreme, defenders of the status quo want things to be more like the taxi business. Market entry is strictly controlled to prevent oversupply and economic hardship for all stakeholders.

I would suggest however, that the truth is in between, more like the post office, which has both competition from the courier business for many of its service areas, as well as an ongoing obligation to service customers in relatively high-cost sectors. This obligation to serve lowers the price for some consumers and may be socially desirable, but it leaves the Post Office with higher rates than might otherwise be the case.

In any of these discussions about models, there are two principles we like to repeat:

Separate the monopolistic and non-monopolistic functions from each other; and

Regulate the monopolistic functions as much as necessary, and regulate the non-monopolistic functions as little as possible. Each will provide benchmarks for the other, and cross-subsidization will be explicit if it exists at all.

I suspect that the final regulatory model for electricity will be a compromise that recognizes certain basic principles:

1. The natural monopoly part of the system is transmission and distribution, and it needs to be separated from the generation so it can be treated separately. This means that the Power Grid, or pool, would be an independent entity responsible for power system reliability, power quality, and servicing at least some of the existing debt. The grid would not be allowed to make investments in generation. You cannot be player and referee at the same time, or in other words, you cannot make a product and, at the same time, be the broker who decides which competitor's product gets bought first.
2. All suppliers should be treated equally. The Ontario Hydro business units should pay for the use of the grid on the same terms as other electricity suppliers.
3. Regulatory decisions are complex and demanding public policy questions. They bear on everything from aboriginal economic development

> to international competitiveness and environment. They are not purely technical or market questions, although market conditions must necessarily be central considerations. Yet, despite the extremely broad context, such decisions still must be rendered in a timely fashion. This of course implies that the regulator, whatever it may be, needs to be well-equipped in terms of its analytical and data-processing capabilities.

An effective electrical regulator is critical because the electricity system is highly important to our economic competitiveness and to the environment. As it does in natural gas, the regulator must have the broad public interest as its mandate. It must deal with many issues beyond price and rate of return. These issues that regulators have to face fit broadly into three categories: Economic, Social Equity, and Environment. I should like to touch briefly on some of these issues. I do not pretend to have the detailed solutions, but I think the range of questions demonstrates why we urgently need a capable regulator.

In terms of the economy, electricity is in many respects an essential service. Society's need for reliable, environmentally sound electricity must be paramount. Security of supply at stable and reasonable prices must take precedence over private struggles for market share or for lowest possible short-term price. Short-term savings are not always the best buy either for the firm or for society. Ontario cannot be competitive in an increasingly open world without an agile, responsive, effective energy regulator. Such a body is necessary to ensure an economically and environmentally efficient electric power system.

One of the reasons that some forms of independent power have a price advantage these days is because the price of natural gas is low and supplies are abundant. Current projections are that this will continue to be the case for many years. But forecasting is a very problematic business. Just ask a meteorologist or a Hydro load planner.

Looking back 25 years is easier than looking forward 25 years. That backward look shows a volatility in gas supply and price that suggests caution. As I said, the economy needs stable electricity prices. Regulators will therefore have to address issues of fuel diversity. We cannot afford to be blind-sided by unexpectedly expensive pipeline capacity or unforseen competition for gas reserves.

Diversity will mean larger numbers of generators, some using new technologies. But that brings problems of its own. One of the things Ontario Hydro has always done very well is keep power quality high. Today's computerized industrial technology will not tolerate unstable power. The grid will have to preserve reliability in quality as well as quantity of power supply.

In a regulated, competitive system, in which the grid is responsible for reliability, it will require the financial capability to discharge that responsibility. Besides the power quality issue, there must be back-up capacity. Some generators will be off-line from time to time for maintenance or equipment failures. Who will pay for coordinating that idle, stand-by capacity? How?

Some NUG interests have argued that the inherent diversity in the cluster of technologies represented by NUG, as well as the statistical reliability of larger numbers of units, means that NUGs provide de facto back-up capacity for each other. But these are questions for another talk.

The question of how to pay for central coordination and back-up services of course brings us to the trickiest element of the new regulatory order – wheeling and other charge for use of the transmission. There is going to have to be some charge for back-up, grid maintenance, and expansion. Those decisions are going to have to be made by an independent body. That body must have enforcement authority. It must consider the public interest as its first priority. It is worth noting that many of these issues have already been addressed within the framework of the Ontario Energy Board's gas regulation.

These charges or fees are also going to have to make some allowance for past mistakes. Hydro has a massive debt. Some of us criticized the policies that led to the creation of that debt over the last 30 years. But no amount of "I told you so," or even financial penalties, however just or satisfying for the critics, or competitors, is going to make that debt go away.

It is also a fact that many interests in the province actively supported the policies that lead to the accumulation of that debt. As recently as three years ago, during the Environmental Assessment Board hearings into Hydro's 25-year plan, some interveners criticized Hydro load forecasts as too low. They wanted even more new generation capacity than Hydro proposed and that, of course, would have meant even more debt.

Blaming and finger pointing is not going to get us anywhere. The debt exists and it must be paid. The question of how to pay the debt embraces, in addition to economic considerations, the second broad topic: social equity.

In a completely free market it is conceivable that Ontario Hydro could be competed into bankruptcy. In that case the province – meaning taxpayers – as guarantor, would have to pay the debt. It is a simple and unavoidable fact that no politician without a death wish could find it acceptable to dump the Hydro debt onto taxpayers. Government may provide some strategic help, but the electricity debt is going to be paid by electricity users one way or another.

With Ontario Hydro having surplus capacity and tottering on the brink of bankruptcy, access to the electricity market must be regulated in some respects. That regulation may take the form of either market rationing or additional charges imposed for transmission access, to prevent the Hydro debt from falling onto the taxpayer's back. This inflicts an unfair burden on independent power producers, but we acknowledge and accept for the time being that these are the political facts of life we face. (Isn't it funny how NUGs are being called on to help out Ontario Hydro at times like this?)

Another social equity issue is the regulation of regional pricing. In Ontario we have made a decision that everyone pays pretty much the same price for electricity. In some U.S. states there can be wide discrepancies in pricing. Electricity in New York City is much more expensive than in Niagara Falls. In Ontario we have made a largely political decision rather than a pure market decision when it comes to prices across the province. Should a regulator in a competitive market preserve that equity situation? If not, how far should the regulator go in allowing price differences between regions of the province?

Business opportunities for NUGs will also be a regional challenge. The system might need 300 megawatts and need it in the east side of the province. What would be the regulatory attitude to 300 megawatts of low-cost power in the west side of the province? That is not where it is needed. Could this be interpreted as discrimination? Would it be in conflict with the obligation to treat proponents equally and fairly? How would we judge?

Finally there is the issue of the environment. A regulator cannot ignore the environmental cost of power. As Maurice Strong has said, "It is fully consistent with the principles of market economics, that the price of a product should reflect its full environmental cost." It is also consistent with international trade policy that all products, regardless of their origin, be subjected to the same regulations, both economic and environmental.

Independent power's market share will increase as Ontario Hydro's thermal generators wear out. The political inevitability is that bigger market share means more public accountability, especially on issues like the environment. Independent power will become more subject to environmental scrutiny through the regulatory process. That will make environmental impact an element of competition for market share between NUGs.

It is possible to displace some dirty power from Hydro with clean power from generation that uses renewable energy or is very highly efficient cogeneration or tri-generation, for example. The regulator could encourage green power by lowering its cost for access to the grid to reflect environmental and social cost savings elsewhere in society. Conversely, dirty power, whatever its source, would pay a premium for such access.

The amount of such premiums is partly science and partly politics, a topic I should like to get into at another time.

Independent power producers have great faith in Ontario. We think that it is possible to overcome, with competiveness, problems that Ontario has suffered in the last couple of years without rate shock or environmental damage. Nor do we believe that Ontarians have to choose between clean power and economic power. The economy and the environment are synergistic. However, the only way to take advantage of that synergy is to manage it conscientiously. And such management will undoubtedly involve the quantification of environmental costs and benefits associated with various power sources, both foreign and domestic. Fortunately, getting this information right makes it possible to reduce the use of other forms of regulation, because the resulting economic instruments allow the market to be the inspector, the judge, and the enforcer. There will ultimately be savings to society in reduced costs of regulation.

The pressures we face are not unique to Ontario. The same technological and investment opportunities and environmental challenges face our economic competitors. Jurisdictions with the most agile and best developed regulatory system will have an important competitive advantage.

In fact, it is safe to say that the most successful economies in North America will be those who install the most effective and agile regulators first. There is major economic advantage to be gained by putting in place a regulator that is a leader in North America.

DAVID GOULDING

It is clear that while Ontario Hydro can offer advice and provide some degree of influence on the future direction of the electricity industry in Ontario, it is the government that will determine the course to be taken. In this context, Hydro has recognized the customers' demand for increased choice and the increasingly competitive thrust in the business and is preparing for such inevitabilities. It is our intention to help ensure a viable electricity supply industry in Ontario whether this be public or private.

Having studied the various electricity restructurings that are taking place around the world, I have observed that there is no "one size fits all," no panacea, and that the structures favoured by interested parties too often owe their proposal more to self-interest rather than the benefit of society. For example, would all of those who wish to see Ontario Hydro generators compete on "a level playing field" be willing also to risk stranding assets by tearing up lucrative long-term contracts, or forego franchises and operate on a strictly competitive basis?

In Hydro we know that we must continue to change, and we have recognized that such change must be in the interests of the province and our customers. The Electricity Exchange Business Unit, operational since January 1995, is a focal point in that change, and it will continue to evolve as we further rationalize the current functions of energy merchant and marketplace provider/operator.

The development of an internal competitive market for Ontario Hydro generation is an initial step in driving a "value" based culture: if the turbine isn't spinning you likely are not earning revenue. We do recognize that this is still only a surrogate for open competition, but we also recognize the need to manage Hydro's debt and our relationships with our competition and other markets during the transition to an uncertain but likely wider North American marketplace. In attempting to ensure that the people of Ontario are key beneficiaries of the evolution of the industry, there are the relationships between choice, competition, and stranded debt to be considered in both the form and timeline of transition.

In my view, the issue is further complicated, and more opportunities are provided, by the convergence of previously distinct businesses that will no doubt take place over the next few years. The rapid growth of capabilities in the information domain prepares the stage for alliances and partnerships in electricity, gas, telecommunications, banking, and many other fields. Just as I view the Electricity Exchange Business Unit as an information-based, risk management business, I also expect information and risk management to be key in future structures. In this context, I suggest that the current debates on the merits of micro versus macro regulation need to be moved to a new paradigm. Future regulation will need to accommodate, hopefully on a non-intrusive basis, the linkages between these currently somewhat separate domains. The ability to "self-deal" and the possession of market power will become increasingly complex as the various market segments become intertwined via both partnerships and a comprehensive information network.

The vast majority of restructuring taking place around the world is occurring for the sake of cost, competition, and choice. Furthermore, in most cases it is taking place during a period of significant surplus generating capacity and supply capability. Such changes are inevitable and can result in benefits to society. However, the management of the obligation to serve and reliability of supply must be given their proper profile and not subverted by commercial interests. Restructuring for the sake of reliability does not usually occur until after the blackout.

FRANK MATHEWSON

Given the announced policies of the current government in Ontario, privatization and the breakup of the behemoth Ontario Hydro is not some academic dream but a distinct political reality. And none too soon, many would add. It is therefore a matter of practical relevance to consider the organizational structure of the fall-out from this exercise. This is the point made by the authors, who consider various scenarios that may emerge and the economic principles that should be applied to maximize wealth (i.e., GDP) in the generation, sale, and distribution of electrical energy in Ontario.

I am not acquainted in detail with the developing literature on privatization and the sets of rules that various authors have proposed to deal with specific facts and situations that may arise. My comments, therefore, are driven by general economic principles. But this is good and not bad: before too much detail obscures issues, it is important to set out general principles that ought to guide the development of the private supply of electrical energy. This is also the goal of the authors of this chapter.

As they point out, we are not operating in a vacuum in Ontario as this province will not be the first jurisdiction to privatize electricity – there is considerable experience in both the U.K. and the U.S. with respect to the competitive supply of electricity – and there is Canadian experience with respect to the introduction of competitive principles in the supply of natural gas.

As Booth and Halpern realize, when the privatization gun goes off, we will inherit a system of generation and distribution driven by whatever set of political forces built the then-current Ontario Hydro structure. For the purpose of establishing the appropriate set of operating rules that ought to be implemented, pointing fingers at those guilty for past mistakes seems pointless.

There are some historical facts, however, that do warrant consideration: first, the current system is a mixture of hydro, thermal, and nuclear generation. As I understand things, all of the existing feasible hydro sites have long ago been exhausted. The mix between thermal and nuclear is alive with issues of relative reliability, cost, and environmental purity and liability. Second, the current fact seems to be that subject to reliability, and at least at current prices, the province has excess generating capacity. And that capacity constitutes a fixed and sunk cost. Of course, the other inherited feature of the system is the transmission and distribution network, which is also a fixed and sunk cost.

Let us suppose that privatization is implemented. And let us imagine that there is no transition path, but we leap to a final private structure.

The search is for a set of rules that yield the greatest net benefits. Any privatization exercise must start with the existing generating and transmission facilities of Ontario Hydro as these are sunk costs. This is the short run. The immediate issues are (1) how to operate these existing facilities in a cost minimizing fashion, and (2) how to price the output so that economic wealth is promoted. There is also a long-run issue: the system as it is operated should provide clear signals about investment or de-investment in all of the facilities.

The consensus seems to be that, similar to natural gas, there is a natural monopoly in transmission and distribution, but there is competitive potential for generation. Booth and Halpern analyze this possibility. There are currently some independent generators, and we imagine the further sale of existing Hydro generating facilities. Liability issues for nuclear capacity cloud the issue of the privatization of these facilities. Once there is competitive generation, all generators of electrical energy should be able to sell power to any buyers, including buying groups, under whatever terms are negotiated. As Booth and Halpern point out, similar to natural gas, we would expect to see sales of electrical energy under both long-term contracts and on a spot market.

If the transmission network is an "essential facility" (to use the legal term flowing from *U.S. v. Terminal Railroad Association of St. Louis* 224 U.S. 383, 32 S.Ct. 507, 56 L.Ed. 810), then this system must be operated to provide access to all users of power. These users could be large industrial users, local municipal purchasers, or coalitions of buyers organized by agents into buying groups.

The next issue is to design the reward mechanism for the operator of this transmission and distribution network to provide open access to the system. For example, with a system where there is a capacity constraint, there can be an issue concerning the frequency with which spot market purchasers can access the system. In natural gas in Canada, for example, there was recent pressure to restrict to one point in each month the ability of spot market brokers to reserve capacity on the transmission system. Spot market brokers provide the economic benefit of sales at lower costs when the system suddenly and unexpectedly at a moment of time has excess product (capacity in the case of electricity). Restricting the ability of these brokers to move their product will limit the competitiveness of the system.

Booth and Halpern separate their analysis into two parts: no congestion and congestion on the transmission grid. Wealth enhancement would ask for pricing rules that in the presence of congestion allocate electrical energy to those demanders who value the product the most *and* give clear signals where and when additional transmission investment should occur. Along this line, the authors suggest that perhaps the transmission issues including access would be less relevant if generation were to move closer to demand. In this regard, natural gas and electricity (non-hydro) production are different.

Some small independent generators in Ontario already operate on this principle, for example, at least historically, in company pulp and paper towns where the employer produces electrical energy for its own industrial uses and sells power to its employee/residential customers. A critical issue, however, with respect to thermal and nuclear involves environmental and liability issues. Whatever environmental standards are imposed on thermal generation, I believe that no one is talking about zero emissions. Aside from the externality issues, say between, Canadian and U.S. thermal plants, it may still be socially efficient not to locate thermal plants on top of large population centres. An analogous issue for nuclear liability also applies.

The authors address the issue of how to solve the dilemma associated with the monstrous debt guaranteed by the Province with respect to the past building excesses of Ontario Hydro. Facilities may in fact be sunk costs; managing the debt is not. Furthermore, there are issues of political saleability as well. Frankly, I do not know the answer to these problems, but like Booth and Halpern I recognize their existence.

Lawyers like to use the phrase "at the end of the day." Well, at the end of this day privatization is here. As the authors indicate, judged from where we are, there is plenty of opportunity for the skilful application of economic principles.

The Regulation of Trade in Electricity: A Canadian Perspective

ROBERT HOWSE
GERALD HECKMAN

In this chapter we attempt to provide a comprehensive overview of trade in electricity and its regulation from a Canadian perspective. We are grateful for comments and criticisms on an earlier draft of this paper by participants in the University of Toronto Electric Power Project, as well as Michal Gal. Section I describes the technical and economic basis of trade in electricity in the North American context. Section II examines the existing domestic regulatory framework for international trade in electricity in both the United States and Canada, at both the federal and state levels.

The international legal framework for trade in electricity between Canada and the United States is considered in Section III, with reference to both GATT/WTO and NAFTA. This part of the chapter reviews the consistency of existing domestic regulation of electricity trade with GATT/WTO and NAFTA rules, and also speculates on the manner in which these rules will affect trade in electricity under circumstances where at least some jurisdictions in North America have moved substantially towards either wholesale and/or retail competition. Section IV considers trade in electricity within Canada across provincial borders, and discusses the constitutional rules affecting such trade and its regulation. The issue of the role of the federal government in assuring an integrated market in electricity is addressed, and some of the implications of pro-competitive regulatory reform for interprovincial trade are also outlined. Finally, in Section V we attempt to provide a comparative perspective on trade in electricity in North America by briefly examining trade in electricity and its regulation within the European Union, including the progress of efforts to create a single European electricity market.

I. THE TECHNICAL AND ECONOMIC BASIS OF EXISTING TRADE IN ELECTRICITY

An Electric Power System Primer[1]

In order to understand how electrical power is traded, it is useful to review the typical means by which power is produced, transmitted, and distributed within a single (domestic) power system.

Figure 1 illustrates a single power system. A generating station (A) produces 60 Hz alternating current (AC) with a voltage typically between 12 and 30 kilovolts. This voltage is usually increased or "stepped" to a transmission voltage of 115, 230 or 500 kilovolts using a "step-up" power transformer. Increasing the transmission line voltage increases the amount of energy which can be economically transported.[2] A high voltage transmission line (B) (usually an alternating current or AC line) carries the electric energy from the generating station to the distribution systems. Transformer stations (C) step-down the high transmission voltage to one which can be carried to rural or municipal distribution stations (E) or to large industrial users. This lower output voltage carried by the transmission lines (D) to municipalities is typically 27.6 to 44 kilovolts. Distribution stations step down this voltage to distribution voltages between 4.8 and 12.5 kilovolts. Pole-type transformers finally step down this voltage to levels of 120, 240, and 600 volts, which can be used by residential or commercial users (F).

Figure 2 represents an international electric power system. This system allows for trade of electric power through the interconnection of single electric power systems of the kind illustrated in figure 1. The circles represent "control areas," geographic areas in which a power system (such as that represented in figure 1) is operated and monitored by a control centre. Every utility in North America is part of a control area. This enables the coordinated operation of the interconnected

Figure 1
Simple Electric Power System

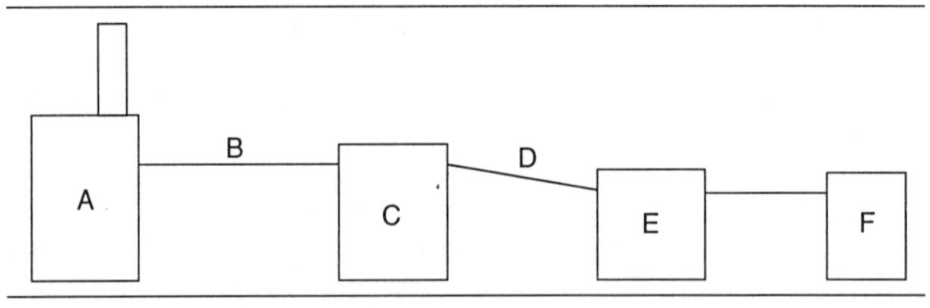

Figure 2
Electric Power System/Transmission System

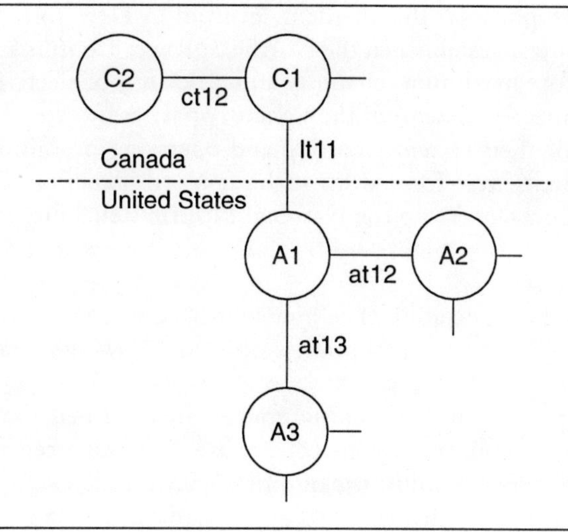

power system. The group of transmission lines (ct12, it11, at12, at13) connecting neighbouring utilities forms the transmission system. Control areas directly connected one to the other form a synchronous system. Five synchronous systems supply all of North America.[3]

Operation of an Interconnected System

RELIABILITY

The concept of reliability comprises both adequacy (the availability of sufficient generating capacity and energy resources to meet the normal needs of customers) and security (the ability of the system to withstand losses of major components (generators, transmission lines) and continue to provide proper service).[4] To provide reliable service, the power system must have sufficient generating capacity to meet peak electricity demand, taking into account scheduled and unscheduled outages in generation or transmission. Transmission and generation "reserves" are needed to ensure reliability.

The interconnection of power systems puts high demands on the reliability and security of individual systems. Failures occurring in one system will have repercussions throughout the entire interconnected system. This was clearly demonstrated on November 9, 1965 when the failure of a relay during a large power transfer between an Ontario Hydro generating station and New York's Niagara-Mohawk Power caused a catastrophic cascade of failures throughout the power system from southern Ontario to New York City leaving thirty million people without electrical power.[5] In response to this incident, utilities in New York, New England, and Ontario established the Northeast Power Coordinating Council to "promote maximum reliability and efficiency of electric service in the interconnected systems of the signatory parties by extending the coordination of their system planning and operating procedures."[6] Reliability Councils were formed throughout North America. Today, nine regional councils make up the National Electric Reliability Council.

CONTROL AREAS

At the heart of the operation of an interconnected system are control areas, defined as a utility or group of utilities with the following characteristics:[7]

- the total power flow between the control area and neighbouring control areas is monitored by the control area operating centre; and
- automatic controls adjust the output of generators inside the control areas in such a way that the total power transferred across the borders of the control area is equal to the net sales or purchases transacted between the utilities inside and outside the control area. This is called the control of the tie line loadings.

In other words, the power system within each control area is able to match its load demand with its generation and to make controlled purchases and sales of power through its interconnections. Short-term operating problems resulting in a mismatch of load and generation are corrected automatically because of the interconnected nature of the system. If any one system is deficient in generation, power flows from systems in neighbouring control areas to reduce the deficiency. Similarly, an excess production of power is absorbed by neighbouring systems. Interconnection thus increases the reliability of individual member systems. It is important to remember that individual systems cannot abuse this "automatic" correction feature and are responsible for maintaining a sufficient generating capacity to meet the load demand or to arrange for purchases of available capacity from other systems.[8]

TRANSMISSION

Electricity is a unique "commodity." It cannot be stored; production and consumption of electricity must be simultaneous. The flow of electricity through a transmission network obeys physical laws called Kirchoff's Laws.[9] Inter-utility exchanges of electricity in a synchronous system are physically implemented by decreasing the generation of the purchasing utility and increasing that of the selling utility. A flow of electricity will occur over the interconnecting lines from a system with excess generation to one with deficient generation.

The capacity of the interconnecting transmission lines to carry the electricity depends on many factors, including the size and type of conductors, the length of lines, and the location of generators and loads. The limited capacity of transmission lines imposes a constraint on the quantity of electrical power which can be transferred from one power system to another. This means that a power system must keep sufficient generating reserves, because it cannot expect to import as much electrical power as it needs from a neighbouring system due to transmission constraints. If sufficient reserves are not maintained, system reliability is impaired.

Transmission is said to have two main functions. The first is to link power systems in order to allow the generators of one system to back up those of another in case of outages; this is known as transmission's "capacity role." Transmission's second function, known as its "energy role," is to allow the economic dispatch of energy between power systems.[10] The cost of producing an additional unit of energy (incremental production cost) is a function of the additional fuel cost and the additional operating costs incurred by operating the generating unit at the higher level of production. Economic dispatch "consists of supplying the system's total power requirements at any time by loading each available generating unit to the point where the incremental cost is the same as that of all other generation."[11] Referring to figure 2, if power system A1 finds that producing additional power using its oil-fired generators is more expensive than simply importing hydroelectricity produced by C1 at lower incremental cost, such an exchange will be arranged. The magnitude of this exchange will be limited by the transmission capacity of transmission line it11. Limited transmission capacity causing deviations from economic dispatch will increase the cost of electricity.

In the years between 1980 and 1990, the "capacity role" of the transmission network in the United States has lost ground to the "energy role" due to an increase in generating reserves. This has caused heavy loading of some tie lines. Due to increasing load, declining generation reserves, and decreasing transmission construction, it has been predicted that the transmission network will shift from its present "energy role" back to a "capacity role" within the next decade. The ability of utilities to practise economic dispatch will be impaired, because they will have to maintain higher transmission reserves on the inter-area tie lines to ensure reliability.[12]

The Economic Basis for Existing Trade in Electricity

Existing trade in electricity is for the most part structured as trade between vertically integrated utilities, which usually own and operate the means of transmission and distribution of electric power within a particular jurisdiction, typically on a monopoly or quasi-monopoly basis but increasingly (especially in the U.S.) subject to the right of access to transmission facilities for "wheeling" of power from non-utility generators to wholesale customers.

A vertically integrated utility that is either profit-maximizing or has a regulatory mandate to minimize costs will import electricity so long as imports are competitively priced with its own generation or domestic electricity otherwise available to it. If the marginal cost of the imported electricity, including the transmission costs, is lower than the marginal cost to the utility of producing electricity from its own generating capacity, the utility will purchase the imported electricity. A number of factors may influence the relative cost of domestically produced versus imported electricity. Among the most important of these are seasonal and daily variations in electricity demand, reliability requirements, the mix of different generating technologies available to the utilities, and the time required to build new generating facilities.[13]

SEASONAL AND DAILY VARIATIONS IN ELECTRICITY DEMAND[14]

The electricity demand (or load) on an electric utility may vary over the course of a year due to seasonal weather variations or seasonal industrial activity in the utility's service area. While Canadian utilities face higher electricity demands in the winter (winter peak loads), many U.S. utilities have summer peak loads. Instead of putting into operation its costlier generators to meet the peak demand for electricity, a utility will import the cheaper excess capacity produced by utilities which are in their off-peak periods. Such transactions are called "summer-winter" diversity exchanges of power. Utilities in different time zones may also profit from the fact that their daily peak loads occur at different times through "daily time zone" diversity exchanges of power. International daily diversity exchanges seldom occur, as utilities trading on either side of the U.S.-Canada border are usually in the same time zone.[15]

RELIABILITY REQUIREMENTS

As normal weather conditions such as cold snaps, combined with reduced supply due to generation problems or low water reservoir levels, may cause the total demand for electricity to exceed the capacity of a utility, in order to maintain a high degree of reliability (ability to meet demand consistently), the utility will purchase the electricity it requires from a neighbouring foreign utility.[16]

MIX OF DIFFERENT GENERATING TECHNOLOGIES

A utility will operate those generators which produce the most inexpensive electricity (typically hydroelectric or nuclear generators) continuously throughout the year to provide its base load power, while expensive oil-fired generators will only be used during peak load periods. If sufficient capacity is available from foreign utilities, the utility may choose to import the electricity it needs to meet its peak load demand instead of using its more expensive peak load generators. As seen in the preceding chapter, the only factor limiting such transactions is the transmission capacity of the network interconnections with the exporting utilities.

NEW GENERATING FACILITIES AND EXCESS CAPACITY

It is advantageous for utilities with excess generating capacity to export electricity during low demand periods. Excess generating capacity exists for three main reasons. First, many generator technologies are subject to economies of scale. For example, the larger the capacity of a hydroelectric generator, the smaller the cost per unit of hydroelectricity produced. Hydroelectric resources will thus be developed to the fullest, creating generating facilities with more than enough capacity to meet the utility's load requirements. Secondly, the load-shape of a utility (with a high and narrow peak load, for example) may necessitate a high capacity for a short period of time, resulting in excess capacity for the duration of the year. Thirdly, generating facilities are generally built in anticipation of an increasing domestic demand for electricity. Excess capacity will be created during the period between the completion of the generator and the predicted increase in demand.

The Economic Basis for Trade in Electricity Under Conditions of Wholesale or Retail Competition in Domestic Markets

The above discussion suggests several reasons why international trade may be economically rational even under conditions of monopoly or near-monopoly supply within domestic (including federal sub-unit) markets. Under a scenario of competition within such markets, however, liberalized trade in electricity may be viewed as an extension of the benefits of domestic competition beyond national boundaries. Under such a scenario, trade may not only be economically rational between domestic power systems, for example, for reliability or cost avoidance reasons, but it should also occur between a generator or distributor in one domestic jurisdiction and a user of power located in another domestic jurisdiction wherever the foreigner is able to offer the best price/service mix for the customer. The point is well put by Boscheck, discussing the movement towards a single energy market in the European Union: "...there is no obvious reason why a European electricity market should not be able to follow the U.K. reform experience and benefit from the same type of price reductions, increases in contract options, and improvement in profitability of all parties involved."[17]

II. THE REGULATORY FRAMEWORK FOR ELECTRICITY TRADE BETWEEN CANADA AND THE UNITED STATES

A Brief History of Canada-U.S. Electricity Trade and Its Regulation

The first recorded electrical transmission line between the United States and Canada was constructed in 1901 by the Niagara Power Company Ltd., a Canadian subsidiary of U.S.-owned Niagara Falls Power Company. The 12 kilovolt line carried electricity from a power plant on the Canadian side of the falls to supply the Buffalo, New York area. By 1903, three hydroelectric plants built on the Canadian side of Niagara Falls were exporting most of their power to the United States.[18] Trade in electricity between both countries continued to expand. Although many of the U.S.-Canada interconnections were in the nature of "trans-border accommodations," by which utilities in one country supplied cross-border industrial customers or towns not yet connected to the distribution networks of domestic utilities, more and more transmission lines designed to export excess electric capacity were built.

In 1907, Parliament enacted the Electricity and Fluids Exportation (EFE) Act, which required electricity exporters to obtain a licence from the Government of Canada. This legislation was motivated by concerns over the long-term contracts by Canadian utilities to export power to the United States, and reflected the fear that Canada would be unable to repatriate this power in times of need.[19] The objective of the EFE Act was to ensure that power exported to the United States was surplus to Canadian needs.[20] In the United States, the President regulated electricity exports and the construction of export facilities until 1935, when the Federal Power Act was enacted. Section 202e of the FPA provided that no person could export electric energy without first having obtained an order from the Federal Power Commission (now the Federal Energy Regulatory Commission). The President retained the power to approve the construction and maintenance of international transmission facilities through the granting of permits. This power was delegated to the Federal Power Commission in 1953 and later to the Secretary of Energy.[21] The presidential permits were typically granted as a matter of course.[22] Neither country regulated the importation of electric power.

In 1955, the EFE Act was amended to include the regulation of the importation and exportation of natural gas. In addition to the licence requirement, the Exportation of Power and Fluids and Importation of Gas Act provided that the price charged Canadians for electricity would not be higher than that charged foreign buyers. In 1959, in recognition of the importance of the growing continental energy trade, a separate board was created under the National Energy Board Act to regulate energy exports. Section 83(a) of the NEB Act provided that exported power had to be surplus to foreseeable Canadian requirements, while s. 83(b) specified that the price to be charged by the applicant for an export licence had to be just and reasonable in relation to the public interest. The objective of s. 83(b) appears to have been to prevent Canadian electricity from being sold more cheaply to foreign purchasers than to Canadian consumers.[23] The NEB Act also required certification of international electric transmission lines by the NEB and extended the maximum term of export licences to 25 years.

In October 1963, the Liberal government's Statement of National Power Policy marked a major change in Canadian policy with respect to the export of electricity. Henceforth, the development of large, low-cost power sources as well as exports of electricity through increased interconnection with U.S. power systems were to be encouraged. In addition, the sale of large blocks of firm power to the U.S. would be licensed subject to the requirements of the NEB Act, in order to make the development of remote, large-scale electric power projects viable for Canadian utilities. The National Power Policy reflected the Canadian government's confidence that because U.S. utilities were no longer dependent on a few sources of power, even firm exports of Canadian power could safely be repatriated in times of domestic need.[24]

By the late 1960s the Canadian and U.S. electric power industries had evolved into large-scale interconnected power networks comprising a total of five synchronous power systems in North America.[25] While most utilities in the United States were investor-owned, all Canadian provinces, with the exception of Alberta and Prince Edward Island were served by provincially owned electric utilities.[26] Canadian and U.S. utilities formed regional reliability councils to improve the coordination of all bulk electricity supply systems in both countries.[27] In addition to coordinating the operation of their interconnected systems to improve reliability, Canadian and U.S. utilities formed power pools with the mandate to "minimize joint generation and operation costs through the planning and coordination of operations of the electric power systems in the pool."[28] Total electric power production costs within the pool could be minimized through the practice of economic dispatch of power among the member utilities. The integration of Canadian utilities in a North American system was crucial to the expansion of economic transactions of electric power: network interconnections provided the facilities for these transactions, and synchronized operation of utilities greatly facilitated the construction of additional interconnecting transmission lines.[29]

The Oil Crisis of the early 1970s created an important discrepancy between the cost of electric generation from the oil-fired generators of U.S. utilities and the production costs of Canadian hydroelectricity and nuclear power. The lower than expected domestic demand for electricity in Canada provided Canadian utilities with great amounts of inexpensive surplus generating capacity. From a fairly constant level of 5000 Gigawatt-hours throughout the 1960s, Canadian electricity exports doubled in the early 1970s and then climbed steadily to levels over 40000 Gigawatt-hours in the late 1980s.

In 1986, the Federal Minister of Energy, Mines and Resources asked the NEB to review all reforms which "would simplify and reduce regulation of electricity exports and international transmission lines."[30] The elimination of overlapping provincial and federal regulation, the streamlining of the licence application and certification process, and the extension of the maximum term for energy exports were to be considered in the review. These reviews by the NEB culminated in an amendment to the National Energy Board Act and the establishment of a new regulatory framework, which will be explored in detail in the next section.

A Snapshot of U.S.-Canada Trade in Electricity

This section presents data from the NEB on Canadian exports of electricity to the United States. Table 1 shows estimated exports and imports from each Canadian province for 1994.[31]

The largest U.S. importers of Canadian power in 1994 were New York state (29%) and the New England states (26%). The total quantity of power exported in 1994 represented slightly over 8% of the total Canadian electricity supply (542.8 Terawatt-hours). Exports in 1994 continued an upward trend from 1991 values, as shown below.

The increase in exports is due for the most part to increasing U.S. demand for electricity and to improving hydraulic conditions. Hydraulic

Table 1
Electricity Exports to the U.S. by Canadian Provinces

Province	*Exports (Gigawatt-hours)*	*Imports (Gigawatt-hours)*
New Brunswick	2 338	40
Québec	16 835	453
Ontario	12 663	185
Manitoba	8 854	158
Saskatchewan	21	79
British Columbia	4 110	23
Total	44 821	941

Figure 3
Total Canadian Electricity Exports: 1965-1994[32]

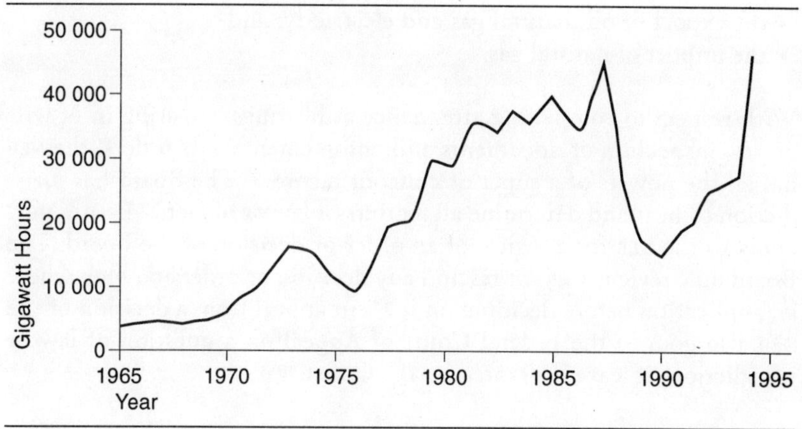

conditions have extremely important effects on Canadian generation capacity (and thus export capacity); 68% of Canadian electricity exported in 1994 was hydro power. Electricity from coal-fired generators and nuclear generators accounted for 19% and 5% of exports respectively.

The Canadian Regulatory Framework for Trade in Electricity

FEDERAL REGULATION

The Constitutional Basis of Federal Regulation

The jurisdiction of the Parliament of Canada to legislate in respect of the international trade in electricity is grounded in section 91(2) of the Constitution Act, 1867,[33] which gives Parliament jurisdiction over the regulation of trade and commerce. Federal jurisdiction over the regulation of international power lines is based in section 92(10) of the Constitution Act, 1867, which grants Parliament jurisdiction over interprovincial works and undertakings.

The National Energy Board

The National Energy Board Act[34] established the National Energy Board (NEB), an independent federal regulatory tribunal with jurisdiction to grant authorizations for:[35]

- the construction and operation of interprovincial and international oil and gas pipelines, international power lines and designated interprovincial power lines;

- the setting of tolls and tariffs for oil and gas pipelines under its jurisdiction;
- the export of oil, natural gas and electricity; and
- the import of natural gas.

With respect to compelling attendance at hearings, swearing in of witnesses, inspection of documents and enforcement of its orders, the NEB has all the powers of a superior court of record.[36] The Board has jurisdiction to hear and determine all matters of law or of fact.[37] Parties may apply to the NEB for a review of an order or decision of the Board. The Board may review, vary, or rescind any decision or order and may rehear an application before deciding on it.[38] An appeal from a decision of the NEB also goes to the Federal Court of Appeal on a question of law or jurisdiction, if leave is granted by the Court.[39]

Regulation of Electricity Exports and Transmission Lines
Division II of Part VI of the NEB Act as well as the National Energy Board Electricity Regulations set out the framework by which applications for the export of electricity and the construction of international electric transmission lines are reviewed. This framework is explained in the revised Memorandum of Guidance (MOG 1993) issued by the NEB on July 7 1993 and summarized below in figure 4.

Electricity Exports
Section 119.03 of the amended NEB Act provides that an application for electricity exports will be authorized by the issuance of a permit without a public hearing unless the Governor in Council upon the recommendation of the Board designates the application under section 119.07 for licencing procedures. Section 119.092 specifies that the maximum duration of such a permit is 30 years. In deciding whether to recommend designation to the GOC, section 119.06(2) of the NEB Act directs the Board to

> seek to avoid the duplication of measures taken in respect of exportation by the applicant and the government of the province from which the electricity is exported, and shall have regard to all considerations that appear to it to be relevant, including
>
> a. the effect of the exportation of the electricity on provinces other than that from which electricity is to be exported;
> b. the impact of the exportation on the environment;
> c. whether the applicant has
>   i) informed those who have declared an interest in buying electricity for consumption in Canada of the quantities and classes of service available for sale, and

Figure 4
NEB Application Processing Procedure

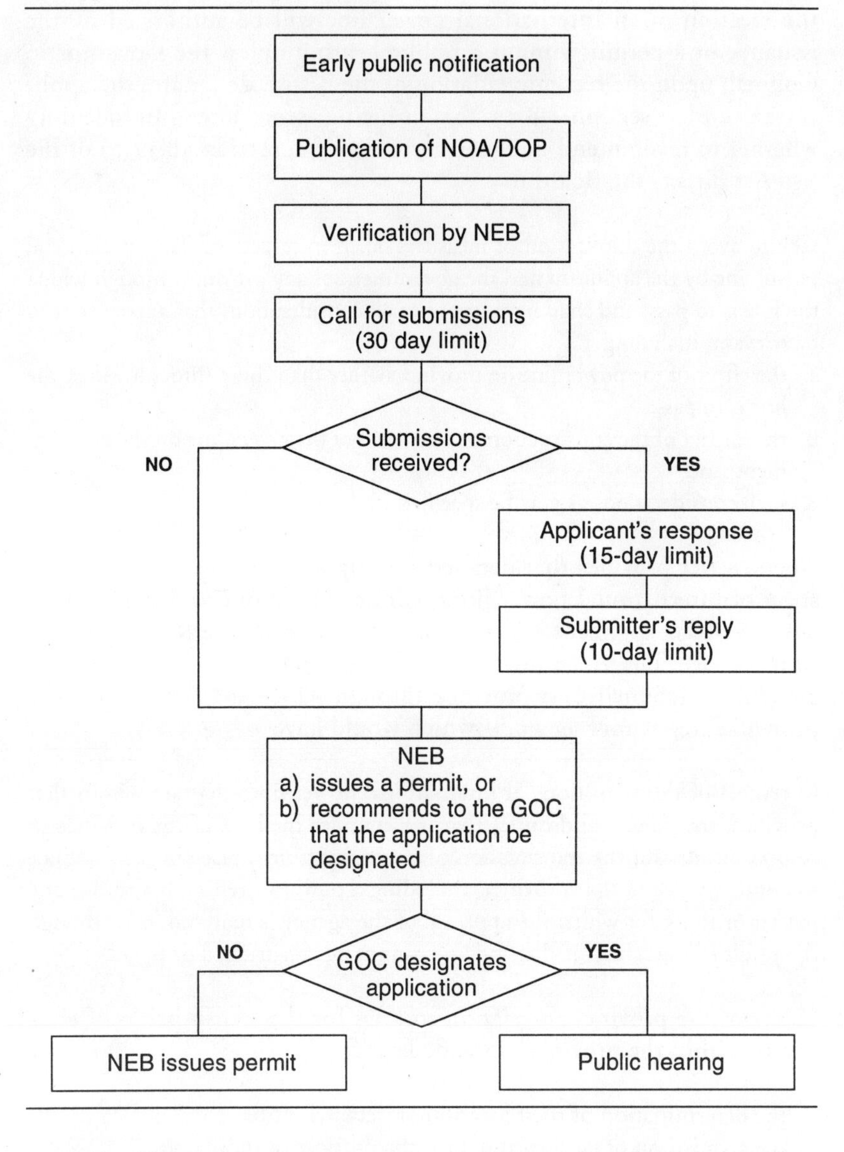

ii) given an opportunity to purchase electricity on terms and conditions as favourable as the terms and conditions specified in the application to those who, within a reasonable time after being so informed, demonstrate an intention to buy electricity for consumption in Canada;[40] and

d. such considerations as may be specified in the regulations.

International Power Lines

Section 58.11 of the amended NEB Act provides that an application for the erection of an international power line will be authorized by the issuance of a permit without a public hearing unless the Governor in Council, upon the recommendation of the Board, designates the application under section 58.15 for licencing procedures. In deciding whether to recommend designation to the GOC, section 58.14(2) of the NEB Act directs the Board to

> seek to avoid the duplication of measures taken in respect of the international power line by the applicant and the government of any province through which the line is to pass, and shall have regard to all considerations that appear to it to be relevant, including
> a. the effect of the power line on provinces other than those through which the line is to pass;
> b. the impact of the construction or operation of the power line on the environment; and
> c. such considerations as may be specified in the regulations.

The NEB Act provides that detailed routing and land acquisition in respect of international power lines will be carried out under provincial law unless an applicant elects to have federal law (the NEB Act provisions) apply. The Act provides for the designation by the lieutenant governor-in-council of a province through which the line passes of a provincial regulatory agency[41] which would have

> in respect of those portions of international power lines that are within that province, the powers and duties that it has under the laws of the province in respect of lines for the transmission of electricity from a place in the province to another place in that province, including a power to refuse to approve any matter or thing for which the approval of the agency is required, even though the result of the refusal is that the line cannot be constructed or operated.[42]

"Laws of the province in relation to lines for the transmission of electricity within the province" include laws for

> a. the determination of their location or detailed route;
> b. the acquisition of land required for the purpose of those lines;
> c. assessments of their impact on the environment;
> d. the protection of the environment against, and the mitigation of the effects on the environment of, those lines; or
> e. their construction and operation and the procedure to be followed in abandoning their operation.[43]

The applicant may also elect under section 58.23 for the provisions of the National Energy Act to apply to the proposed line and not the laws of the province.

The NEB examines the adequacy of the application and of the information submitted by the applicants in light of the submissions of intervenors and of the relevant considerations outlined in sections 58.14(2) and 119.06(2) of the NEB Act. Based on this examination, the Board may either grant the requested authorization or permit or recommend to the Governor-in-Council that the application should be designated under s. 58.14(1) or s. 119.06(1) of the Act. If the Governor in Council makes the order designating the proposed power line or electricity export for a licensing or certification process, the Board holds a full public hearing, and has regard to all considerations that appear to it relevant pursuant to s. 58.16 or s. 119.08 of the Act. If the GOC does not make such an order, a permit or authorization is granted by the Board.

Interpretation of the Amended NEB Act by the NEB: A Case Study

We now briefly review a recent decision of the NEB to grant permits to Hydro-Québec to export firm and interruptible energy.[44] The decision shows how the Board fulfils its obligation, set out in section 119.06(2) of the NEB Act, to review the environmental impact of the export, its effect on other provinces and the provision of fair market access.

In May 1994, Hydro-Québec requested a 30-year permit authorizing the export of electricity. Exports under contracts of a maximum period of five years to any customer whether directly linked to the applicant or not would be authorized by the permit. In examining the adequacy of the application, the NEB reviewed environmental considerations, effects of the exports on other provinces, fair market access, and several other considerations.

Environmental Considerations. The NEB considered the environmental impacts of the possible construction of additional production facilities, of the operation of production facilities, and of the transmission facilities.

The Board first noted that its assessment of the impact of the electricity export on the environment pursuant to s. 119.06(2) of the NEB Act was governed by the Supreme Court of Canada's decision in *The Grand Council of the Crees (of Québec) and the Cree Regional Authority v. The Attorney General of Canada et al.*[45] The Supreme Court held that the Federal Court of Appeal "had erred in limiting the scope of the Board's environmental inquiry to the effects on the environment of the transmission of power by a line of wire across the border."[46] The Court reasoned that if the construction of new facilities was required to serve the demands of an export contract, the environmental effects of the facilities were export-related, and it was thus "appropriate for the NEB to consider the source of the electrical power to be exported, and the environmental costs that are associated with the generation of that power."[47]

Several interveners warned the NEB that providing Hydro-Québec with the requested permit for the 30-year term would reduce capacity available for domestic power consumption and lead to the construction of new generation facilities. The Board surmised that because of the substantial capital investment and lengthy lead time required to construct new facilities, such facilities would only be built in response to additional domestic load or long-term firm export commitments, and not on the basis of short-term sales.

Pursuant to s. 199.06(2) of the NEB Act, the Board decided not to include a review of the effects of the operation of production facilities in order to avoid duplication of measures taken by the Government of Québec. The operation of the facilities complied with provincial environmental requirements, and Québec had jurisdiction to consider the environmental impact when the facilities were built.

As Hydro-Québec proposed to use existing transmission facilities, all of which already had the necessary permits and certificates and were operating within federal and provincial standards and guidelines, the NEB did not include in the scope of its assessment the operation of the transmission facilities.

The Board concluded that it was satisfied "there could be no potentially adverse effects on the environment of the exportation of electricity under the requested permit."[48]

Effects on Provinces Other than Québec. Hydro-Québec submitted to the NEB that its own analysis had identified no potential negative impacts on the other utilities' systems, and that it complied with standards set by the Northeast Power Coordination Council. Ontario Hydro, New Brunswick Power, and Cornwall Electric appeared to concur, stating that they expected no reliability problems. Nevertheless, the Board accepted the concern raised by the Grand Council of the Crees (of Québec) and the Cree Regional Authority that Hydro-Québec's estimate of the maximum quantity of firm power to be simultaneously exported (4825 Megawatts) exceeded the present 4300 Megawatt capability of Hydro-Québec's existing lines. In order to guarantee that the proposed exports would have no unacceptable effects on the reliability of systems in neighbouring provinces, the NEB decided to limit the simultaneous transfer capability allowed by the permit to 4300 MW.

Fair Market Access. According to the Board, fair market access "is meant to afford to Canadian purchasers who have demonstrated an intention to buy electricity for consumption in Canada an opportunity to purchase electricity on terms and conditions, including price, as favourable as those offered to an export customer."[49] The export permit requested by Hydro-Québec would allow exports on terms and conditions that were unknown at the time of the granting of the permit. The Board recognized that Hydro-Québec needed the authorization to engage in the changing electricity export market. Providing Hydro-Québec with the flexibility to define terms and conditions in the future was essential for the utility to secure maximum economic and reliability benefits from its future exports. In order to ensure compliance with the Act, however, the Board decided to include as a condition in the permit that Hydro-Québec would ensure that prospective Canadian buyers would be provided fair market access.[50]

Recognizing in effect that it could not constantly look over Hydro-Québec's shoulders to ensure that the utility was providing fair market access, the Board established a mechanism by which Hydro-Québec has to keep Canadian purchasers informed of the terms of its export contracts. Canadians who determine from this information that they were not offered fair market access will then have an opportunity to complain to the Board.[51]

PROVINCIAL REGULATION

The Constitutional Basis for Provincial Regulation

The jurisdiction of provincial legislatures to legislate in respect of the intraprovincial generation and transmission of electricity is affirmed in section 92A(1)(c) of the Constitution Act 1982 which states that the provinces have exclusive jurisdiction over the "development, conservation and management of sites and facilities in the province for the generation and production of electrical energy." In addition, provincial jurisdiction over the regulation of electricity as a specific industry or trade may also be grounded in exclusive provincial jurisdiction over "property and civil rights in the province." (92(13)).[52]

In the case of *Fulton* v. *Energy Resources Conservation Board*[53] the Supreme Court held that provincial jurisdiction over intraprovincial works and undertakings" (s. 92 (10)) encompassed jurisdiction with respect to intraprovincial generation and transmission facilities, even though the facilities at issue were to be interconnected with the power system of another province. However, Peter Hogg makes the following important observation about this case: "[the facility in question was intended to] deliver 99 per cent of its output to customers within the province" and the interconnection was to enable trade only in exceptional "emergency" circumstances.[54] According to Hogg, "if a more substantial operational connection had existed between the two systems [of Alberta and British Columbia], the systems would have been classified as a single interprovincial undertaking and would have come within federal jurisdiction under s. 92 (10)(a)."[55] This has the important implication that, as interconnection becomes a basis for facilitating significant trade in electric power, either interprovincially or internationally, *intraprovincial* transmission facilities will increasingly be susceptible to being characterized as "interprovincial undertakings" and therewith the general regulatory field will shift to the federal government. Since federal jurisdiction over "interprovincial undertakings" is an exclusive head of power, provincial regulation (apart from certain provincial laws of general application) will then become ultra vires of the constitution, *even if the federal government is reluctant to regulate.* This principle was set out with clarity in the *AGT* case,[56] where intraprovincial telephone companies were held to be interprovincial undertakings within the meaning of 92(10), since they provided the means by which international and interprovincial telephone service was delivered to customers within the province.

It is arguable, however, that the explicit grant of provincial jurisdiction under 92A(1)(c) precludes a shift of general regulatory authority from the provinces to the federal government[57] on the basis that the significance of interconnection makes intraprovincial electric facilities interprovincial works and undertakings. Indeed, this view was expressed in an opinion signed by three justices of the Supreme Court in a recent Supreme Court decision which concerned the scope of federal legislative authority over nuclear energy pursuant to the declaratory power, 92 (10)(c).[58] According to LaForest J., 92A(1)(c) was actually *intended* to guard against the danger of "the possibility of the transformation of these enterprises into purely federal undertakings by reason of their connection or extension beyond the province. Section 92A ensures the province the management, including the regulation of labour relations, of the sites and facilities for the generation and production of electrical energy that would otherwise be threatened by s. 92(10)(a)."[59]

Even if these dicta accurately reflect the general approach that a majority of the Supreme Court would take if it were to adjudicate the issue of the interrelationship between 92(1)(a) and 92A(1)(c), as LaForest himself notes, the regulatory environment in which 92A(1)(c) was born assumed vertical integration of the generation, transmission, and distribution functions within provincial utilities. It is, of course, now recognized that the unbundling of these functions is possible, and indeed such unbundling may be an important part of the shift to competition. So even if 92A(1)(c) precludes provincial generation facilities from being characterized as interprovincial works and undertakings, it is far from clear that 92A(1)(c) would act to prevent unbundled transmission and distribution facilities from being considered interprovincial undertakings on the basis of interconnection. LaForest J.'s dicta were focused on generation facilities, since the language of 92(1)(c) refers to "sites and facilities in the province for the generation and production of electrical energy." From a technical standpoint, production is largely synonymous with generation, and is clearly distinguishable from transmission and distribution.[60]

Apart from the possibility that *transmission* and *distribution* may be matters of exclusive federal jurisdiction under 92(10)(a), the federal government is granted the power over trade and commerce, including international trade and commerce under s. 91(2) of the Constitution Act, 1867, and this power has been held to preclude otherwise valid provincial legislation that interferes with international trade and commerce, even in some cases where the interference does not take the form of an explicit border restriction on imports or exports.[61] As well, the Supreme Court has held that the general territorial limitation on provincial jurisdiction renders unconstitutional provincial legislation that, in its pith and substance, purports to destroy or impair contractual rights outside the province.[62]

As will be seen in the discussion which follows, several provinces purport to regulate directly international trade in electricity. It is quite possible that, where this regulation had the effect of interfering with international trade, it could be found ultra vires the province. Prior to the 1990 overhaul of the federal statutory framework for regulation of international trade in electricity, the federal regulator had the statutory responsibility to ensure that any exported power was surplus to Canadian needs and that the price charged was in the "public interest." The 1990 legislative framework did not simply abolish these criteria, but removed their application from the mandate of the NEB by means of a concept of deference to provincial determinations on these matters. It is difficult to discern whether this concept of deference was merely based on the assumption that the provinces have constitutional authority to regulate exports of electricity beyond Canadian borders, or whether what was intended was an actual delegation of federal authority to the provinces. In Canadian constitutional law, such delegations have come to be viewed as constitutionally permissible.[63]

A Sampling of Provincial Regulatory Practices

The following is a selective sample of provincial regulatory approaches:[64]

Ontario. Ontario Hydro is required to submit all export contracts to the Ontario cabinet for approval. Financial and operational impacts of exports are reviewed at rate hearings of the Ontario Energy Board. Environmental assessments are carried out under the authority of the Minister of the Environment. Under the Power Corporation Act of Ontario, Ontario Hydro may enter into export contracts, subject to cabinet approval, only where the "supply of power is surplus to reasonable foreseeable power requirements of Ontario and other customers in Canada" and where "the price to be charged for that supply of power will recover the appropriate share of the costs incurred in Ontario and be more than the price charged to customers in Canada for comparable service."[65] Furthermore, the Board of Ontario Hydro "shall ensure that the requirements for power of Ontario customers and any requirements for power under any contracts with other customers in Canada are met before meeting the requirements for power of any customer outside of Canada."[66]

Québec. Cabinet approval of all exports is required. However, there are no statutory criteria for such approval, the legislation merely stating that the government "may on such conditions as it may determine, authorize any contract for the exportation of electric power from Québec."[67]

British Columbia. British Columbia law requires that an Energy Removal Certificate be issued for the removal from the province of an energy resource produced, manufactured, or generated within the province.[68] There are no statutory criteria for the granting or denial of a certificate, although the statute does describe the general purpose of export licensing as "to ensure the efficient use of energy resources and to ensure that present and future requirements of the Province may be met..."[69] The Minister may either issue a certificate or submit the matter to the Utilities Commission for a final determination of whether a certificate should be issued.

The U.S. Regulatory Framework for Trade in Electricity

FEDERAL REGULATION

Electricity Exports

Exports of electricity from the United States are regulated and require authorization under section 202(e) of the Federal Power Act.[70] Section 202(e) provides that the Federal Power Commission (which has been replaced in this function by the U.S. Department of Energy) will not issue an order authorizing an application for the transmission of electric energy to a foreign country if

> after opportunity for hearing, it finds that the proposed transmission would impair the sufficiency of electric supply within the United States or would impede or tend to impede the coordination in the public interest of facilities subject to the jurisdiction of the Commission.

Construction and Operation of International Power Lines

Applicants proposing to build and operate international power lines must first obtain a Presidential Permit from the Department of Energy

(DOE). Permits are granted if the construction and operation of the facilities are consistent with the public interest. In order to determine this, the DOE considers the line's effects on the environment and on system reliability. Environmental review is required by virtue of The National Environmental Policy Act (1969).[71] As Prodan notes, the policy of the DOE is to consider the environmental impact only of the construction and use of the transmission line itself, and not the environmental impact of the generation of the energy to be transmitted, or any environmental impact outside the United States.[72] The Departments of State and Defense review the application for its effects on trade policy and national security.[73]

Electricity Imports

Imports are not regulated per se by the U.S. Federal Government. However, if the proposed imports include plans to construct or operate international transmission facilities which have not been approved by a Presidential Permit, the applicant must apply to the DOE for a permit.

STATE REGULATION

State public utility commissions (PUCs) regulate investor-owned utilities which produce and distribute electricity. PUCs attain their objective of ensuring "that utility customers receive reliable service at reasonable cost" by "reviewing the rates they charge customers and by authorizing (or not authorizing) construction or expansion of facilities or establishment of services."[74] Typically, PUCs will allow the construction of new facilities for the importation or generation of electric power only if the applicant utility shows that the new facilities are required by the present or future public convenience and necessity. Because applicable regulations differ from state to state, this section will examine the role of state regulators in U.S./Canada trade in electricity through selected representative examples or case studies.

Canadian Electricity Exports to the State of Maine[75]

Before a utility's proposal to import electricity from a foreign utility can go forward in the State of Maine, it must clear a number of regulatory hurdles.

First, statutory law and regulations of the Public Utilities Commission require electrical utilities to exercise least-cost planning.[76] Simply put, least-cost planning requires a utility to examine different alternative plans to produce the required electricity including the construction and operation of generators, the purchase of electricity from other sources, energy conservation programs, and load management and to follow the plan with the least costs.

The Small Power Production and Cogeneration Act[77] provides for the sale of power from qualifying facilities (QFs), which are usually small power producers and cogenerators, to utilities without prior approval from the Maine PUC. The price of the electric power must be smaller than the avoided costs of the utility as calculated by the utility. The importance of this statute is that a utility exercising least-cost planning must consider the purchase of power from QFs as a possible alternative plan.

The Maine Energy Policy Act[78] requires the Maine PUC to give preference first to conservation and demand management measures and then to the purchase of energy from QFs if all available alternatives are "equivalent." Thus, a utility seeking approval from the Maine PUC to purchase Canadian electricity must show that this energy alternative is superior to the alternatives of conservation, load management, and purchase from QFs.[79]

Utilities must obtain a Certificate of Public Convenience and Necessity from the Maine PUC before they can proceed with the purchase of generating capacity, transmission capacity, or energy.[80] In order to obtain the Certificate, utilities must not only establish before the PUC that the power purchase is necessary, but they must also show that the purchase is consistent with their overall least-cost plan.

In 1987, Central Maine Power Co. (CMP) proposed a large purchase of hydroelectric power from Hydro-Québec.[81] CMP failed to obtain the required Certificate of Public Convenience and Necessity from the Maine PUC because it had not explored the alternatives to the purchase thoroughly and could thus not demonstrate that its proposal was superior to these alternatives. Central to the PUC's decision to deny CMP's request was the utility's refusal to negotiate with small power producers and cogenerators (QFs) to determine price and availability for an equivalent purchase of power.

Trade in Electric Power with the State of Vermont

The Vermont Public Services Board (VPSB), which reviews all purchase contracts entered into by state utilities, has vigorously reviewed proposals to purchase energy from Canadian sources. In its review of a long-term contract for the purchase of firm energy from Hydro-Québec by Vermont utilities, the VPSB "claimed jurisdiction over environmental (and other) consequences of projects located outside of the state to the extent that they affect *the general good of the state*" [emphasis added].[82] In extensive evidentiary hearings, the VPSB heard evidence on the environmental and economic impacts of the proposed purchase. The Board ultimately approved the contract.

Trade in Electric Power with the State of New York
Regulatory hurdles applying to the purchase of Canadian electricity may be lower in the State of New York than in the States of Maine or Vermont. Prodan notes that the New York Power Authority is authorized by state statute to import power for resale to utilities without approval by the New York Public Service Commission and observes that public pressure, not regulation, appears to have been responsible for the cancellation of a $19 billion contract to purchase electricity from Hydro-Québec.[83]

III. THE INTERNATIONAL TRADE LAW FRAMEWORK: THE GENERAL AGREEMENT ON TARIFFS AND TRADE (GATT) AND THE NORTH AMERICAN FREE TRADE AGREEMENT (NAFTA)

GATT and NAFTA are the key international law instruments that govern trade in electricity between the United States and Canada. The general provisions of the GATT[84] that directly affect trade in electricity are parallelled by largely identical provisions of the NAFTA. In addition, the energy chapter of the NAFTA contains specific provisions that apply to trade in energy, including hydroelectric energy. Further, GATT provisions on state-trading enterprises and NAFTA provisions on investment, monopolies, and competition may affect trade in electricity, particularly under some pro-competitive reform scenarios that involve at least partial vertical de-integration.

The 1947 GATT (including subsequent amendments and interpretative understandings) applies only to trade in goods. The General Agreement on Trade in Services (GATS), negotiated in the Uruguay Round, comes within the GATT/WTO institutional umbrella, and has been in effect since January 1, 1995. NAFTA has separate provisions that apply to trade in goods and to trade in services. A threshold issue is whether electrical energy is to be deemed a good or a service.

Many of the characteristics of services that feature in economists' definitions of what a service is as opposed to a good, seem to apply to electric power – these characteristics include intangibility or invisibility, a high degree of simultaneity between production and consumptions, and non-durability.[85] More importantly, many of the impediments to trade in electricity are of the kind addressed in GATS and in the NAFTA Services chapter – i.e., they are related to domestic restrictions on competition as well as lack of coordination and harmony between domestic regulatory regimes more generally.

Neither the GATS nor the NAFTA Services chapter contain any explicit definition of services that would obviously determine whether electric energy is to be considered a service or a good. What little academic commentary that exists on the matter seems to assume that electrical energy is a good. The strongest support for this view is that electrical energy is contained within the Harmonized Commodity Description and Coding System as classification 2716, thereby suggesting some measure of international agreement that it should be treated as a good. Secondly, the provisions of the NAFTA Energy chapter, which explicitly apply to electrical energy, refer back frequently to provisions of the NAFTA and the GATT that apply to trade in goods. Finally, neither the GATS nor the NAFTA Services chapter contains any obvious reference to electrical energy, or any sectoral commitments that apply to it. The issue of whether the treatment of electrical energy as a good or service is more appropriate within the trade law framework may have to be at least partly re-examined if there is a strong movement towards vertical de-integration, for even if electrical energy as such is now treated as a commodity, it is far from clear that provision of access to transmission lines or distribution systems should not be considered as a tradeable service, rather than simply as the means for trade in the "commodity" of electrical energy.

Although federal sub-units in Canada and the United States are not parties to GATT/WTO or to NAFTA, national governments are, in most instances, required (at least to the extent constitutionally possible) to ensure compliance of state and provincial governments with these agreements.

Existing Export Control Regimes and GATT Art. XI/NAFTA Art. 309

Art. XI (1) of the 1947 GATT bans import and export "prohibitions or restrictions other than duties, taxes, or other charges, whether made effective through quotas, import or export licenses or other measures." Art. XI (2) contains several exceptions to this general ban, two of which may be relevant to the case of import and export restrictions on electricity. Thus, Art. XI (2)(a) permits "export prohibitions or restrictions temporarily applied to prevent or relieve critical shortages of foodstuffs or other products essential to the exporting contracting party" and Art. XI (2)(b) permits import and export prohibitions or restrictions "necessary for the application of standards or regulations for the classification, grading or marketing of commodities in international trade." The provisions of Art. XI are incorporated into NAFTA through NAFTA Art. 309.

At first glance, it appears that the federal export control regimes for electrical energy in both Canada and the United States are in violation of Art. XI (1). The Canadian regime confers on an administrative tribunal, the National Energy Board, the authority to grant or deny a permit or licence for the export of electrical energy, based upon several statutory criteria, such as the effects of the proposed exports on the environment and whether potential Canadian purchasers have been given an opportunity to purchase the electricity in question on equally favourable terms. "Export licences" constitute one of the forms of prohibition or restriction on exports or sale for export specifically mentioned in Art. XI as impermissible. Both the burden imposed by the existence of a licensing scheme as well as the possibility of denial of a permit or licence on the basis of substantive legislative or regulatory criteria likely constitute violations of Art. XI (1). Similarly, in the American case, the requirement of authorization by the Federal Power Commission (FPC), as well as the mandatory denial of an authorizing order where the transmission would impede either the sufficiency of electricity supply within the United States or the coordination of facilities, also likely constitute violations of Art. XI (1).

It is possible that, to some extent, both of these licensing schemes could be justified, at least in part, under Art. XI (2) (a). Thus, it could be argued that control of exports by licensing is a necessary means to the prevention of a "critical shortage" of electrical power within the export restricting state. Art. XI (2)(b) might also be invoked on the basis that export licensing is necessary to sustain "standards" with respect to system reliability. In as much as environmental concerns justify export licensing schemes, Arts. XX(b) and XX(g) of the GATT could be invoked. Art XX(b) permits otherwise GATT-inconsisent measures that are "necessary to protect human, animal or plant life or health."

Art XX(g) authorizes otherwise GATT-inconsistent measures relating to the conservation of exhaustible natural resources if such measures are made effective in conjunction with restrictions on domestic production or consumption." GATT Panels and Canada-U.S. Free Trade Agreement Panels interpreting these GATT provisions have taken a quite narrow view of these exceptions. Thus, under XX(b), to show that a trade restriction is "necessary to protect human, animal or plant life or health," a contracting party must show that the measure in question is the least trade-restrictive means available to achieve the environmental goal in question.[86] The issue would be whether alternative policy instruments, such as measures to encourage domestic conservation of energy or environmentally-friendly forms of generation, or domestic "green" taxes on the consumption of energy, could achieve environmental goals while restricting trade less than an export licensing scheme. With respect to Art. XX(g), the jurisprudence has moved in the direction of the equivalent of a least restrictive means test as well, despite the fact that the language "related to" seems to imply a less stringent standard of justification than the wording "necessary" in Art. XX (b).[87] In any case, Art. XX (g) requires that trade restrictions be taken in conjunction with domestic restrictions on production or consumption, and export licensing is not part of any federal scheme of restrictions on the production or consumption of exhaustible natural resources.

Provincial schemes such as those of B.C. and Québec that do not contain any precise statutory criteria to determine when a licence may or may not be granted would be very difficult to justify under any of the exceptions to Art. XI, since there is no basis for ensuring that exports would only be prohibited or restricted where the strict rationales for the GATT exemptions applied.

As noted earlier in this paper, in both Canada and the United States, federal law requires that permission be obtained to construct and operate international transmission lines, which are essential to the export and import of electricity. Whether these schemes, or a denial of permission under these schemes, would constitute a violation of Art. XI(1) is debatable. On the one hand, a significant number of large-scale export transactions depend upon the construction of such lines, and therefore a government-imposed impediment to their being constructed would in effect be a prohibition or restriction on imports or exports. On the other hand, since exports or imports are not directly being prohibited or restricted, there is a counter-argument that such regimes should be regarded as internal regulations or requirements within the meaning of Art. III of the GATT (which applies only to imports, however) rather than as prohibitions on imports or exports within the meaning of Art. XI. In the *Salmon Herring and Landing Requirement* case,[88] an FTA Panel held that Canadian measures that did not directly impede exports were nevertheless a restriction or prohibition within the meaning of Art. XI, because they imposed a particular burden on export transactions that was not imposed on comparable domestic transactions. In extending the ambit of Art. XI to measures that did not on their face purport to restrict the flow of exports, the Panel placed much emphasis on the fact that the exact wording of Art. XI extended to prohibitions and restrictions not only on exports as such but also on "sale for export." Federal schemes for authorization of construction of international transmission lines may, of course, be justifiable under, for instance, Art. XI (2)(b), where reliability concerns can be invoked under the "standards" language of this subsection.

National Treatment: GATT Art. III (4/NAFTA Art. 301)

Art. III (4) of the 1947 GATT requires that: "The products of the territory of any contracting party imported into the territory of any other contracting party shall be accorded treatment no less favourable than that accorded to like products of national origin in respect of all laws, regulations and requirements affecting internal sale, offering for sale, purchase, transportation, distribution or use." This provision is incorporated into NAFTA by virtue of NAFTA Art. 301.

National Treatment will probably have a very significant impact both on trade liberalization in electricity and on regulatory reform within individual jurisdictions. Of course, where domestic law mandates a fully vertically integrated monopoly within a particular jurisdiction, would-be foreign competitors cannot complain of discrimination on the basis of Art. III(4) because would-be domestic competitors are equally being denied market access. In other words, foreigners are being treated as badly as their domestic counterparts, but no worse. However, under National Treatment, any competitive rights that are granted to domestic producers or suppliers must be granted to producers or suppliers of other GATT/WTO members as well. Thus, for example, if a jurisdiction decided to require that a utility provide transmission access to its own independent power producers at a regulated price, this right of access would have to be provided on no less favourable terms to independent power producers of other GATT/WTO members. Similarly, where government regulation requires that a utility purchase a certain quantity of its energy from non-utility generators (NUGs), such a requirement must, arguably, apply without discrimination to NUGs within the jurisdiction and NUGs of other GATT/WTO members.

Thus, National Treatment means that a decision to move towards competition domestically necessarily entails opening up the domestic market to *international* competition as well. On the surface, then, National Treatment appears to provide a clear and relatively straightforward link between the process of domestic pro-competitive regulatory reform and trade liberalization. Nevertheless, a problem may arise where, due to substantial variation in the pace and extent of regulatory reform, the National Treatment standard results in highly asymmetrical degrees of market access between different jurisdictions. Suppose that jurisdiction A maintains a fully vertically integrated monopoly, whereas jurisdiction B has moved to competition in generation. The monopoly in jurisdiction A will, pursuant to National Treatment, have a right to compete for generation contracts in jurisdiction B; however, generators in jurisdiction B will have no reciprocal right to compete for contracts in A, given that A still maintains a monopoly. This lack of reciprocity, although the natural outcome of the application of a National Treatment standard under circumstances of regulatory diversity, can be expected to offend deeply held beliefs about "fair trade" and may lead to significant trade irritants where a foreign monopolist actually represents a significant competitor in a particular market. This will put pressure on laggard jurisdictions to move towards pro-competitive regulation. However, as noted, non-reciprocity in market access is entirely legal under GATT and the corresponding provisions of NAFTA, and therefore could not be used, as a matter of law, to justify trade retaliation.

State Trading Enterprises: GATT Art. XVII

Art. XVII of 1947 applies to state enterprises; publicly-owned and/or operated utilities clearly fall into this category. Art. XVII(b) requires that, when such state enterprises engage in purchases and sales "involving either imports or exports," they "make any such purchases or sales solely in accordance with commercial considerations, including price, quality, availability, marketability, transportation, and other conditions of purchase or sale, and shall afford the enterprises of the other contracting parties adequate opportunity, in accordance with customary business practice, to participate in such purchases and sales." There is considerable debate as to whether Art. XVII actually imposes a kind of National Treatment obligation on state enterprises in their business decisions.[89] However, this may become a moot issue, as a recent Panel decision suggested that Art. III(4) may directly apply to the practices of state-trading enterprises "at least when the monopoly of importation and monopoly of the distribution in domestic markets" are combined (although it was not necessary to apply Art. III(4) in this case because a violation of Art. XI had already been found).[90] Of course, to apply Art. III(4) to the internal purchases and sales within a vertically integrated monopoly would almost be tantamount to depriving the monopoly of a significant part of its raison d'être, namely a monopoly or quasi monopoly on generation. However, where a degree of deintegration has occurred, for example with competition in generation, but where publicly owned utilities maintain distribution monopolies, or where a grid or pool is publicly owned, National Treatment of generating sources from other GATT/WTO members would be required.

NAFTA Ch. 15: Competition Policy, Monopolies, and State Enterprises

The provisions of Ch. 15 of NAFTA go considerably further than those of GATT Art. XVII. Ch. 15 requires that Parties to NAFTA ensure, through government regulation, that private monopolies act "solely in accordance with commercial considerations" in purchase and sale of the monopoly good or service in the relevant market, and furthermore, that such private monopolies not discriminate against investments, goods and services of other NAFTA Parties. Of considerable potential significance in the context of pro-competitive regulatory reform is the requirement that the government of each Party ensure that private monopolies act "in a manner not inconsistent with the Party's obligations under this Agreement" whenever they exercise any delegated administrative or regulatory authority. Under an "unbundling" scenario, where a grid or pool was privately owned (for example by a cooperative of utilities), this provision would ensure, at a minimum, that the prices charged foreign producers for access to the grid, and other conditions of access, did not constitute violations of National Treatment or the equivalent of GATT Art. XI restrictions or prohibitions on imports.

NAFTA Ch. VI: Energy and Basic Petrochemicals

The Energy chapter of NAFTA restricts the extent to which export restrictions in violation of Art. XI(1) of the GATT can be justified either under Art. XI(2)(a) or under several provisions of Art. XX, including conservation of exhaustible natural resources. According to Art. 605, restrictions that would otherwise be justifiable under these GATT provisions may only be permitted under NAFTA where the proportion of total supply exported by the trade restricting Party to another Party is not decreased. In other words, in case of a shortage, a Party may not favour domestic interests over those of other NAFTA Parties.

Art. 603.5 states that each Party may administer a system of import and export licensing for energy or basic petroleum goods provided that "...such system is operated in a manner consistent with the provisions of this Agreement..." This provision would appear, at a minimum, to counter the interpretation, which was suggested above to be plausible on the basis of GATT jurisprudence, that electricity export licensing schemes could in themselves be deemed a violation of GATT Art. XI/NAFTA Art. 309. However, the language "in a manner consistent with the provisions of this Agreement" takes away much of the force of 603.5, since even though the very existence of a licensing scheme may not be considered inconsistent with NAFTA, any system that places a substantial burden on export or import transactions is still likely to run afoul of GATT Art. XI/NAFTA Art. 309.

Ontario's statutory criteria for granting approval of Ontario Hydro export transactions most clearly run afoul of the provisions of this Chapter. As discussed above, the Ontario legislation provides that absolute priority must be given to Ontario and other Canadian customers over foreign customers, and moreover, that energy must be sold at a higher-than-domestic price to foreign customers. Indeed, the Ontario legislation even anticipates the possibility of a conflict with the Canada-U.S. FTA, the predecessor to NAFTA that contains much the same provisions on energy and states explicitly that the statute is to apply even in the face of any conflict with the FTA.[91]

NAFTA *Ch. 11: Investment*

The Investment provisions of NAFTA would have relevance in the case of the privatization of publicly-owned utilities. Art. 1102 elaborates a National Treatment standard with respect to, inter alia, "acquisition" of investments. The implication is that investors of another Party must be afforded an equal opportunity to purchase privatized assets – privatization entails therefore the possibility of allowing foreign ownership and control. It is to be emphasized here that any restrictions on the sale of privatized electric utility assets to investors of other NAFTA Parties would constitute new measures, rather than the continuation of existing measures, and therefore would not qualify for grandparenting under Art. 1108. Finally, it should be noted that the National Security exemption in Art. 2102, which refers to measures deemed necessary for the protection of national security interests, might justify some restrictions on foreign ownership of nuclear assets, where this could undermine national control of nuclear technology with military applications, or domestic or international nuclear non-proliferation policies.

The Limits of the International Trade Law Framework: Outstanding Issues

We have already alluded to one significant potential difficulty in the existing framework: National Treatment in the absence of comparable regulatory regimes can lead to asymmetrical market access. Another important limit on the trade law framework is that it provides neither rules nor institutions to govern international cooperation, whether between utilities, grids, regional groupings, or regulators, to insure the fair and secure operation of the common infrastructure that sustains trade in electricity. In an environment where all trade transactions, or almost all, are bilateral arrangements between vertically integrated utilities, such rules and institutions are of limited importance – the benefits of building and maintaining reliable connection are internalized to a relatively small number of major traders. Bilateral agreements between importing and exporting utilities and the exercise of some systems control functions by regional reliability groupings have allowed the facilitation of considerable inter-utility trade in the past. However, in a much more competitive, decentralized environment, the interconnected trans-boundary power system will increasingly resemble a genuine international commons. Under these circumstances, it is far from evident how basic decisions will be made that affect the use and maintenance of this commons. Certainly, self-regulation of the commons by utilities themselves would pose serious competition policy issues, since the interests of utilities, where they still exist as vertically integrated entities, may well be in conflict with those of other stakeholders in the commons – distributors, generators, wholesale or even retail customers who may gain from expanded trade in electric power.

In the United States, as part of the general federal regulatory thrust towards requiring that state utilities provide transmission access for wheeling to utilities of other states, the Federal Energy Regulatory Commission (FERC) enacted a policy providing that Regional Transmission Groups (RTGs) – composed of various state utilities as well as other transmission users (such as independent power producers) – "would be responsible for, among other things, planning and prioritizing the transmission needs of their members, determining the rates, terms and conditions of transmission service(s) provided by member utilities, and resolving transmission-related disputes between and among members."[92] It is noteworthy that the FERC requires that all RTG agreements contain an obligation that members provide all types of transmission service to other members, even if this involves expanding transmission facilities.[93]

One solution to the problem of governing the Canada/U.S. trans-boundary power system commons would be to integrate Canadian utilities and independent power producers into the RTGs. Indeed, in a recent report, the Canadian NEB notes that a number of Canadian utilities are engaged in trans-boundary regional planning activities with a view to the creation of RTGs.[94] At first glance, integration of Canadian utilities and other bulk users of transmission services into the U.S. RTGs would seem a problematic solution from the perspective of Canadian sovereignty, since the basic rules of the game for the solution of conflicts or disputes would be set by a domestic American regulatory authority in Washington.[95] However, despite extensive policy reviews by the NEB after a 1988 Report[96] recommended that the Canadian federal government take a leadership role in developing a national approach to trans-boundary transmission access, no such national approach has emerged.

Under these circumstances, the federal government may have only itself to blame if the basic ground rules end up being written in Washington. As the NEB notes, one implication of integration of Canadian utilities into the RTGs is that these utilities may well be required to provide themselves transmission access through allowing wheeling of power into their own Canadian networks.[97] To some extent, then, this solution would address the reciprocal market-access concern raised earlier in this section of the paper. Thus, one can easily imagine the pressure for pro-competitive regulatory reform in Canada being driven by the necessity for Canadian utilities to maintain access to increasingly competitive American markets. It is to be noted that a requirement that Canadian producers join RTGs to maintain access to American transmission facilities may well not of itself violate any provision of the GATT or NAFTA, provided that such membership or participation does not place a burden on Canadian producers disproportionate to that which it imposes on their American counterparts, and that Canadian producers enjoy the same rights and obligations as American producers.[98] In sum, the combination of American regulatory activism, Canadian regulatory inertia, international trade law rules, and Canadian interest in continued access to American markets may bring about an integrated Canada/U.S. market in bulk power, even in the absence of some kind of explicit binational code or agreement on trans-boundary market access.

Domestic Trade Remedy Laws

A further set of difficulties is likely to surround the application of countervailing duty and dumping laws to trade in electric power, as well as related fair trade claims, like those related to divergent environmental standards. The GATT authorizes the imposition of countervailing duties against subsidized imports (Art. VI), with the Uruguay Round Subsidies Agreement providing more detailed rules on the circumstances under which such duties may be imposed, which include a requirement of injury to the domestic industry of the importing country.[99] The United States is, notoriously, the world's sole user, for all intensive purposes, of countervailing duties. In a competitive environment, where Canadian utilities are actively competing with American producers to supply electric power in the United States, government policies, mostly at the provincial level, that appear to confer competitive benefits on these utilities may give rise to countervailing duty actions. For instance, government guarantees of a firm's debt has long been regarded as a subsidy under U.S. countervailing duty law, on the basis that such quantities reduce the firm's cost of borrowing below what it would be in a normal competitive market. Terms of access to government-owned land or resources for power generation will also be subject to scrutiny, just as was the case with access to trees on Crown lands in the recent softwood lumber dispute between the United States and Canada.

Dumping is defined as the sale of goods in export markets at a price below that which prevails in domestic markets. The GATT permits the imposition of duties on dumped products, where dumping is found to be causing material injury to the domestic industry in the importing state.[100] Contracts for the sale and purchase of electricity exhibit complex variations in the mix of price and service, depending on the quantity of energy to be supplied, the kinds of guarantees of security of supply and so forth. For this reason alone, increased competition for markets between foreign and domestic suppliers will likely give rise to dumping actions by the latter against the former. Where a utility is a monopoly in its domestic market, but is actively competing in open foreign markets there is a significant possibility that it will end up charging higher prices for comparable domestic sales than those it charges foreign customers. By penalizing this kind of conduct anti-dumping actions may, perhaps ironically, benefit consumer interests in the *exporting* country, even if they harm consumer interests in the importing country.

Under conditions of increasing competition, the link between trade and environment will likely engender considerable debate. Already, as the example of the James Bay project in Québec demonstrates, this is likely to be one of the most sensitive issues in the future. Under competitive conditions, liberalized trade allows a shift of production of electricity to jurisdictions with lower environmental standards or lower environmental costs.[101] This is not to suggest that competition, domestically or internationally, is necessarily at odds with environmentalist goals. Where environmental costs of energy production are fully internalized through taxes and charges, a competitive market may well lead to an improvement in environmental welfare, with choices among generating sources reflecting the relative environmental costs.[102] Nevertheless, where trade actually results in an increasing percentage of total power production occurring in jurisdictions where environmental standards are inadequate to internalize environmental costs, and especially where some of these costs spill over jurisdictional boundaries, there is a legitimate cause for concern. However, in addition to this kind of legitimate environmentalist concern, there will also likely be increased complaints from producers in jurisdictions with higher environmental standards that it is unfair that they compete with imports from jurisdictions where producers are faced with lower environmental compliance costs. There is reason for considerable scepticism concerning this kind of "fair trade" claim.[103] At the least, one can say that there is a wide range of government policies that affect, positively and negatively, international competitiveness of particular industries – this makes it normatively largely incoherent to claim that, abstracting from all other policy differences, one particular regulatory differential renders trade "unfair."

IV. INTERNAL CANADIAN TRADE IN ELECTRIC POWER

Interconnections[104]

Provincial electric systems are synchronously interconnected through AC ties to U.S. systems or to each other, with the exception of the Québec/Labrador and Newfoundland systems. The Québec/Labrador system is connected to other Canadian systems and to the U.S. systems via DC or radial ties. The Newfoundland system is isolated from other systems. Almost all Canadian utilities are members of regional reliability councils making up the North American Reliability Council (NARC). Notably, the total transfer capacity of the interprovincial interconnections is approximately 10000 Megawatts (including the 5300 MW Québec-Labrador interconnection), while that for international interconnections is 13000 MW. According to the NEB, this difference "results from relative market potential, relative proximity to loads, and greater economic attractiveness of international export markets".[105] Most Canadian utilities have interconnection agreements with utilities in neighbouring provinces. These agreements typically provide for mutual assistance during emergencies, coordination of operations, reliability exchanges and diversity exchanges of surplus power. Such cooperation can produce significant benefits.

A Snapshot of Interprovincial Trade in Electric Power

In 1988, interprovincial transfers of electricity amounted to 49201 GWh, a significantly greater amount than imports and exports from and to the United States taken together; however, more than half of the total of interprovincial exchanges was accounted for by sale of power from Labrador to Québec.[106] Almost all of the interprovincial exchanges consist in transfers between adjoining provinces as opposed to transfers of power wheeled through the transmission facilities of intermediary provinces.

The Regulation of Interprovincial Exchanges

S. 92A (2) of the Constitution Act 1982 confers on the provinces the jurisdiction to "make laws in relation to the export from the province to another part of Canada" of, inter alia, "production from facilities in the province for the generation of electrical energy." No jurisdiction is conferred on the provinces, however, to regulate imports. Further, and of potential significance, s. 92A (2) actually prohibits the provinces from "discrimination in prices or in supplies exported to another part of Canada." The federal government, by virtue of its interprovincial trade and commerce power, has jurisdiction over all interprovincial trade, both imports and exports, concurrent with the limited jurisdiction over exports conferred on the provinces by 92A (2). S. 92A (3) explicitly states the principle of federal paramountcy – i.e., in the case of conflict between provincial and federal regulation the federal law shall prevail.

Successive federal governments have attempted to achieve an integrated power market in Canada through spontaneous inter-utility coordination, as well as through interprovincial and federal-provincial negotiations. Despite the fact that notions of an integrated "Canadian grid" go back to the Diefenbaker era, every effort at arriving at a national framework through "cooperative federalism" has faltered on the inability of the provinces to agree among themselves.[107] The latest such impasse is represented in the failure of the provinces and the federal government to agree to provisions on trade in energy in the Agreement on Internal Trade, concluded in July 1994. Indeed, until an Energy chapter can be negotiated, none of the general trade liberalizing provisions of the Agreement are to be applied (Art. 1811.3). The negotiations on an Energy chapter were supposed to conclude "no later than June 30, 1995." We are unaware of any substantial progress on this front.

In recent studies undertaken for the National Energy Board, it was found on the basis of economic modelling that enhanced inter-utility cooperation among Canadian utilities and between utilities in Canada and the United States, assuming new generation projects started commercial operation in the year 2000, would bring long-term benefits totalling over $23 billion (projected year-2000 dollars).[108] These benefits from diversity sales and long-term firm sales of electricity would require transmission access and the wheeling of electric power. Transmission access is "the right or opportunity of electricity generating entities to use transmission facilities owned by others," while wheeling is defined as "the authorized use of the transmission facilities of an intermediate entity by two other entities whose transmission facilities are not directly interconnected, in order to sell, purchase, or exchange electricity between them."[109]

Options to achieve greater inter-utility trade considered in the NEB's review were:[110]

The continuation of voluntary cooperation between utilities (allowing individual utilities to develop their own policies on transmission access and wheeling).

Voluntary cooperation with monitoring by a federal agency, which would report regularly on the progress made in enhancing inter-utility cooperation and in developing transmission access and wheeling policies.

The establishment of voluntary regional planning entities, which would do for the development and coordination of economic improvements and efficient trade what regional reliability councils and NERC do for reliability concerns. Utilities would be free to develop individual access and wheeling policies, or could establish regional policies. The operation of these regional access and wheeling policies could be supervised by the regional planning entities, which could also offer dispute resolution mechanisms to members.

The establishment of regional planning entities with mandated federal power which would be similar to the previous option, except that a federal agency with jurisdiction to resolve disputes over international electricity trade and interprovincial and international wheeling would provide the dispute resolution mechanism.

As could have been expected, the responses of utilities and provincial governments to these options, and particularly the last option involving mandated federal powers, was divided between those provinces and utilities which have direct access to external markets and those who do not. The first group preferred the status quo, where utilities could develop individual policies on access and wheeling and enter into voluntary cooperation agreements. Those utilities who lacked access to external markets were open to the option of mandated solutions to disputes with other utilities and provinces over wheeling and access.

The federal government has failed to act on any of the options. One positive sign, however, is that, based upon the recommendation of the Board of Trustees, in 1993 membership in Regional Reliability Councils was opened up to any entity with an interest in generation or transmission including independent power producers, power marketers, and power brokers. So far over 40 different independent power producers and at least one broker have become members of Regional Reliability Councils.[111] If this trend continues, it would bode well for the development of a credible institutional capacity for self-regulation of access to transmission and wheeling across Canadian jurisdictions through the Regional Reliability Councils and their national umbrella organization, the National Energy Reliability Council. However, given the divergent and sometimes contradictory stakes that the various actors have in access issues, it is hardly realistic to expect a full-blown self-regulatory framework or agreement to be forthcoming without guidance from Ottawa. As was suggested in the previous section of this paper, if federal inaction continues much longer, what may determine the resolution of wheeling and transmission issues throughout North America will be the agreements struck within Regional Generating Groups, on the basis of ground rules from a U.S. regulatory agency, the FERC. As Canadian utilities and others with a stake in wheeling and transmission access in the U.S. increasingly integrate themselves in the RTGs, they will likely be required to give reciprocal access to their own networks to other members of the RTGs, whether Canadian or American. The regulatory path for trade between, say, British Columbia and New Brunswick may well run through Washington not Ottawa!

V. A COMPARATIVE PERSPECTIVE: EUROPEAN TRADE IN ELECTRICITY

A Brief Description of the Electricity Industry in Europe

INDUSTRY STRUCTURE

The electricity supply industry (ESI) in Europe, owing to its different origins in each country, is characterized by a diversity of institutional forms. These vary from highly centralized, state-owned utilities to more decentralized structures comprising utilities owned by both private and public interests.[112] Apart from the United Kingdom, monopolistic structures dominate the electricity sector in the European Community.[113] Monopolies are either state-established or created regionally through private agreements between suppliers and regional governments.[114] In France for example, state-owned Electricité de France (EDF) owns the bulk of the electricity supply industry. By virtue of the 1946 Nationalisation Law, EDF has monopoly rights over the generation of electricity and expansion of generation by other producers as well as over the importation and exportation of electricity.[115] In contrast, the German electricity industry is composed of many supplier companies which provide electricity to municipalities under exclusive agreements. Despite structural differences between European countries, electric supply industries "have tended towards concentration and integration on regional and national lines."[116]

INTERCONNECTION OF NATIONAL SYSTEMS

As in North America, national transmission systems in Europe are closely interconnected and form three main synchronous systems which are connected by DC transmission links:[117]

- "Western Europe": The Union for the Co-ordination of Production and Transport of Electricity (UCPTE), founded in 1951, comprises the utilities of Austria, Belgium, France, Germany, Italy, Luxembourg, the Netherlands, Switzerland, Greece, Portugal, Spain, and Yugoslavia.
- Scandinavia: The Nordel comprises the utilities of Denmark, Norway, Sweden, Finland, and Iceland (the last of which cannot for geographical reasons participate in actual interchanges of electricity).
- Central and "Eastern Europe": The CDO/IPS (Central Dispatching Office of the Interconnected Power Systems) comprises the utilities of Hungary, Poland, the former Czechoslovakia, Bulgaria, Rumania, and the former Soviet Union's southern systems.[118]

The role of UCPTE and Nordel is very similar to that of North American Reliability Councils and is to "aim to secure the optimal use of equipment for the production and transmission of electricity, whether existing or to be built" and "facilitate co-operation between utilities by organising the exchange of information, giving advice and issuing recommendations."[119] The United Kingdom and Ireland form separate systems. The United Kingdom is connected to the UCPTE through a 270 kilovolt DC link with a 2000 Megawatt capacity.

AN OVERVIEW OF THE EVOLUTION OF EUROPEAN INTER-UTILITY TRADE IN ELECTRICITY

Patterns of inter-utility trade in electricity similar to those in North America are found between Western European countries. The economic basis for such trade is in most ways identical to that in North America. The IEA describes the European and North American electricity trade as follows:[120]

Transfers of electricity between utilities in neighbouring regions have been common for many years. Exchanges based on differences in natural production costs between regions are economically efficient, and fluctuations in load can be balanced by exchanges with neighbouring utilities with different load fluctuations.[121] Such exchanges reduce the overall reserve margins needed by diversifying the potential sources of supply. Surplus capacity in a neighbouring region can result not only from simple differences in load timing but also from differences in climate, economic structure, or the timing of forced and scheduled unit outages.[122]

Until the 1980s, the electricity trade between European utilities largely consisted of "balanced" diversity exchanges and reliability exchanges.[123] The nature of the electricity trade changed in the late 1970s, when certain utilities became net exporters and others net importers of electric power. This came about as a result of the development by certain utilities of excess generating capacity combined with a good operating performance of their existing capacity.[124] The leader of the exporting utilities is Electricité de France, which undertook a vast and successful expansion of its nuclear electricity generation capacity in the 1970s and early 1980s. EDF has increased its exports of low-cost nuclear electricity to Italy and Belgium, but it has been unable to expand trade with Germany, where undertakings by the electric utilities to support the domestic coal industry have limited power imports. Utilities keen on exporting their surplus capacity of low-cost electricity and large industrial consumers of electricity seeking to purchase this electricity have emerged as driving forces behind reforms which would see the creation of a single European market in energy.

TRADE BETWEEN WESTERN AND EASTERN EUROPE

Trade in electricity between the countries of Eastern and Western Europe has been limited in the past. Electricity transfers have been effected using DC links and "back to back" converter stations, which can be operated independently from general operations of the eastern and western networks but which offer limited transfer capability.[125] Direct interconnection between western and eastern systems has not been possible because of the "fundamentally different operational philosophies" applied within the CDO/IPS and UCPTE systems.[126] Various working groups within the European electric industry are now examining the feasibility of the synchronous connection of sections of the East European network with the UCPTE system. The justification for such an interconnection can be found in the desire for increased energy exchanges with Western Europe, and in the endeavours by East European countries to raise the quality of power supplies to West European standards.[127] A further reason may also be the use of Western capital for investments in East European countries, and the associated commitments to supply energy to the West. Such interconnections raise important technical issues. In 1985, the International Energy Agency reported that Eastern and Western networks were not of comparable quality. Eastern generating capacity was barely sufficient to meet domestic peak electricity demand, causing frequent power disruptions and frequency variations unacceptable by Western standards. Even if the interconnection were technically feasible, many Eastern European countries depend on electricity imports from the former Soviet Union and would require some way of retaining access to the United Power System of the former Soviet Union, further complicating the integration of the two systems and expanded trade between East and West.[128]

A Brief Overview of Electricity Trade in Seven European Countries

In this section electricity trade figures for France, Germany, Italy, Sweden, Norway, Finland, and the United Kingdom are examined. Unfortunately, no detailed information concerning the regulatory framework governing the import and export of electricity for these countries was located. Any available general information was included, however.[129]

In 1985, the International Energy Agency noted:[130]

In Europe, there is no requirement for licenses for electricity exports. There is however, a tendency to enter into future electricity exchanges on the basis of short-term cooperation – apart from historically established long-term exchanges and exceptional participation in foreign generating capacities.

The absence of licence requirements for the export of electricity in 1985 may possibly be explained by the fact that the evolution of European inter-utility trade in electricity (described above) from balanced short-term reliability and diversity exchanges between neighbouring utilities to increased exchanges between net importers and net exporters is a very recent phenomenon.[131] The IEA noted in 1985 that[132]

Short-term electricity exchanges are of particular importance and have been flexibly developed to take advantage of considerable differences in marginal electricity production costs. The tendency towards a policy of short-term exchanges reflects both the desire of utilities to retain control over their longer-term electricity supply and the reluctance of regulators in several countries to license generating capacities which are not intended to supply the home market.

Licensing requirements now exist in certain European jurisdictions.

FRANCE

State-owned Electricité de France (EDF) holds monopoly rights for the import and export of electricity. In the face of pressure from the European Commission to remove these monopoly rights, which it claims are in breach of the Treaty of Rome, the French Government set up a working group to study possible changes to the industry's regulation and structure. The working group's report (the Mandil report) made two main recommendations:[133]

- Removal of EDF's monopoly rights over generation and expansion of generation by other producers, particularly from renewables and cogeneration. EDF would retain operational responsibility for dispatching on the basis of merit order from all French power plants, and for long-term capacity planning, under the control of the public authorities. Competitive bidding open to all producers or potential producers would be used for new capacity requirements. EDF would be obliged to buy power from authorised independent producers using cogeneration, small hydro, wind and solar units, regardless of their ranking in the merit order. All tariffs for EDF power purchases would be set by the public authorities.
- Removal of EDF's import/export rights; the Government would reserve the right to authorise imports by large, energy-intensive industries. Other end-users and independent distribution companies would not be allowed to import directly, since this would not be compatible with France's principle of geographic uniformity of tariffs. Exports

Figure 5
Electricity Trade: France

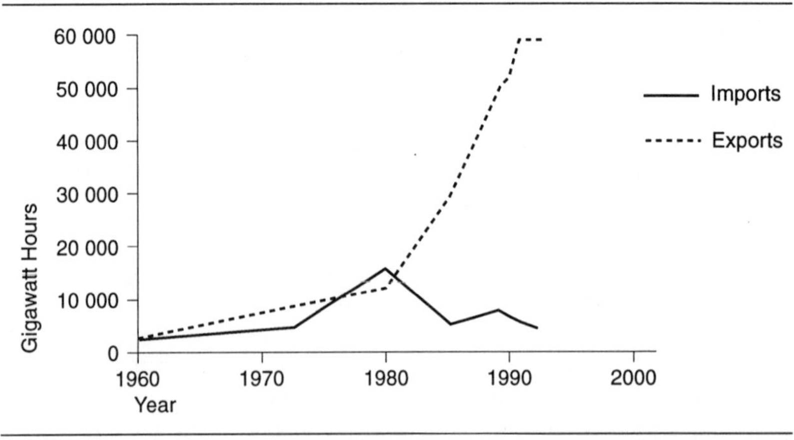

would be authorised by the Government on condition that national requirements are met first.

GERMANY

As mentioned above, although many companies are involved in the German electricity supply industry, most companies are granted exclusive supply rights by the municipalities they service, precluding competition. This is done through a regulatory framework based on concession and demarcation contracts. Concessions are contracts between municipalities and utilities providing for the granting of exclusive supply rights. Demarcation agreements, concluded between energy transmission companies, define the territories in which they agree not to compete.[134] Such agreements are allowed under German anti-trust law under an exception for grid-based energy forms, which include natural gas and electricity.[135] German electric utilities must use domestic coal.

Figure 6
Electricity Trade: Germany

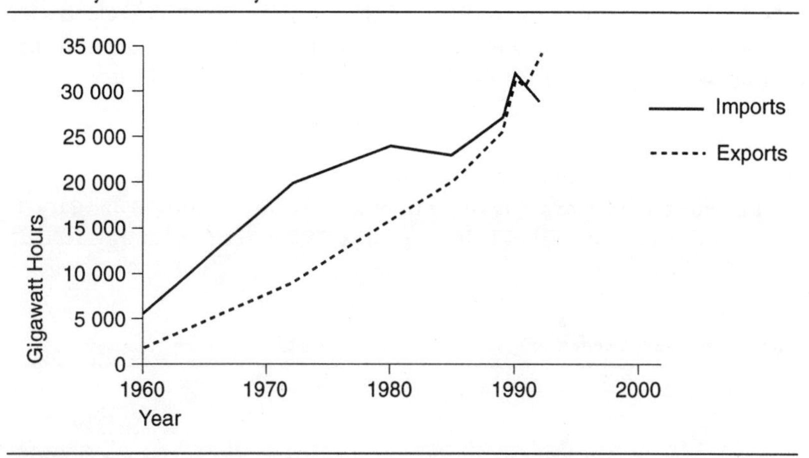

ITALY

A significant proportion of Italy's electricity requirements are satisfied by imports, mostly from France. Italy's net imports of electricity were of 35 TWh, while its gross domestic generation was of 227 TWh. The Italian government is preparing to sell ENEL, the state-owned electricity company, to the private sector. ENEL was established in 1962 to provide for the production, importation and exportation, transmission, transformation, and distribution and sale of electricity in Italy.[136] Under a planned concession

Figure 7
Electricity Trade: Italy

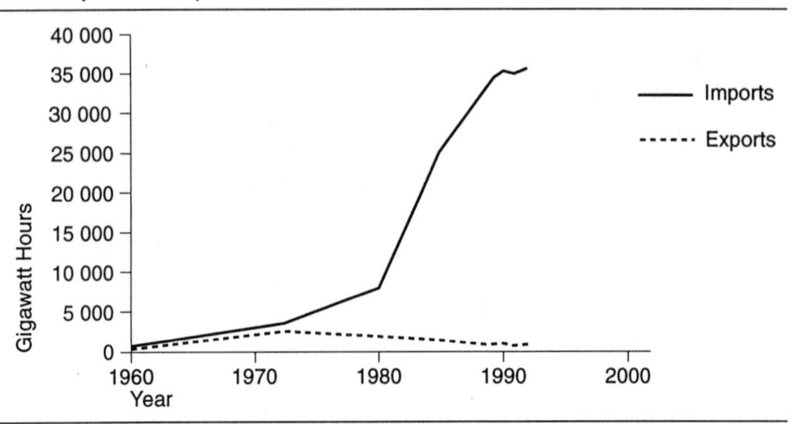

agreement, ENEL would retain monopoly rights over electricity transmission, exports and imports and, to a limited extent, distribution.[137]

The European Commission intends to challenge ENEL's electricity export/import monopoly before the European Court on the ground that it violates the Treaty of Rome.

SWEDEN

The International Energy Agency notes the following:[138]

The electricity market is to be opened to competition on January 1, 1995. A government bill in February 1994, "Competitive Energy Trade,"

Figure 8
Electricity Trade: Sweden

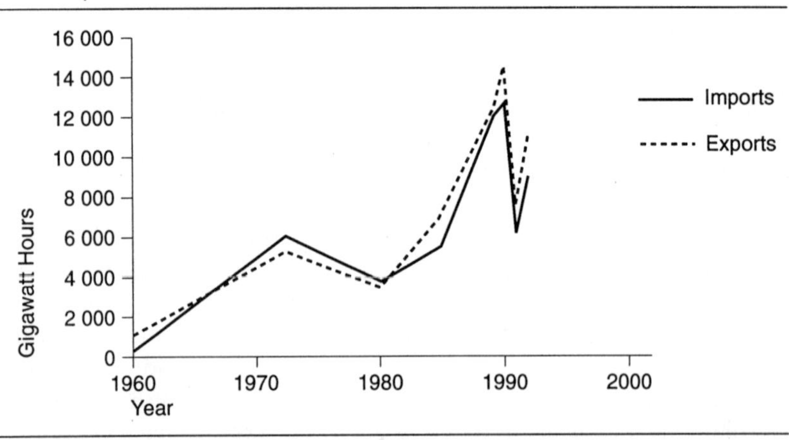

based on previous parliamentary decisions on principles and a Law Committee report, proposes changes in the Electricity Act of 1902. The objectives are to increase electricity trade, give producers and consumers choice, monitor and control anti-competitive behaviour and operate all electricity networks as separate economic activities with transparent cost accounting.

NORWAY

The IEA summarizes the Norwegian regulatory framework governing electricity trade as follows:[139]

Power exports and imports require government authorization; concession-free exports of firm power are allowed for up to five years, with the aggregate amount of power under these terms not to exceed 5 TWh a year (controlled through a concession-trading scheme). Through a recent report to the Storting (No. 46), the Government expressed a concern that significant exports of firm power over long periods are undesirable due to concerns about energy security and the environment. Instead, the Government prefers power exchanges – exports of peak power and imports of non-peak power. The report discusses the possibilities and benefits associated with long-term power exchanges, and concludes that such exchanges would contribute to resource management and energy security objectives.

In approving long-term power export agreements, the government has placed substantial emphasis on the economic benefit to Norway of the agreement and has both accepted and rejected 25-year power exchanges with German utilities on this basis.

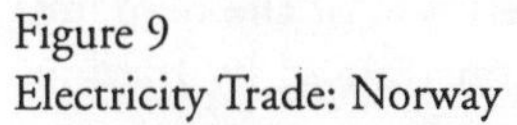

Figure 9
Electricity Trade: Norway

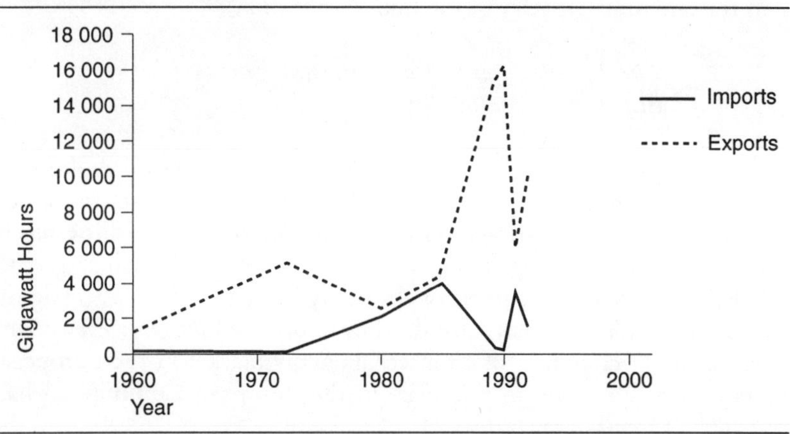

Figure 10
Electricity Trade: Finland

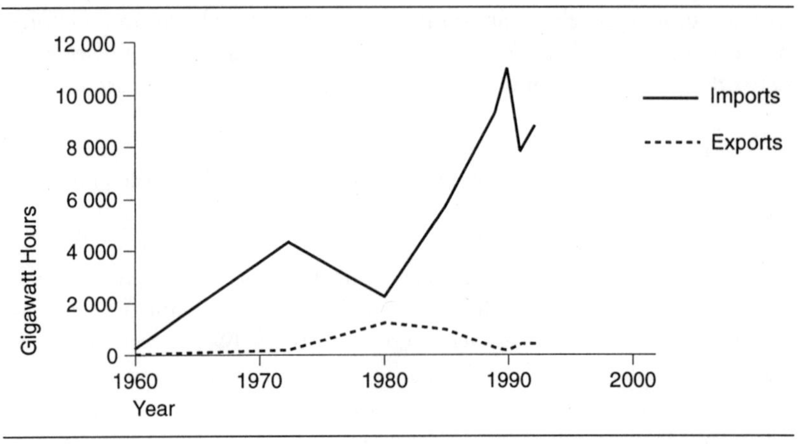

FINLAND

The IEA notes that "the Electricity Market Act, to be submitted to the Parliament in 1994 and expected to come into force in 1995, will encourage more competition and abolish unnecessary regulation of electricity production and sales (the parts of the market where competition is considered possible), and remove several licence requirements for power plant construction and electricity exports."[140]

UNITED KINGDOM

Although its electricity export levels are minimal, the United Kingdom imports over 16000 GWh annually (approximately 5% of the demand) from France. The level of electricity imports is constrained by the limited transmission capacity of the U.K.-France link.[141]

The Development of an Internal Energy Market in the European Community

As seen on pages 138-9, European trade in electricity has been limited until very recently to balanced diversity and reliability exchanges between the utilities of various countries. The emergence of utilities keen on exporting their surplus capacity of low-cost electricity (such as EDF) and large industrial consumers of electricity ready to enforce their rights to purchase this electricity are driving reforms which will eventually lead to the development of an internal energy market in the European Community. To establish this market, the European Commission has pursued a three-phase strategy:[142]

The first phase concerns the cross-border transit of electricity and gas through large grids. The second phase envisages a true opening of the markets by establishing common rules for electricity production or gas storage, as well as for transmission or transport and distribution. The third and final phase will then be defined in the light of the necessities for further market liberalization which remain after the second phase has been implemented.

The Commission has proposed to implement this strategy through the adoption of directives pursuant to Article 100a of the EC Treaty.[143] Article 100a "empowers the [European] Council together with the European Parliament to ensure the establishment and functioning of the internal market by adopting measures of secondary EC law in order to bring about the approximation of the Member States' legal systems."[144] Directives, under Article 189(3) of the EC Treaty, are binding on Member States as to the result to be achieved, but allow state authorities the choice of method to reach this result. Recognizing the strategic importance of the energy sector to Member States, the Commission has declined to use Article 90(3) of the EC Treaty which allows it, by way of a Commission directive, to "liberalize the sector's legal framework without having to achieve the European Parliament's and the Council's consent."[145] It has been observed that the European Commission's role to mediate and achieve consensus between Member States and sectoral interests has eroded its ability to initiate and implement policies in respect of the single European energy market.[146]

Phase one of the European Commission's strategy was implemented through directives which provided that Member States facilitate the transfer of electricity and natural gas through large high voltage and high-pressure lines respectively.[147] Another directive ensures price transparency in the electricity and gas sectors by obliging Member States to ensure that utilities operating on their territory disclose the prices they charge end users.[148] The purpose of phase two is to establish common rules for the internal market in electricity and gas. The Commission's original proposal comprised three main thrusts:[149]

- Removal of exclusive rights: rights to construct new electric generation and transmission facilities must be granted by Member States in a transparent and non-discriminatory way. This addresses the practice of many governments to give state-owned utilities monopolies over new generation, transmission, and export/imports. The same principle would apply to the operation, purchase, or sale of existing facilities.
- Unbundling: separation of generation, transmission, and distribution functions for both accounting and management purposes. The Proposed Directive provides relatively detailed guidelines for the operation and

management of transmission and distribution systems after unbundling, with a view both to ensuring access on fair competitive terms and at the same to assuring the reliability of the systems.
- Third-party access: large-scale industrial users should have a right of access to the existing transmission grid providing they offer adequate compensation.

In response to the European Parliament's suggestion of substantial changes to the Commission's proposed Directives implementing its proposal, the Commission presented an amended proposal, which by and large maintained the removal of exclusive rights, abandoned the unbundling of management, and abandoned in large part the obligatory third-party access in favour of a right to negotiate access in good faith with the transmission grid operator. The proposals, as of 1994, had not been adopted.[150]

Conclusion

When contemplating the bold (but yet unrealized) proposal for a Single European Energy Market advanced by the European Commission in 1992 North American observers may feel a little overwhelmed. There is simply no institutional structure in North America to facilitate, from the top down as it were, the design and implementation of a comprehensive regulatory framework for free trade and competition in electrical energy throughout Canada and the United States. Yet, ever so slowly, we on this continent are crawling from a system based on vertically integrated monopoly suppliers and highly regulated trade between them towards a competitive, integrated market. The peculiar combination and interaction of pro-competitive reform initiatives in a few jurisdictions, regulatory activism in Washington and inertia in Ottawa, of international trade law rules and the imperative to sell surplus capacity beyond one's own borders, all these forces acting together, may even get us to a competitive integrated market before the Europeans.

NOTES

1 See generally Ontario Hydro, Forestry Manual, Section 06 – Electrical Theory, (1991) 4; see also Canada, National Energy Board, "Inter-Utility Review: Transmission Access and Wheeling," (1992) 2-1. (Hereinafter "NEB Wheeling Review.")
2 Ontario Municipal Electric Association, Electric Power in Ontario, 55. (Hereinafter "Electric Power in Ontario.")
3 "NEB Wheeling Review," 2-2.
4 Ibid., 2-3.

5 "Electric Power in Ontario," 53.
6 Hydro Ontario, "System Interconnections," 7. Memorandum to the Royal Commission on Electric Power Planning with respect to the Public Information Hearings, 1976.
7 "NEB Wheeling Review," 2-7.
8 "System Interconnection," 4.
9 "NEB Wheeling Review," 2-5.
10 Ibid., 2-14.
11 Ibid., 2-6.
12 Ibid., 2-14.
13 Department of Energy (Energy Information Administration – Office of Coal, Nuclear, Electric, and Alternate Fuels), "U.S. Canadian Electricity Trade," November 1982, 3.
14 "System Interconnections," 2.
15 Ibid., 8.
16 For an example of such exchanges between Ontario Hydro and American utilities, see M. Perlgut, "Electricity Across the border: The U.S.-Canadian Experience," The Canadian-American Committee, 1978, 6.
17 Ralf Boscheck, "Deregulating European Electricity Supply: Issues and Implications," *Long Range Planning* 27 (1994):111, 118.
18 Perlgut, 10. It was only in 1910 that the Hydro-Electric Power Commission of Ontario (now Ontario Hydro, the provincial utility) built a transmission line from the Niagara Power Co. generator to supply seven Ontario cities, including Berlin (Kitchener) and Toronto, with electricity from the Canadian side of Niagara Falls.
19 These fears were apparently not without foundation. The heavy reliance of U.S. utilities on a limited number of generation sources in the 1920s and 1930s prompted President Franklin Roosevelt to state that the cancellation of an electric power export licence would be deemed "an unfriendly act" by Canada; see, National Energy Board, "Regulation of Electricity Exports," 6.
20 "Regulation of Electricity Exports," 5.
21 United States Department of Energy; Energy, Mines and Resources Canada, "Canada/United States Electricity Exchanges," May 1979, 63. The Presidential permits are now granted by the Department of Energy.
22 Perlgut, 13.
23 National Energy Board, "Regulation of Electricity Exports," 8.
24 Ibid., 8-9.
25 See "NEB Wheeling Review," 1-4. While all utilities within a synchronous power system are directly connected to each other by AC transmission lines, utilities in different synchronous systems must be connected by special DC transmission systems. The five synchronous systems are: the Western System (comprising Alberta, British Columbia, and most states west of the Montana/North Dakota border); the Eastern System (comprising Saskatchewan, Manitoba, Ontario, the Maritimes, and all other States except Texas); Québec and Labrador, Texas, and Mexico. The Island of Newfoundland is not connected to any other system.

26 There are several reasons for provincial ownership of electrical utilities. The massive investments required for the development of hydroelectric generating capacity, the natural monopoly nature of certain components of the electric power industry, and the socio-economic objectives of the provincial governments in extending electricity service to the population militated in favour of the creation of public provincial utilities; see "Regulation of Electricity Exports," 4, and Perlgut, 20.
27 "Canada/United States Electricity Exchanges," 35.
28 "Regulation of Electricity Exports," 3.
29 "U.S.-Canadian Electricity Trade," 11.
30 Letter of the Honourable Marcel Masse to Mr. R. Priddle, Chairman of the National Energy Board (2 September, 1986). (Hereinafter "Marcel Masse Letter.")
31 Canada, National Energy Board, "1994 Annual Report," 16-18. (Hereinafter "1994 NEB Annual Report.")
32 Data collected from National Energy Board annual reports from 1964 to 1994.
33 Constitution Act 1867 (U.K.), 30 & 31 Vict.,c.3.
34 National Energy Board Act, R.S.C., 1985, c. N-7 (am. 1990, c. 7). (Hereinafter NEB Act.)
35 Canada, National Energy Board, "1994 Annual Report," 3.
36 NEB Act, 34, sections 11 and 17.
37 Ibid., s. 12(2).
38 Ibid., s. 21(1).
39 Ibid., s. 22(1).
40 The process described in subparagraph (c) is called "the provision of fair market access."
41 NEB Act, 34, section 58.17.
42 Ibid., section 58.21.
43 Ibid., section 58.19.
44 In the Matter of Hydro-Québec: application dated 24 May, 1994 for a permit to export firm and interruptible electricity (December 1994), (NEB) (Hereinafter Hydro-Québec Application.)
45 1 S.C.R. 159 (1994). (Hereinafter Cree Regional Authority.)
46 Ibid., 191.
47 Ibid., 192.
48 Hydro-Québec Application, 10.
49 Ibid., 11.
50 Ibid., 26.
51 Ibid. The following conditions were appended to the export permit: Hydro-Québec shall: (i) for exports of less than one month duration, subsequent to the commencement of an export, inform all accessible Canadian purchasers, on request, of the terms and conditions under which a particular export was made; (ii) for exports of one month or more in duration, file with the Board, within

fifteen consecutive days of execution, a copy of any specific contractual arrangements associated with the export and, upon request, serve a copy thereof on the requesting Canadian purchaser.

52 Citizen's Insurance Co. v. Parsons (1881) 7 App. Cas. 96.
53 S.C.R. 153 (1981).
54 P. Hogg, *Constitutional Law of Canada,* 3d ed. (Toronto: Carswell, 1992), 728.
55 Ibid.
56 Alta. Govt. Telephones v. CRTC (1989) 2 S.C.R 225.
57 It is to be noted that in this section of the paper federal powers are considered only to the extent that they may limit the constitutional scope of provincial regulation.
58 Ontario Hydro v. Ontario (L.R.B.) (1993) 3 S.C.R 327.
59 Ibid., 378.
60 See B.S. Black and R.J. Pierce, Jr., "The Choice Between Markets and Central Planning in Regulating the U.S. Electricity Industry," *Columbia L. Rev.* 93 (1993):1339, 1343.2.
61 Central Canada Potash Co. v. Saskatchewan, (1979) 1 S.C.R 42.
62 Re Upper Churchill Water Rights, (1984) 1 S.C.R 297.
63 In the case of PEI Potato Marketing Board v. Willis, (1952) 2 S.C.R 392, the Supreme Court of Canada held that the delegation of federal powers over interprovincial and international marketing of potatoes to a provincial administrative agency was constitutional. However, this decision is rather inexplicably at odds with the slightly earlier Nova Scotia Inter-Delegation case, A-G. N.S. v. A-G. Can, (1951) S.C.R 31, which held that the delegation of legislative powers to a provincial government was unconstitutional.
64 See "Regulation of Electricity Exports," 66, for the more complete sampling; because the cited report was written in 1987, it is possible that the information relating to some of the provincial legislation is obsolete.
65 Power Corporation Act, R.S.O. 1990, c. P.18, s. 85 (2), (a) and (b).
66 Ibid., s. 85(3).
67 R.S.Q. 1990, c. 6, 6.1.
68 B.C. Statutes 1980, Utilities Commission Act, c. 60. ss. 22-25.
69 Ibid., s. 22 (1).
70 16 U.S.C. § 824a.
71 42 U.S.C. § 4321-27 (1988).
72 Prodan, see n. 68, 460.
73 "U.S.-Canadian Electricity Trade," 17.
74 Lise H. Jordan, "The Role of State Regulators in United States/Canadian Energy Trade," *Journal of Energy and Natural Resources Law* 10 (1992):4, 380.
75 See generally Pamela Prodan, "The Legal Framework for Hydro-Québec Imports," *Tulsa Law Journal* 28 (1993):435.
76 Ibid., 451.
77 ME. REV. STAT. ANN. TIT. 35-A, § 3302 (West 1988); see also Prodan, 452.
78 Title 35-A, § 3191; see also Prodan, 454.

79 There are questions as to the authority of state regulators to review the prudence of utilities in choosing to enter into international power purchase agreements instead of purchasing power from an alternate source; see Jordan, 386.
80 ME. REV. STAT. ANN. TIT. 35-A, § 3133 (West 1987); see also Prodan, 454.
81 A complete account of the demise of CMP's proposal as well as a treatment of the role of environmental considerations in the regulatory process governing electricity exports is found in Prodan, see note 75.
82 Jordan, 383.
83 Prodan, 454.
84 GATT here refers to the 1947 General Agreement, as modified by various amendments and interpretive understandings, as well as any relevant provisions of the various Uruguay Round Agreements. All of these instruments now come under the institutional umbrella of the World Trade Organization (WTO) and are frequently now referred to as a single regime, either the WTO or the GATT/WTO.
85 See M.J. Trebilcock and R. Howse, *The Regulation of International Trade,* (London and New York: Routledge, 1995), 217. See also P. Nicolaides, "The Nature of Services," in *The Uruguay Round: Services in the World Economy,* ed. K.P. Sauvant (Washington and New York: World Bank and United Nations Centre for Transnational Corporations, 1990).
86 "Thailand: Restrictions On Importation of and Internal Taxes on Cigarettes," BISD 37th Supp. (1990):200; "United States – Restrictions on Imports of Tuna," 30 I.L.M. (1991):1594, not adopted by GATT Council; "In the Matter of Canada's Landing Requirement for Pacific Coast Salmon and Herring," Final Report of the Panel, 16 October, 1989 (Canada-U.S. Free Trade Agreement).
87 "Salmon and Herring Landing Requirements."
88 Ibid.
89 See J.H. Jackson, *The World Trading System: Law and Policy of International Economic Relations,* (Cambridge, MA: MIT Press, 1990), 284.
90 "Canada-Import, Distribution and Sale of Alcoholic Drinks by Canadian Provincial Marketing Agencies," BISD, 35th Supp., 37, 90.
91 Power Corporation Act, R.S.O. 1990, c. P.18, s. 85 (4).
92 L.S. Portasik, "The Transition to Fully Competitive Bulk Power Markets: Federal Regulatory Developments in the Electric Power Industry," *Energy Law Journal* 15 (1994):365-71.
93 Ibid., 372.
94 National Energy Board, "Review of Inter-Utility Trade in Electricity," (January 1994):18-19.
95 Ibid., 19-20.
96 Energy Options Advisory Committee, "Energy and Canadians: Into the Twentieth Century."
97 National Energy Board, "Review of Inter-Utility Trade in Electricity," 19.
98 In the Puerto Rican Milk case, a Canada-U.S. FTA Panel considered whether a requirement that Québec milk producers join an American regulatory scheme for milk quality as a condition of continued access to the Puerto Rican market

violated provisions of the FTA, the predecessor agreement to NAFTA. The Panel abstained from making a finding as to whether National Treatment or the provision parallel to Art. XI of the GATT were violated by this requirement. The Panel did find, largely on the basis of the extraordinary obstacles that Puerto Rico and the American authorities had placed in front of Québec satisfying American regulatory requirements, that reasonable expectations of market access had been undermined. "In the Matter of Puerto Rico Regulations on the Import, Distribution and Sale of U.H.T. Milk from Québec," Final Report of the Panel, 3 June, 1993.

99 The basic legal framework is outlined in Trebilcock and Howse, *The Regulation of International Trade,* ch. 6.

100 The legal framework as it has evolved from the 1947 GATT through to the Uruguay Round Agreement is outlined in Trebilcock and Howse, *The Regulation of International Trade,* ch. 5.

101 See Prodan, 447.

102 See generally, Black and Pierce. As they suggest, full internalization of environmental costs through these instruments is easier said than done. Nevertheless, it remains the appropriate and most coherent goal of regulatory policy.

103 R. Howse and M.J. Trebilcock, "The Fair Trade-Free Trade Debate: Trade, Environment and Labour Rights," Law and Economics Working Paper, Faculty of Law, University of Toronto, 1994. See also R. Stewart, "Environmental Regulation and International Competitiveness," *Yale Law Journal* 102 (1993):2039.

104 Canada, National Energy Board, "Inter-Utility Trade Review: Inter-utility Co-operation" (1992), 5-1 and 5-27.

105 Ibid., 5-6.

106 NEB, "Wheeling Review."

107 The history is well summarized in National Energy Board, "Inter-Utility Trade Review: Transmission Access and Wheeling," 1-3; and National Energy Board, "Inter-Utility Trade Review: Inter-Utility Co-operation," (1992):1-12.

108 National Energy Board, "Review of Inter-Utility Trade in Electricity," 4 January,1994.

109 Ibid., 5.

110 Ibid., 7-9.

111 Conversation with Eugene F. Gorzelnik, Director of Communications, NERC, 30 May, 1995.

112 Energy and Environmental Programme, Royal Institute of International Affairs and Science Policy Research Unit, University of Sussex, "A Single European Market in Energy," (1989):35. (Hereinafter "Sussex report.")

113 The monopolistic structure and public ownership of national electric industries can be explained by the fact that they are commonly regarded by the state governments as natural monopolies of strategic economic and infrastructural importance, viable industrial policy tools and significant contributors to public budgets; see Ralf Boscheck, "Deregulating European Supply: Issues and Implications," *Long Range Planning* 111 27 (1994):5.

114 Rüdiger Dohms, "The development of a Competitive Internal Energy Market in the European Community," *Connecticut Journal of International Law* 9 (1994):805.
115 International Energy Agency, Energy Policies of IEA Countries – 1993 Review, (Paris: OECD/IEA, 1994), 232. (Hereinafter Energy Policies of IEA Countries.)
116 Sussex report, 35.
117 International Energy Agency, *Electricity in IEA Countries: Issues and Outlook* (Paris: OECD/IEA 1985), 87. (Hereinafter "Electricity in IEA countries.") See also International Energy Agency, Seminar on East-West Energy Trade, Proceedings, 162, Vienna, 3-4 October, 1991. (Hereinafter "East-West Energy Trade.")
118 The IPS normally operated in parallel with the United Power System (UPS) of the former USSR, although it could be run independently under certain circumstances; see "East-West Energy Trade," 164. The electric system of the former GDR, a former member of CDO/IPS is being integrated into the UCPTE system. The interconnection of the electricity grid of the former GDR to the UCPTE system is not expected to come into operation before mid-1995: see "Energy Policies of IEA Countries," 253.
119 Ibid.
120 International Energy Agency, *Electricity Information* 1993 *(Paris: OECD/IEA* 1994), 47. (Hereinafter "Electricity Information 1993.")
121 Such exchanges are referred to as diversity exchanges of electricity; see Heckman, *Canada-*U.S. *Trade in Electricity,* 9.
122 Ibid., 9-10.
123 "Sussex report," 39; Dohms, 806; Boscheck, 111.
124 "Sussex report," 39.
125 "Electricity in IEA countries," 93. Because it lacks a primary fuel base and has inadequate base load capacity, Austria imports electricity from Eastern Europe through back-to-back converter stations in the winter, and exports electricity back in the summer.
126 "East-West Energy Trade," 210. While the operation of the CDO/IPS system is managed by a central entity, the UCPTE system operation, based on the "self-management" of control areas, is decentralized.
127 "East-West Energy Trade," 210.
128 Ibid., 169.
129 For import/export data shown in the charts, see *Electricity Information 1993*, 48-50.
130 "Electricity in IEA countries," 103.
131 It will be recalled that the regulation of electricity exports from Canada to the U.S. was prompted by concerns over long-term contracts by Canadian utilities to the U.S. and the fear that Canada would not be able to repatriate this power in times of need.
132 "Electricity in IEA countries," 90.
133 "Energy Policies of IEA Countries," 232.
134 Ibid., 247.

135 The exclusivity clauses in supply contracts are being challenged by the German Federal Cartel Office on the basis of E.U. legislation; Ibid, 253.
136 "Electricity in IEA countries," 276.
137 "Energy Policies of IEA Countries," 325.
138 Ibid., 425.
139 Ibid., 387.
140 Ibid., 218.
141 Ibid., 476.
142 Dohms, 809.
143 Treaty Establishing the European Economic Community as amended in 1992 by the Maastricht Treaty. (Hereinafter the "EC Treaty.")
144 Dohms, 807.
145 Ibid., 809.
146 Stephen Padgett, "The single European Energy Market: the Politics of Realization," (1992) *Journal of Common Market Studies,* Vol. XXX, 1 (March 1992):53-54.
147 Directive of 29 October 1990 on the transit of electricity through transmission grids (90/547/EEC), OJ NO. L 313, 13.11.90.
148 Directive concerning a Community procedure to improve the transparency of gas and electricity prices charged to industrial end-users 90/337/EEC), OJ NO. L 185, 17.7.90.
149 Commission of the European Communities, Proposal for a Council Directive concerning common rules for the internal market in electricity (92/C65/04), OJ NO. C 65, 14.3.92.
150 Ibid., 813.

Comments by

KENT L. EDWARDS
YVES MÉNARD
ADONIS YATCHEW

KENT L. EDWARDS

"Regulation of Trade in Electricity: A Canadian Perspective" by Howse and Heckman is an excellent overview of the current regulatory framework, and impressions emerge based on the premise that the paper is a foundation work for further study.

Looking back is a little like reviewing the "as built" plans of the House that Jack Built. It served in the past, but somehow it does not have much comfort for today's needs. We can either knock it down and rebuild from the ground up, or attempt to remodel the existing structure, preserving that which can be saved.

This is only difference in degree. It seems that however resolved, change is needed.

Allow me to make two assertions shaping my outlook, based on observation within the Ontario scene.

First, Regulation in the broader context is anything requiring somebody else's approval. Regulation defines the boundaries of autonomy.

Second, Destiny is determined by fundamental laws and need. Applied to trade in electricity, regulation will ultimately adapt to the marketplace. Attempts to shape the marketplace through imposition of law will be transitory at best and at worst will only delay the inevitable.

On these premises, there are three key elements shaping trade in electricity today that we must look to for future regulation. Resolution of these issues will determine whether to fix the house or rebuild it. These elements are of course:

- Competition
- Privatization
- Regulation

Here I define regulation more broadly to include the natural or market regulation that in turn impacts on the more narrow regulation of trade in electricity.

COMPETITION

Competition needs a broad enough definition to include not only competition within the electric industry, but also competition to the electricity industry.

It also means genuine competition that has "winners" and "losers."

We have witnessed the deregulation of the gas industry with plummeting gas prices resulting. This factor, among others in Ontario has realigned the energy market. Natural gas is taking a larger market share at the expense of electricity. Electricity has not responded to the market forces. The fallout has been the over-forecasting of electricity needs and the present glut of electricity. The monopoly stance of the industry cannot be supported in the face of this competition. The monopoly rates are not competitive and cannot be tolerated. Unfortunately, a monopoly carries significant risk in this environment. There are no competitors (within the electric industry) that can be put out of business, absorb the losses, to help weed out inefficiency, and to restore a new more competitive industry (in this case, with the natural gas industry). We need, therefore, to consider new regulations to create and ensure real competition. This leads me to the issue of privatization.

PRIVATIZATION

Discussion on privatization generally centres on the virtues of the efficiencies of the private sector and the relative inefficiency of the public sector. This is not the real issue of privatization. The real issue is risk. A private monopoly is inherently more ruthless in its charges than a public monopoly, thus demanding heavy regulations to protect consumers. The only risks to a private monopoly are attracting unwanted additional regulation or external competition to the industry (the same as with a public monopoly).

However, bringing competition within an electricity industry blended with privatization creates a major shift in the risk set.

With privatization an inefficient and private generator can go out of business – be bankrupted if you like. The industry continues as inefficient operations disappear, or are purchased at low cost by survivors. The risk of losses is removed from the electric customer and placed on the private investor, the shareholder. Inefficient technologies cease to operate in favour of better technology (a good example is the closing of coal-fired generation in the U.K. after privatization there).

In Ontario, the apparent competition within Ontario Hydro business units is a pale comparison to real competition. Although some salary levels and jobs may be affected for the "winners" and "losers," in the end the customer's money is redistributed from so called "winners" to "losers." Inefficient technologies are continued as the customers are the shareholders and cannot avoid the loss. The only real competitive threat in Ontario remains the external market forces outside the electricity monopoly (such as Natural Gas and sometimes WUC). However, this is a commendable first step.

A competitive and privatized electricity market will require industry regulation only to ensure competition and privatization. Market forces will otherwise prevail.

REGULATION

Despite its requirement for reciprocal policies on issues such as competition and access, I suspect more fundamental regulation will prevail. Environmental regulation and public pressure better known as "NIMBY" or "Not In My Back Yard" will continue to exert greater restrictions on electricity trade than NAFTA. State and provincial regulation is already directed at looking after the local needs first by prohibiting dumping and demanding so-called "least-cost planning."

Significant new transmission rights of way are not expected. New generation is difficult enough to site for local consumption. Resistance to new generation to serve distant markets is becoming insurmountable. Barring a technology breakthrough to allow transmission corridors to carry significantly more power, we can expect transmission tie lines in future to be increasingly dedicated only to system reliability. This has a number of consequences:

- Generation will be increasingly sized and sited to suit local loads.
- Large distance wheeling is improbable as it requires excess load over generation at the source end and excess generation over load at the receiving end to avoid overloading tie line capacity. Electricity systems contemplating wheeling will be wise to factor distance into the wheeling charge formula, whether wholesale or retail, to encourage locating generation as close to load as is practical for regulating the need to transmit energy (vs. "postage stamp" rate).

SUMMARY

To conclude, I submit that the next step in regulatory evolution is to compare the future trends of competition and privatization with the present electric industry regulatory environment. Only then does it become abundantly apparent what regulatory changes are needed to let these future trends occur. While attempts to delay the inevitable will possibly extend the life of some existing monopolies and provide a source of funding for a body of lawyers, the electric customer will pay. Instead, it is time to change the focus of industry regulation to open the electricity market to the competition and privatization best serving the energy consumer.

YVES MÉNARD

I would like to commend the authors for having taken the trouble to try to explain what distinguishes electricity from other products or services, on the one hand, and what an electric power system is, on the other hand.

I think it is a truism to affirm that only a good comprehension of these realities will bring about efficient and realistic regulation of the trade in electricity. After all, I would argue that rather than strive to make reality conform to the law, we should strive to make the law conform to reality.

I would also like to commend the authors for having managed to achieve the "comprehensive overview of trade in electricity and its regulation from a Canadian perspective" announced at the beginning of their paper. This is no mean task and achievement.

However, as also announced at the beginning, their contribution was intended as a basis for further studies and not to provide strong policy prescriptions.

I will try to address this in my commentaries.

Of course, I do not pretend to be able to set down detailed policy prescriptions in the short space allotted me, and neither do I pretend having the competence to do so, even if I was allotted sufficient space.

I would like, in the following space, to suggest avenues that, I hope, could be useful in conceiving new regulatory frameworks, both at the federal and at the provincial levels, for the trade in electricity.

My commentaries will thus address two main topics, which I will refer to as the "Tomato Dilemma," or "Is Electricity a Good or a Service," and the "Californian Bungalow Syndrome," or the Canadian tendency to import concepts from elsewhere, most particularly from the United States.

Also, in keeping with the commendable effort of the authors to explain the nature of electricity and of electrical networks, I would like to propose another way of looking at those realities, which could also be useful in conceiving new regulatory frameworks. I call that last part "The Flat Earth Paradigm," or the "Spider's Web Analogy."

First, we will try to decide whether electricity is a good or a service.

The "Tomato Dilemma," or "Is Electricity a Good or a Service?"

The authors point out that a threshold issue as regards the trading of electricity between the United States and Canada is whether electrical energy is deemed a good or a service for the purposes of the International Trade Law Framework (i.e., GATT and NAFTA).

I believe this question also has some bearing on the way the domestic trade in electricity, both interprovincial and intraprovincial, and both at the wholesale and retail levels, should, or should not, be regulated.

To begin with, the same way a tomato can be both a fruit and a vegetable, depending on whether you are a botanist or a cook, I would hold that electricity can both be a good and a service. It all depends on whether you are a "captive" consumer or not, and on the purpose for which you are buying this electricity.

Basically, electricity in itself can be characterized as a product, or good, that is manufactured from natural resources, i.e., water, gas, oil, etc., the same way plastic is a product made from petroleum.

For many of its uses, however, there exists no useful substitute for electricity. For instance, I do not know many people interested in using gas can openers or coal toasters or, even, in switching to gas lighting.

As pointed out, electricity is also a unique "commodity" in that it cannot be stored easily and economically, thus making the need for production and consumption to be simultaneous.

Consequently, on the one hand for "captive" uses and "captive" consumers, i.e., for those uses for which no practical and economical substitutes exist, by consumers within the Utility's concession area who cannot practically shop around for suppliers, electricity could be deemed an essential service, albeit rendered by a manufactured good.

This creates an obligation to serve on the beneficiary of an electricity distribution concession or franchise, i.e., an obligation to plan and to invest so as to ensure that at all times this essential service is rendered. Conversely, this gives the utility a right to be able to recuperate necessary investments thus made, whether through costs or incentive-based rates.

It also creates an obligation on governments to ensure that these "captive" consumers will not be charged undue prices by the utility for those "captive" uses, and that the utility will recuperate only those investments it "prudently" made.

On the other hand, for other purposes, electricity should be considered a manufactured product or good, the subject of trade, both within and outside Canada, and a source of significant economic spin-offs for its whole economy.

Future regulations,[1] in keeping with the oft-avowed trend toward less intrusive, more efficient government, should then be focused where they are really needed, i.e., in setting down rates and conditions for "captive" uses by "captive" consumers and ensuring that no cross-financing takes place at the expense of those "captive" consumers.

For the aforementioned other purposes, however, and starting with trading between utilities both within and outside Canada (i.e., wholesale trading and wheeling), I submit the electricity trade should be deregulated as much as is possible, if not completely.

The Import of Concepts from Elsewhere or the "Californian Bungalow Syndrome"

Whether through natural but, of course, misplaced modesty, or through a tendency to be self-deprecatory, Canadians often are under the impression that, if it's being done elsewhere, most particularly in the U.S., the U.K. or, for Québécois, in France, it's probably better and should therefore be done here.

This is what I call the "Californian Bungalow Syndrome"; we go visit California, fall in love with those beautiful bungalows and have one built here, neglecting the fact that it is cold here, and for long periods of time, that it snows a lot, and that courtyard tiles are murderous when it snows, etc.

In other words, we tend to neglect the fact that what is being done elsewhere is done according to *their* needs and takes into account *their* realities, and that what we do here should be done according to *our* needs and take into account *our* realities.

The privatization debate is going on full steam world wide, and the demonopolization debate is going on full steam in the U.S.

Many of the countries where privatization has occurred did not have a choice.

Argentina is a case in point.

A utility such as SEGBA (Servicios Eléctricos del Gran Buenos Aires S.A.) could not but be privatized back in 1992. With an economy just recovering from years of horrendous inflation (over 200% per year) and government intervention and micromanagement, it was almost impossible to judge the utility's financial performance; the service it was offering was of low quality, "non-technical" losses were very high (around 20%), many of the people employed by it were employed under unofficial "make-work" projects, etc., etc., etc. In short, its managers and its personnel, even while competent, were powerless to change the situation and could not be held accountable anymore.

At that point, you privatize, and at the same time demonopolize, and start with a clean slate.

In the U.S., on the other hand, most of the industry already belongs to private interests. You therefore do not need to privatize but having noticed that the existence of vertically integrated monopolies, even if small monopolies, constitutes a barrier to a more efficient use of existing resources and equipment, you force a functional demonopolization. This functional demonopolization then becomes a tool for bringing about an optimized use of resources, of production units and of transmission networks, for reducing rate disparities and, ultimately, for favouring lower uniform rates over extended territories. Ah yes, one other tool to reach this objective is by facilitating the creation of Regional Transmission Groups, whose operations would be coordinated through Central Dispatching Centres.

So in effect, you functionally break down small monopolies and, by favouring subsequent regroupings through Regional Transmission Groups and coordination through Central Dispatching, you aim for more efficient operations and scale economies.

At the risk of sounding self-complacent, it would seem to me that the objectives pursued through this functional demonopolization have already been achieved in Canada.

We use, through Central Dispatching, the lowest-costing production units at all times, we have an optimized use of transmission networks over extended territories, and we enjoy low, indeed some of the lowest in the industrialized world, uniform rates over these same extended territories.

We also manage to supply our "captive" consumers a highly reliable product to fulfil their "captive" uses.

This has been achieved through nationalization of those networks, rather than through privatization *but* the results are there all the same!

What we then should be wary of is to have U.S. solutions imposed on us, or to impose them on ourselves, to the ultimate detriment of our rate payers, a.k.a. "captive" customers.

I suggest that it now becomes urgent that we develop Canadian solutions that, while reflecting important global trends, will not merely be an importation of solutions that have found favour elsewhere.

These Canadian solutions will most importantly have to recognize the structural differences that exist in the Canadian electrical utility industry, most notably, the presence of large integrated provincial power pools and public ownership of the majority of the industry's assets.

The Flat Earth Paradigm or the Spider's Web Analogy

A paradigm is a way through which we apprehend reality. As such it can be helpful, or, more often, it can be a hindrance.

For instance, up to a few centuries ago most European people were convinced the Earth was flat. For day-to-day activities this had no practical consequences. However, when people wanted to sail across the ocean it caused them no small anxieties, the least of which being that they thought that if they sailed too far they would fall into an abyss and dragons would eat them!

But from the moment this paradigm changed and they started viewing the earth as being round, new possibilities emerged, one being that they could go *around* the Earth. It may have been extremely difficult, but it became *possible!*

It is somewhat the same with electric networks and electricity.

People, whether they be engineers, economists, lawyers or lay, tend to think of electric networks as highways over which quantities of electricity are moved from one point to another. This is useful for most day-to-day operations and serves nicely as a general mathematical model.

This is also the current paradigm and, I believe, a hindrance when it comes time to think of new commercial opportunities in relation to the trade in electricity.

In fact, as physicists will tell you, electricity, on an alternating current system, does not really move from one point to another. What you find on such alternating current systems are electromagnetic fields that vibrate.

At the risk of sounding facetious, I propose that, rather than thinking in terms of quantities of electricity moving from one point to another over a highway, we should try to think in terms of a spider's web with "shakers" and "shakees" disposed across the web.

The transmission network would be the web, the "shakers" would be the generating units and the "shakees" would be the loads.[2]

This would then be the paradigm I propose.

This new paradigm is useful in that it helps to dissociate the physical, or "wires," aspect of an electric network from the commercial aspect.

In fact, on an integrated network, where, or from whom, your electricity is supplied, i.e.,what or who shakes you, is of no real concern to you. Electrons, or electromagnetic waves, are not identified or identifiable and as long as you are pleasurably "shaked," everything is peachy!

This is where the usefulness of Central Dispatching, both at the commercial and technical levels, comes in. It ensures that at all moments and at all places on the "web," "shakers" and "shakees" are balanced on the one hand, and that operating profits generated, notably by optimized use of production units, or "stacking,"[3] are duly apportioned on the other hand.

With this paradigm, it appears to me that the commercial possibilities of the Electricity trade grow appreciably. For instance, it then becomes possible for Hydro-Québec to sell electricity to Southern California Edison. It is not *probable*, at least in the short term,[4] but it is *possible*.

Again, I submit that future regulation should focus on those controls necessary to ensure that the smooth technical operation of networks is ensured and that, at all times, "captive" consumers are supplied a highly reliable product, at the lowest cost, to fulfil their "captive" uses.

Conclusion

We are now at a crossroads as regards the trade in electricity. The old ways of doing business and of looking both at that trade in electricity and at the way it should be regulated are being questioned, if not squarely put aside, in most of the industrialized world.

In Canada, as mentioned in my above comments, we have the advantage of already enjoying many of the benefits sought by other countries that are either privatizing, or demonopolizing, or both.

I submit it is now opportune for us to rethink the Canadian regulatory frameworks of the Trade in Electricity with a *Canadian* perspective, focusing those regulatory frameworks where they are really needed so as to preserve what we already have achieved on the one hand, and to take full advantage of the opportunities of the new North American energy Market on the other hand.

NOTES

1 I am talking here, of course, only of the regulation of the *trade* in electricity. Regulation of the *means* of production, transmission, and distribution is an altogether different matter and not the subject of the present comments. Suffice it to say that once those means (i.e., generating stations, transmission lines, etc.) have been duly authorized, and for as long as they are operated within duly authorized environmentally acceptable limits, the trade in the electricity they produce should be regulated as little as possible.

2 This works also for a radial circuit, i.e., one with generation at one end, load at another, and no interconnected circuits in between. Only, then, you have to view the circuit as a vibrating antenna. However, this way of viewing the network, in the case of a radial circuit, possesses no advantage over the "highway" way for the purpose of separating the "wires" operation from the commercial operation.

3 In an integrated network, the act of putting on line at any given time, sequentially, the unit least costly to operate.
4 That is until, and if, a more fully integrated and coordinated North American Transmission System, with stronger ties between the five synchronous systems that now compose it, comes into existence.

ADONIS YATCHEW

This is an interesting and useful presentation in that, in conjunction with certain other publications (and I have in mind here some of the NEB publications), it provides extensive relevant legal background on trade in electricity. What is particularly pleasant is that despite the fact that it is a document written on legal matters and by legal scholars, there is a notable absence of "heretofores" and "notwithstandings"in the text.

I will begin my comments by focusing on GATT Article III (4), which has been incorporated into NAFTA Art. 301. This article requires that imports be accorded the same treatment as domestic products. As the authors correctly point out, the provision will have important impacts on the way trade liberalization takes place, the restructuring that is implemented, and regulatory changes that are introduced.

What I would like to see is a more detailed discussion of the likely implications of this article and other legislation for specific restructuring scenarios. In sketching such an analysis, it is probably reasonable to assume that the current surplus in generating capacity in Ontario will persist for an extended period of time. (Domestic demand growth is unlikely to absorb existing capacity prior to the turn of the century, barring early retirements of nuclear units at Pickering or Bruce.)

Suppose now that transmission access is permitted at the wholesale level and that a significant number of, say, industrial companies or municipal utilities seek to purchase their supplies south of the border. If, as has been suggested by various parties, a surcharge is imposed on transmission charges to cover off the costs of stranded assets, then the level of the surcharge will increase with the loss of load. An increase in surcharge could lead to further loss in load and so on, though the loss in load is constrained by the availability of cross-border transmission capacity.

In any event, the "similar treatment" provision of NAFTA would seem to require that transmission charges for similar service be the same for parties supplied by provincial sources as well as those purchasing electricity extra-provincially. Put another way, those leaving the Ontario pool could not be charged a differential transmission fee.

Let me come at this issue from a different perspective, one that is well established in the U.S. debate on stranded assets. Existing electricity industry assets were constructed for the benefit of provincial users – present and future – on the broad expectation that the builder/owner would be permitted reasonable recovery of costs over the lifetime of the assets without inequitable redistribution of these costs to captive customers or to taxpayers. (In the U.S. there is the related question as to how much of the risk of loss in value should be transferred to the shareholder when the structure of the industry changes as a result of legislation such as the Energy Policy Act of 1992.) The net implication is that extra-provincial purchases by a wholesale entity that intends to remain in the province could arguably be held responsible for at least some portion of the assets that it strands.

The question then for our legal minds is whether under such circumstances, attempting to impose differential transmission charges related to asset stranding on such customers could be successfully contested under NAFTA Art. 301.

So far I have talked about transmission access. I would be curious to find out whether from a legal point of view, uneconomic stranding could be better avoided by alternative models of restructuring, such as a structure where the grid acts as a monopsonistic buyer, acquiring supply on a competitive basis and, perhaps during the period of excess capacity, is supervised by a regulator. Summarizing then, if there are going to be stranded assets, do alternative restructuring models constrain, from a legal point of view, one's ability to deal with them? I suspect they do.

Let me preface my next remarks by the following perhaps obvious observation. Firms, be they private or public, resemble homeostatic organisms in a number of ways. The view that pervades standard economic texts, i.e., that firms are profit maximizers or cost minimizers within a certain set of constraints, is far too narrow. Even analysis of strategic behaviour using game theory that focuses principally on *economic* choice variables is inadequate. In fact, the survival instinct translates into strategic behaviour in political and legal settings as well.

One of the implications of these statements is that the passage of trade-enabling legislation is endogenous to the interests of various groups, some better organized, others less so. (In Ontario, these include Ontario Hydro, its collective bargaining units, the MEUs, industrial customers, private power producers, commercial and residential customers.) Indeed, politicians routinely engage in a kind of interest group calculus, in order to guide their sponsorship, support, or opposition to legislation. When legislation involves international trade, the compromises required to balance interests become that much more complex.

Now, a major portion of the Howse and Heckman chapter discusses the recent European experience with respect to establishing international trade in electricity. At last report, the movement towards institutionalizing an open electricity market in Europe has stalled. It seems that the winners in a free European electricity market have yet to persuade the losers. Interest groups have been able to severely retard continent-wide restructuring.

What I would like to see is some analysis, admittedly speculative at this point, relating the recommendations of the European Commission to the passage of legislation by member countries – why certain legislative steps have been taken, why others have been avoided. Furthermore, for our purposes it would be particularly useful to assess the implications of the European legislative experience for North America in general and Ontario in particular.

Let me begin my final area of comments by noting one stubborn empirical fact. In the United Kingdom, Norway, Sweden, and New Zealand one of the first steps in restructuring involved separation of monopolistic segments of the industry from those amenable to competition. In particular, all of these countries created separate transmission companies. In the U.S., competition was institutionalized by legislating access to the grid (Energy Policy Act 1992). Why the different approaches? It would seem that ownership played an important role. In all of the former countries, transmission was publicly owned, while in the U.S. the preponderance of such assets were part of private vertically integrated utilities.

This observation underscores the point that private property rights play a central role in determining the path along which the industry evolves. Evidently, vertical separation of transmission in the U.S. would have involved complex legal proceedings. (Only recently have there been indications that some U.S. utilities may be required to vertically deintegrate.)

In this context, I would make the point that the creation of *international* private property rights – through privatization or through contractual agreements of purchase and sale that would be present under transmission access – will shape and perhaps constrain further restructuring or evolution.

This statement is not intended to undermine the virtues of private ownership, but merely to point out that before one proceeds to create private property rights, it is important to undertake all necessary and logically prior restructuring steps. Two critical steps that fall into this category are the separation of transmission from generation and the establishment of a mechanism that can deal competently with the stranded assets issue.

Hydro Restructuring and the Regulation of Conventional Pollutants

DONALD N. DEWEES

I. INTRODUCTION

Restructuring of electricity generation in Ontario may result in competition between Ontario Hydro and non-utility generators (NUGs) to install new generating capacity in the future.[1] Some scenarios involve dismantling of Ontario Hydro and the privatization of the parts including individual generating stations. Here we will consider a more modest scenario in which the thermal division of Ontario Hydro continues as a public entity, but NUGs are allowed to enter the generation market without the approval of Ontario Hydro. In fact NUGs have already installed some capacity in Ontario in recent years. Some NUGs, such as pulp and paper mills, generate their own steam and have turned to cogeneration of steam and electricity to improve efficiency and to reduce their total energy costs. Other steam producers wish to produce electricity and sell it to Hydro for distribution on the grid to other customers. Interest in NUG generation has increased in recent years with the emergence of efficient combined cycle gas turbines (CCGT) that generate steam with the waste heat from the gas turbine and then produce electricity from both the gas turbine and from a steam turbine. CCGT installation might be of interest to an electricity consumer or to a NUG that wished only to sell electricity to the grid. I assume that new thermal generating capacity in Ontario may be fueled by coal, oil, gas, biomass, or some portion of the municipal waste stream.

The construction of any new thermal generating station will require a Certificate of Approval from the Ontario Ministry of Environment and Energy (MOEE) and will be subject to some environmental regulations, possibly including environmental assessment. For an extensive review of environmental regulation and electricity generation technology, see SNC (1992). This paper considers whether existing regulations give rise to a level playing field as between Ontario Hydro and NUGs as they compete to provide additional electrical generating capacity, and if not, what regulations might do so. The motivation for the study is the concern that existing regulations, which often treat different industries differently, might be more strict for one type of generator than for another that causes similar environmental risks. I will refer to regulations that create a level playing field as "efficient" regulations, and those which do not as "discriminatory" regulations. Discriminatory regulations are potentially a major problem if there is likely to be close competition between Hydro and NUGs. The cost of air pollution controls can represent a significant proportion of the capital and operating cost of a well-controlled coal-fired generating station.

The identification of efficient or discriminatory regulations requires the specification of an objective against which existing or proposed regulations may be measured: I will use the concept of economic efficiency as the basis for comparison. It has long been established that efficient pollution control requires that the polluter pay for the marginal external harm caused by his emissions.(Baumol and Oates 1988, 45). This will cause abatement until the marginal cost of abatement equals the payment for one marginal unit of additional emission; if this equals the marginal harm of that emission then marginal costs will equal marginal benefits of abatement, and efficiency is achieved. If pollution control is achieved through command-and-control regulation, then in addition to optimal regulation the source should pay for the marginal external harm caused by an increase in electricity generation. (Freeman, Burtraw, Harrington, and Krupnick, 1992.) This will ensure that the price of electricity includes the residual harm being done by post-abatement emissions. It is this latter element that has led to the incorporation of "environmental adders" into the rates of some U.S. utilities. If the policy instrument is marketable pollution permits (MPPs) then, under certain assumptions, holding those permits will impose the external cost on the polluter and no adder is required. MPPs exist in some circumstances in the U.S. but not in Ontario, so our analysis of existing regulations will consider only command-and-control regulation.

One might use some concept such as fairness or equity as the basis for assessing the existence of bias in environmental regulations. Unfortunately, the definition of these terms is not unambiguous. Fairness might require that a plant of a given type in a given location using a given fuel should be treated the same regardless of ownership, but this would result from any efficiency rule as well. What about two plants that differ greatly in the cost of abatement: does fairness require that they be limited to identical pollution density in the stack gases; that they be limited to identical quantities of pollution per kilowatt-hour (kWh) generated; that they be required to install abatement equipment of equal efficiency even if this results in unequal emissions? Should identical plants in different locations meet identical emission standards, identical ambient air quality standards, or should their emissions be related to the harm that they cause? I have not found a simple and workable definition of equity or fairness that could be used as a basis for evaluating bias, so this study will focus on economic efficiency.

It is true that estimating the physical harm caused by pollution emissions is difficult and uncertain; valuing that physical harm in dollar terms is at least as difficult and uncertain. Yet these problems do not disable the analysis proposed here because the important analysis is a comparison between alternate forms of regulation. We need not be certain that reducing sulphur dioxide (SO_2) emissions causes benefits of $x per kilogram, in order to conclude that a regulation that causes Ontario Hydro to incur marginal SO_2 abatement costs twice those of competing NUGs will inefficiently tilt new investment toward the NUGs.

I assume that new generating capacity may be fuelled by coal, oil, gas, biomass, or some portion of the municipal waste stream. I assume that any new nuclear facilities would be built only by Ontario Hydro. I exclude new hydroelectric investment simply to confine the scope of this paper. The focus is therefore on new thermal combustion generation. I assume that the NUG will likely engage in cogeneration, producing process steam and electricity or operating a CCGT.

The pollutants of interest are those that have been identified with these sources and regulated by the MOEE or identified as a serious public concern. The air pollutants are sulphur oxides (sulphur dioxide and sulphate: SO_2 and SO_4, jointly known as SO_x); particulate matter (PM); nitrogen oxides (NO_x); volatile organic compounds (VOCs); carbon dioxide (CO_2), a greenhouse gas; and some trace toxic contaminants associated with coal and waste combustion: arsenic, cadmium, lead and mercury. See Table 1. The principal air pollution problems are SO_x, NO_x, toxics, and PM from coal-fired utility boilers and SO_x, NO_x, and PM from oil-fired boilers of utilities and NUGs. Wood waste boilers release PM. Municipal waste boilers will give rise to PM and possibly toxics. From gas-fired boilers and CCGTs the concern is VOCs and NO_x. Water pollutants are waste heat, in the case of generating stations using lake or river water for cooling, and oil and grease, in the case of some coal-fired power plants. Where solid fuel is burned, whether coal, biomass, or municipal waste, the ash must be disposed of. This may give rise to solid waste disposal problems and in some cases to air or water pollution associated with the ash piles.

Table 1
Air Pollution Emission Rates (gms/kWh) (Uncontrolled, except as noted)

Source	SO_X	*PM*	NO_X	*VOC*	CO_2	*As*	*Cd*	*Pb*	*Hg*
Steam									
Coal*									
Uncontrolled	6.4	15.3	3.6	0.014	950	.003	.0002	.0014	.00007
FGD/ESP	0.65	0.14	2.76		950				
Residual Oil	2-7	0.4	0.6-2.2	0.02	750	.000086	.00007	.00021	.00001
Natural Gas	<0.1	0	0.73-2	N/A	450				
Wood	0.34	2.5	0.3-0.6	–					
Municipal Waste	1.18	24	0.3-0.46	N/A					
Gas Turbine									
(Combined Cycle)									
Distillate	1.2	0	0.7-0.9	N/A					
IGCC	0.1-0.4	0	0.18-0.36	N/A	450				
Natural Gas	0	0	0.22-1.6	N/A					

Sources: Ottinger, 1990:111, 115, 117, 119, 120, 122; SNC, 1992: Table D.28.
NA: Data not available.
– : Not relevant.
FGD/ESP: Flue gas desulphurization with fabric filter; or electrostatic precipator.
IGCC: Integrated gasification combined cycle – a coal combustion technology.
*Pulverized bituminous coal, dry bottom, 1% sulphur, 9% ash. Emissions depend on a number of features of the boiler, so these figures are indicative only of emissions from one common design.

The efficiency criterion might be applied to the regulation of Ontario Hydro and NUGs at any of three levels of sophistication. First and simplest, for each pollutant of interest arising from a given technology and fuel, do Hydro and the NUG face the same regulatory requirement? Second, recognizing that Hydro and a NUG might choose different locations or a different scale of plant for the installation of new capacity, do the regulatory requirements for each pollutant vary by location in a way that is consistent with variations in marginal benefits and do variations with scale match corresponding effects on marginal benefits? This criterion is more difficult to address because it requires some understanding of how marginal benefits vary by location and by rate of emission. Third, recognizing that Hydro and a NUG might tend to choose different technologies and fuels, thus emitting different pollutants, does the regulation of different pollutants impose marginal costs that are equal to the marginal benefits for each pollutant and location? This test is considerably more difficult than the first two because it requires the estimation of a value of marginal benefits for each of the pollutants and a comparison of those values. Such valuation is highly uncertain and contentious.[2] We will concentrate on the first two questions.

The existing Ontario environmental regulations are reviewed in section II. Section III discusses the likely shape of damage functions for the pollutants and derives from this the efficient objective function for each pollutant. Section IV evaluates the extent to which existing regulations deviate from efficient regulations. Section V suggests alternative forms of regulation that would reduce the bias identified in Section IV.

II. REVIEW OF ONTARIO ENVIRONMENTAL REGULATIONS

Air Pollutants

Any source of air or water pollution in Ontario would be subject to the Environmental Protection Act.[3] Section 14.(1) of the EPA prohibits the discharge of a contaminant into the natural environment that "causes or is likely to cause an adverse effect." "Adverse effect" includes: impairment of the quality of the natural environment for any use that can be made of it; injury or damage to property or to plant or animal life; harm or material discomfort to any person; an adverse effect on the health of any person; and more.[4] This general nuisance prohibition would apply equally to Ontario Hydro or any NUG. It protects the local environment near any source of pollution to the extent that the Crown can prove that the emission of the contaminant causes or is likely to cause harm given the rate of emission and the circumstances of the local environment.

Section 6 of the EPA prohibits the discharge of a contaminant in excess of the amount allowed in any regulation. Regulation 346, General - Air Pollution,[5] includes another general nuisance prohibition in section 6. Section 5 of Regulation 346 prohibits any person from causing the concentration of a contaminant at a point of impingement to exceed the amount allowed in Schedule 1. Schedule 1 lists a number of contaminants and the allowable concentration for each. (See Table 2.) A point of impingement is any point at which the air pollution might cause harm, generally taken to be ground level near the property line. Because the offence is causing the concentration to exceed the limit, this section could be interpreted to require each polluter to take the environment as it finds it. If several polluters contribute to a concentration exceeding the limit, each might be charged even though the emissions of each alone would not exceed the limit. In fact, however, the Ministry determines compliance with this regulation by estimating the concentration in the air using air pollution dispersion models, and considering only the source in question. Thus these limitations are enforceable as emission limits. They apply equally to Ontario Hydro and to any NUG.

Table 2
Utility Air Emissions and Ontario Regulations

	Source		*Ontario Limits (1/2-hour average)*	
Pollutants	*Coal Oil*	*Gas*	*Schedule I (μg/m³)*	*Other* (μg/m³)*
SO_x	✓			830
PM	✓		100	
NO_x	✓	✓	500	
VOC	✓	✓		
CO_2	✓	✓	-	
Arsenic	✓			1
Arsine	✓		10	10
Beryllium	✓		0.03	
Cadmium	✓		5.0	
Chromium	✓		-	5
Copper	✓		100	
Formaldehyde	✓		65	
Lead	✓		10	
Manganese	✓		-	7.5
Mercury	✓		5	
Nickel	✓		5	

*Appendix 10, "Summary of Point of Impingement Standards, Ambient Air Quality Criteria (AAQCs) and Approvals Screening Levels (ASLs)" in Table 1 "General Information: Certificates of Approval, Section 9, Environmental Protection Act," Toronto, MOE, August 1992.

The MOEE has established guidelines for ambient air quality with respect to a much larger number of contaminants than are listed in Schedule I. These include point of impingement standards based on half-hour averages and 24-hour averages. (See Table 2.) In general these guidelines add little to the limits of Schedule I, with respect to electricity generation.

Regulation 346, section 12 limits the emission from incinerators of organic matter having a carbon content, expressed as equivalent methane greater than 100 ppm by volume, and would probably apply to a watts-from-waste plant that might generate electricity. Any cogeneration facility that will incinerate wastes or will generate wastes requires approval under EPA Regulation 347, section 12. New incinerators have been banned under section 12.1 since 1992.

In addition to these regulations and guidelines of general application, there are more specific regulations of air pollution that might arise from thermal electricity generation. The Countdown Acid Rain Program produced four regulations of sulphur dioxide emissions from four corporate sources in Ontario: Algoma,[6] INCO,[7] Falconbridge,[8] and Hydro.[9] The portions of these regulations that apply after 1994 are reproduced in Appendix A. Each specifies a total annual limit on the discharge of sulphur dioxide from the subject corporation. Ontario Hydro is limited to 175,000 tonnes per year of sulphur dioxide. In addition, the Ontario Hydro regulation limits the sum of sulphur dioxide and nitric oxide emissions to 215,000 tonnes, requiring some control of nitric oxide, but allowing a trade-off with sulphur dioxide.

The Countdown regulations for Algoma, INCO, and Falconbridge specify limits for total sulphur dioxide emissions from the sinter complex or smelter complex. If electricity were generated at any of these sites, any emissions of sulphur dioxide arising from that generation activity might or might not have to comply with the total emission limit in the regulation depending on whether the generation facility was considered part of the "complex."

In 1994, the MOEE adopted air pollution emission guidelines for stationary combustion turbines.[10] These would apply to any new combustion turbine generator installed after November, 1994, and to large turbines installed after June, 1994. The guidelines limit NO_x, carbon monoxide, and sulphur dioxide emissions all in relation to the power output of the turbine. The NO_x limit is derived by multiplying a factor by the power output of the combustion turbine and adding the product of another factor times the heat output that is recovered from the combustion turbine.[11] The limit is enforced by measuring the concentration of NO_x in the exhaust gas and calculating the allowable concentration based on the factors in the regulation. This regulation, which is referred to as a new source performance standard, allows considerably greater NO_x emissions per unit of power output for small turbines than for large turbines, presumably because control costs are higher for small turbines. The regulation also provides greater emission limits for liquid fuel turbines than for natural gas turbines. Carbon monoxide is limited to 60 ppm by volume in the exhaust gas and sulphur dioxide is limited to 800 grams per gigajoule of useful energy output. This regulation does not distinguish between Ontario Hydro and NUGS.

There are also specific regulations for particular activities or industries. Regulation 338 Boilers Regulation applies to boilers not operated by Ontario Hydro that are put into operation after 1986 or to which changes have been made, changes that might increase sulphur dioxide emissions.[12] Section 3(1) prohibits the use of fuel oil or coal with a sulphur content exceeding 1% unless the emissions are no more than if the fuel sulphur content did not exceed 1%. Section 3(3) prohibits the use of fuel that would result in a wet sulphate deposition exceeding 0.1 kg/hectare/year from sources north of a line between approximately Collingwood to Perth Road (north of Kingston), up to 50 degrees north latitude.

There are federal guidelines for thermal generating stations. The current version of these guidelines is reproduced in Table 3. The MOEE would apply these guidelines in any application for a new generation source in Ontario. In general these are less restrictive than the applicable Ontario regulations.

CO_2 discharges are not regulated in Ontario. However Ontario has adopted a management strategy for greenhouse gas emissions which states two objectives: to reduce the carbon intensity of energy supplied by Ontario Hydro by 5% between 1990 and 2000; and to stabilize Ontario Hydro's net greenhouse gas emissions at 1990 levels by the year 2000 and to reduce them 10% by 2005.[13] Whether this commitment would survive a restructuring of Ontario Hydro is questionable. In any event, there is no reason to expect NUGs to make a similar commitment.

Other Pollutants

The discharge of water pollution is controlled by section 14 of the Ontario EPA by the Ontario Water Resources Act[14] particularly section 30, by control orders and program approvals with individual sources, and by regulations adopted under MISA, the Municipal Industrial Strategy for Abatement. MISA has developed separate regulations for each of nine major industrial sectors. One of these, Effluent Monitoring and Effluent Limits – Electric Power Generation Sector,[15] applies explicitly to named existing Ontario Hydro generating stations. Section 16 prohibits discharges that exceed daily

Table 3
Federal Thermal Power Generation Emissions Guidelines*

Pollutant	*Fuel*	*Limit (720-hour average)*
NO_2	Coal	258 ng/J
	Oil	129 ng/J
	Gas	86 ng/J
PM	All fuels	43 ng/J
SO_2	Coal, oil:	
	258-2580 ng/J uncontrolled	258 ng/J
	>2580 ng/J uncontrolled	90% control

Notes: ng = nanogram
J = joule

*Thermal Power Generation Emission National Guidelines for New Stationary Sources, P.C. 1990-333, section 4.

and monthly average concentration limits specified in Schedules 2 and 3. Schedules 2 and 3 specify concentration limits for a set of water pollutants for about a dozen different effluent streams from each of the generating stations. In general, the concentration of a pollutant allowed in a specific effluent stream is the same for all generating stations. Table 4 presents a typical entry in Schedule 2, the limits applicable to the R.L. Hearn generating station. Section 16 limits the ph of discharge water to the range 6.0 to 6.9 for any sample and any stream. I understand that compliance with these MISA regulations will be costly for Ontario Hydro, particularly the oil and grease limitations.

What water pollution regulation would be faced by a NUG? There is no generic NUG regulation, so MISA would only apply if the NUG was a member of a listed industry. The pulp and paper industry and the petroleum refining industry are possible NUGs and each is covered by a MISA regulation. Taking the pulp and paper regulation as an example,

Table 4
MISA: Electric Power Generation Sector

		Schedule 2 R.L. Hearn TGS			
	Parameter	*Types of Non-Event Process Effluent Stream*	*Monitoring Frequency*	*Daily Concentraion Limit (mg/L)*	*Monthly Average Concentration (mg/L)*
ATG	*Column 1*	*Column 2*	*Column 3*	*Column 4*	*Column 5*
8	Total Suspended	ATWE	D	70.0	25.0
	Solids	WTPE	D	70.0	25.0
9	Aluminum	ATWE	W	13.0	4.50
		WTPE	W	13.0	4.50
9a	Iron	ATWE	W	2.50	1.0
		WTPE	W	2.50	1.0
25	Oil and Grease	OWSE	W	29.0	13.0

Explanatory notes:
Types of Non-Event Process Effluent Streams:
ATWE = Ash Transport Water
WTPE = Water Treatment Plant
OWSE = Oily Water Separation
ATG = Analytical Test Group
mg/L = Millgrams per Litre
D = Daily monitoring requirement

Source: O. Reg. 215/95 Effluent Monitoring and Effluent Limits – Electric Power Generation Sector, Schedule 2.

section 3 (1) states that the regulation applies to plants listed in a schedule, implying that it applies to the entire plant and therefore to any boiler or generation facility that was an integral part of the plant. Section 14 (1) prohibits daily discharges that exceed the daily plant loading limits set out in Schedule 2 and section 12 (3) prohibits monthly average discharges that exceed the monthly average loading limits in Schedule 2. Sections 14 (5), (6), and (7) limits the allowable concentration of dioxin in the effluent, and section 14 (8) limits the ph of the waste water to the range of 6.0 to 9.5. Section 15 provides that after 1995, the plant may calculate new plant loading limits by determining the ratio of actual production over the last three years to the reference production rate published in Schedule 4 of the regulation, allowing emissions to increase if output increases and requiring it to decrease if output decreases. Section 16 requires that the effluent from any of the streams not cause more than 50% mortality to fish placed in 100% effluent. Table 5 presents a typical entry in Schedule 2, the effluent limits for the Canadian Pacific Forest Product mill in Thunder Bay. The petroleum sector MISA regulation is similar to that of the pulp and paper sector, specifying daily and average monthly plant loading limits for each plant and adjusting these over time based on actual production compared to the reference production rate.[16]

There is a fundamental difference between the MISA regulation that applies to Ontario Hydro and regulations applying to other industries. The Hydro regulation limits the concentration of pollutants in the effluent stream, while the other regulations limit the total discharge of pollutants in kilograms per day and per month, using factors that were calculated based on the output of each plant. On the one hand, each of these regulations limits pollution discharge in relationship to output, so that if production increases, the allowed discharge should increase, presuming that the effluent flows at a hydro plant are proportional to electricity generated. But, on the other hand, Ontario Hydro may discharge more pollutant if it increases the water flow in a waste stream. It is not obvious whether Hydro has much choice about the volume of these waste flows, but if so, it has added flexibility in the amount of pollution it may discharge. Because the other regulations tie the allowable effluent to the output of product, installing some cogeneration capacity would not increase production output at all and would not increase allowable pollutant discharge. Thus any installation of cogeneration capacity by the pulp and paper industry or the petroleum refining industry would require an emission control strategy that did not allow increased water pollution discharges beyond the pre-existing effluent limits. In this respect, the MISA regulations are more restrictive for NUGs than for Hydro.

Table 5
MISA: Pulp and Paper Sector

	Canadian Pacific Forest Products, Thunder Bay			
	Parameter	*Monitoring Frequency*	*Daily Plant Loading Limit (kg/day)*	*Monthly Average Plant Loading Limit (kg/day)*
ATG	*Column 1*	*Column 2*	*Column 3*	*Column 4*
1A	Biochemical Oxygen Demand (5-day)	D	28400	14200
6	Total Phosphorous	W	796	483
8	Total Suspended Solids (TSS)	D	38100	22400
16	Chloroform	W	10.6	5.34
17	Toluene	W	0.611	0.611
20	Phenol	W	1.17	1.17
33	Adsorbable Organic Haldide Phase one	W	4670	3620
	Phase two	W	2800	2170
	Phase three	W	1490	1160
24	2,3,7,8-Tetrachlorodibenzo-para-dioxin	Q		
	2,3,7,8-Tetrachlorodibenzofuran	Q		
	TEQ	Q		

Explanatory notes:
D = Daily monitoring requirement
W = Weekly monitoring requirement
Q = Quarterly monitoring requirement
ATG = Analytical Test Group
kg/day = Kilograms per day
TEQ = Total toxic equivalent of 2,3,7,8 substituted dioxin and furan congeners

Source: O Reg. 760/93 Effluent Monitoring and Effluent Limits – Pulp and Paper Sector.

If the NUG was not in a MISA sector, it would be covered by regulations of general application. And whether or not the NUG was in a MISA sector, it would be covered by the general provisions of the Ontario Water Resources Act.[17] The Act provides in section 30 that any person who discharges any material into any water that impairs the quality of the water is guilty of an offence, while section 28 deems the water to be impaired if the material causes or may cause injury to any person, animal, bird, or other living thing. This legislation would appear to be neutral in its application to Hydro and any NUG. Because the test is whether the discharge may cause harm, it would appear to favour small sources over large sources, simply because it is easier to detect materials in amounts that might cause harm when the volume of the discharge is large.

In the case of a NUG located in an urban area, such as a hospital, university, or other non-industrial commercial operation, it would likely discharge its water waste into the local sewers and would be subject to the municipal sewer by-law, if any. It seems unlikely that this would be more restrictive than the law applicable to discharges directly into the natural environment. Whether there would be some advantage to a small generating plant discharging water into the municipal sewer system is unclear.

Environmental Assessment and Approvals

The Ontario Environmental Assessment Act[18] requires that some proponents of undertakings submit an environmental assessment before proceeding with the undertaking. In general, public bodies are subject to the EAA unless they are exempted from it, and by regulation Ontario Hydro is indeed covered.[19] The proponent of an undertaking must submit an assessment including a description of the undertaking, its rationale, alternatives to the undertaking, a description of the environmental effects of the undertaking, reasonable mitigation measures, and an evaluation of the environmental advantages and disadvantages of the undertaking.[20] The assessment is submitted to the Minister[21] who must arrange for a review of the assessment by the Ministry.[22] When the review is completed, the assessment and the review must be made available for public inspection[23] and the public is entitled to make submissions to the Minister. At this point,the Minister may order that a hearing be conducted[24] and/or order additional research into the undertaking.[25] No undertaking covered by the Act may proceed until it has been approved by the Minister.[26] Undertakings may be brought within the Act or excluded from the Act by regulation.[27]

The EAA does not apply to private sector proponents and projects unless they are specifically designated for coverage. Thus, a generation project pursued by a NUG would not be subject to the assessment procedure unless the proponent was a municipality, since municipalities are explicitly covered by the Act.[28] Any refuse-fired incinerator burning over 100 tons per day of refuse requires approval under the EAA, but this requirement is of little effect since new incinerators have been banned since 1992. In addition, the Federal Environmental Assessment process requires assessment for any power project that falls under federal jurisdiction.[29]

It appears that the environmental assessment requirement falls heavily on Ontario Hydro and often not at all on a NUG building a similar project. The difference may be somewhat less in practice than it appears from the legislation and regulations, since a NUG that proposed to construct a major power project might be required to undergo EA, and Ontario Hydro could be exempted from EA for small standardized projects, or it could employ a class EA for such projects as a group and thereby avoid individual EA for each such project.

There are significant costs associated with environmental assessment. Studies have shown that the direct costs of environmental assessment in general run from less than 1% of the cost of the project to over 10% of the cost (Ahmad 1987, 7; World Bank 1991, 20). The relative cost is lower for large projects because of economies of scale. Ontario Hydro has found direct EA costs running between 0.4% and 4.7% of project costs. But in addition to these direct costs there are costs associated with the delay caused by the EA process. If the proponent makes a significant capital investment prior to undertaking the EA, it must pay additional interest cost on the invested capital until the project is cleared to proceed. If 10% of the project cost was incurred prior to performing the EA, and if the real interest rate was 5%, a one-year delay would impose interest costs of 0.5%. The cost of delay would be larger if the delay was longer or if there was a need to have the project on line at a particular time so that extra costs were incurred as a result of the delay. It has been suggested that the popularity of CCGT generation plants over the last few years arises in part from the short lead time from project conception to having the capacity on line.

To the extent that Ontario Hydro would be required to perform environmental assessments on its projects while NUGs would not be required to assess projects of similar impact, there is a clear bias against Ontario Hydro as a proponent. The practical importance of this point depends on whether the projects that Ontario Hydro would pursue in the future actually have potentially greater environmental impacts than projects of NUGs, and if not whether the NUG projects would be subjected to similar environmental assessment requirements.

III. EFFICIENCY, REGULATION, AND THE SHAPE OF DAMAGE FUNCTIONS

Theory

Economic efficiency, it was argued above, requires that the marginal costs of abatement with respect to the rate of discharge equal the marginal benefits of abatement with respect to the rate of discharge for each source and pollutant and that the source must pay for the marginal external harm arising from the remaining emissions. Assessing the efficiency of regulations requires that we compare the marginal costs of abatement that they cause to the marginal benefits of that abatement.

This in turn requires that we have some idea of the way in which marginal benefits of abatement for a given pollutant vary from one location to another and how they vary with the concentration or discharge rate of the pollutant.

We may consider three attributes of new generation capacity that might vary between Ontario Hydro and a NUG, and consider what effect each should have on allowable emission rates. The location of the new source might be important, as marginal benefits should be higher in areas with high population density or highly sensitive environments than in other areas. Efficient regulations should therefore impose higher marginal costs of abatement in such sensitive areas. The technology of generation or the fuel source might or might not vary between Ontario Hydro and a NUG, but if the regulation is driven by equating marginal benefits and marginal costs it should impose the same marginal costs of abatement with respect to a given pollutant, regardless of the technology or fuel. Finally, the rate of emission of a pollutant might vary because of the size of the source. If size or technology should cause one source to emit so much more than another that it affected the local concentration and therefore the marginal benefits of abatement, then the efficient regulation would impose marginal costs that reflected this difference in marginal benefits. In all of these cases, we can see that there is not a necessary relationship between ownership and the factor that might cause variations in a regulation. Thus, sources of similar location and emission rate should be subject to regulations imposing similar marginal costs.

A simple model of pollution dispersion, the relationship between the emission rate from a source and the concentration of pollution at a receptor, provides useful background for this discussion. In general, the pollution concentration $C(X,Y)$ at a location X,Y depends on the background concentration $C^o(X,Y)$ that would exist without known emissions on the emission rates d_i of all sources that affect location X,Y and on any mechanism, such as the wind or settlement of pollutants, that removes the pollutant from the area. Assume that the rate of removal is proportional to the concentration. The contribution that source i makes to the concentration at X,Y is described by the product of the emission rate d_i and the transfer coefficient $a_i(X,Y)$ which is defined as

$$a_i(X,Y) = \frac{\Delta C(X,Y)}{\Delta d_i} \qquad (1)$$

Then:

$$C(X,Y) = C^o(X,Y) + \sum_{i=1}^{n} a_i(X,Y)d_i \qquad (2)$$

If the atmosphere is perfectly mixed, as we may assume for the greenhouse gases, then $C(X_j, Y_j) = C(X_k, Y_k)$ for all j and k and $a_i(X, Y)$ is the same for all *X*, *Y*, and *i*.

In this case, equation (2) simplifies to:

$$C = C^o + a\sum_{i=1}^{n} d_i \qquad (3)$$

If d_i is a small fraction of the sum of all discharges, then C will not change significantly with changes in d_i. This in turn means that if the damage as a function of concentration is reasonably well-behaved its derivative with respect to d_i will be essentially constant for changes in d_i. Thus marginal harm will be essentially constant. Otherwise variations in *C(X, Y)* may cause variations in the marginal harm and its negative the marginal benefit of abatement. This requires that we understand the relationship between C(X,Y) and marginal benefits.

This analysis will focus on human health effects of air pollutants, although other effects will also be considered. Studies of the benefits of air pollution abatement generally find that, for most of those pollutants, the vast majority of the benefits arise from reducing risks to human health, so a focus on these effects will capture the most important relationships.[30] The remaining benefits are concentrated in the categories of soiling and cleaning and of property values (aesthetics), both of which would be closely correlated with population density. An implication of this concentration of benefits on the human experience is that the benefits of abatement for these air pollutants are roughly proportional to population density.

Consider first a pollutant for which the sources to be regulated contribute only a small fraction of the total relevant emissions. An example would be greenhouse gases, for which Ontario emissions represent less than 1% of total world emissions. The effect of greenhouse gases arises over long periods of time during which the world's atmosphere may be regarded as well mixed. Thus the effect of the discharge of a kilogram of CO_2 does not depend on where it is discharged; within a few years it will be evenly dispersed throughout the atmosphere. Equation (3) applies. In such a case, the damage arising from Ontario's emissions depends on Ontario's contribution to the world stock of CO_2, which is so small that the world concentration is not significantly changed as a result of Ontario emissions. If the benefit-of-abatement function is well-behaved so that the marginal benefits vary slowly with increases in the world stock of CO_2, the marginal benefits of abatement of CO_2 must be essentially independent of changes in the emission rate in Ontario.

Therefore the marginal benefits of abatement of CO_2 in Ontario may be assumed to be constant. This will be true, with respect to Ontario emissions, for all of the greenhouse gases and indeed for any truly global pollutants.

Consider next a persistent pollutant that accumulates in the environment, increasing in concentration over a period of years. Whatever the shape of the function relating marginal changes in the concentration of this pollutant at any point in time to changes in the harm caused, any change in the rate of discharge will simply change the time at which various concentrations are reached. The benefits of abatement in any year are the present value of the stream of benefits over many years in the future, during which time the concentration will presumably increase steadily. The result will be that the marginal benefits of abatement of a persistent pollutant in any year will not change sharply with the rate of discharge in that year. Again, marginal benefits are insensitive to the discharge rate. This should be true for the persistent toxic pollutants, the most important of which are listed in Table 2.

Turning to general air pollutants that cause harm in the year in which they are emitted, we can explore the relationship between the ambient concentration and the marginal benefits by remembering that marginal benefits are the sum of benefits to all members of society and by examining the relationship between the concentration of a pollutant and the harm to the individual. Figure 1 shows several plausible shapes for the individual damage function and the resulting aggregate damage functions. If damage increases in proportion to concentration, the damage function is linear through the origin and marginal harm is constant as shown in Figure 1a. Different individuals may differ in the slope of their damage functions but the aggregate is linear and aggregate marginal harm is constant. This function is often assumed for carcinogens.

Equally plausible is a damage function that is linear above a threshold and zero below that threshold, shown in Figure 1b. Individual marginal damage is zero below the threshold and constant above it. If individuals have such a damage function and if individuals differ in the threshold at which harm begins to occur, then the aggregate damage function will be spread out much more than the individual function. Aggregate marginal harm will rise slowly up to the concentration at which the least sensitive individual is affected, above which it will be constant.

A third damage function is the log-linear function, which begins at zero for zero concentration then rises exponentially up to some maximum. Marginal harm has a similar shape. See Figure 1c. If individuals differ in the exponent of this function then the aggregate will be a composite exponential function and marginal harm will again be constantly increasing.

Figure 1
Damage as a Function of Pollution Concentration

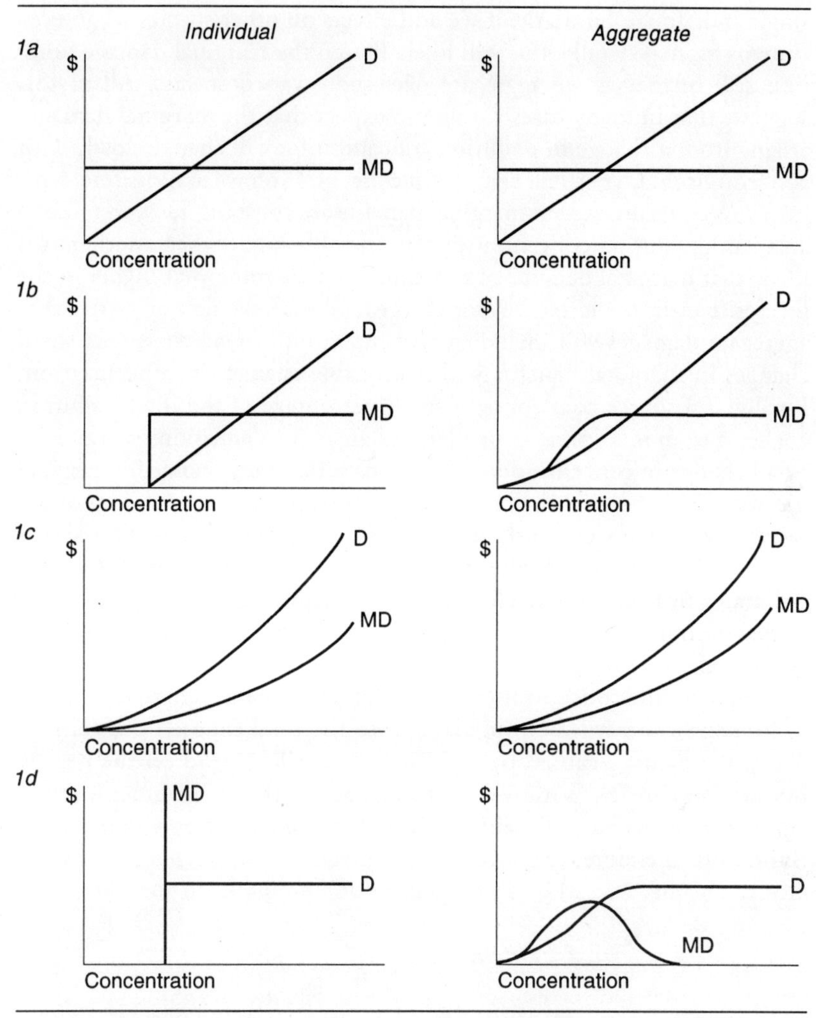

It is sometimes suggested that there may be a critical pollution concentration below which there is no harm and above which serious harm occurs. The extreme version of this function is the step function, shown in Figure 1d. Individual marginal harm is a spike at the concentration at which the harm occurs. Aggregating over individuals with varying sensitivity could yield an aggregate marginal damage curve that was flat over a wide range or which varied slowly over this range, depending on the distribution of sensitivity in the population.

The above discussion shows that as we aggregate over individuals, marginal damage functions tend to become flatter than when we look at a single individual. Similarly, if we add effects on other species or on ecosystems more generally, this will likely flatten the marginal damage function still further as we aggregate over more varied species. All of this suggests that in many cases we should expect that the marginal damages arising from a change in pollution concentration will change slowly with concentration. Over small changes in concentration we may assume, with little loss of accuracy, that marginal benefits are constant. However, if one area has a pollution concentration considerably higher than another, it is likely that marginal benefits of abatement will be somewhat higher in the former than in the latter. So, for the general case, we do not expect that aggregate damages will include a step function; instead we expect small changes in marginal benefits with reasonable changes in concentration. Finally, if a source contributes a small percentage of the total pollution concentration in an area, even a large change in its emissions will cause a small change in concentration, and hence little or no change in marginal benefits. Table 6 shows the proportion of total provincial emissions of the relevant pollutants currently discharged by Ontario Hydro. This shows that moderate changes in Hydro's emissions, or the emissions of the same pollutants by NUGs, should change total Ontario emissions only modestly, so that even if marginal harm is not constant with respect to emissions, it will not change greatly.

Whatever the concentration of a pollutant, the damage that it causes to human health will be proportional to the total exposed population. Thus, the damage caused by a given source will depend on the nearby population density. With respect to damage to the environment, sensitivity to a particular pollutant may vary from one location to another. It is difficult, therefore, to generalize about variations in damage from one area to another caused by a unit release of a pollutant in that area.

Empirical Evidence

The principal concern arising from environmental exposures to carcinogens is an increased risk of cancer. Since dose-response functions are estimated from populations that face high occupational exposures, there is uncertainty about risks at low doses. We do not understand the mechanism by which cancer is caused, at the level of cell biology, with sufficient certainty to know what model to use for extrapolating from the high doses on which epidemiological studies have been based to the low doses of current environmental exposure (Shipp and Allen 1994, 80). Nichols (1984, 131-3) reviews this literature noting that when predicting the effects of low doses, based on observed risks from high

Table 6
Ontario Hydo's Emissions from the Fossil System

Air Pollutant	*1991 Emissions (31.6 TWh) (000 Tonnes)*	*Contribution to Ontario Total Emissions %*	*ECS Study: Emissions at 25 TWh (000 Tonnes)*
SO_2	166.0	16	97.6
NO_x	55.4	16	36.3
CO_2	27000	14	20991
Part.	8.4	N/A	6.9
Mercury	0.002	20-30	0.001
Trace Metals	N/A	N/A	N/A

Source: Ontario Hydro (1993, 73).

doses, the linear (one-hit) model predicts the largest risks at low exposures, while other models produce estimates that are orders of magnitude smaller. It is generally assumed for regulatory purposes that at low exposures the risk is proportional to exposure with no safe threshold, although this may over-estimate risks at low exposures where the risks may be low or even zero (USEPA 1990, 2-41; Shipp and Allen 1994, 85). The linear model yields a linear risk or damage function and a constant marginal damage function. As Nichols notes, if the sub-linear models are correct, then the shape of the benefit function at low doses is irrelevant because the risks are trivially small.

We can review the empirical evidence for four major electric utility pollutants, sulphur dioxide, particulate matter, nitrogen dioxide, and volatile organic compounds (VOCs) to see what shape the evidence supports. Sulphur dioxide is not a carcinogen, but chronic or acute inhalation may give rise to health effects in humans. While the evidence on these effects is not satisfactory, some tentative conclusions have been drawn. The USEPA (1982, 1-105) concludes that respiratory symptoms and pulmonary function decrements may occur in children for exposures above perhaps 0.88 ppm (with some concurrent particulate exposure), but that the risks at annual average ambient sulphur dioxide concentrations prevalent in the United States are likely nonexistent or minimal. DPA (1990, c-155) states that there may be no risk of fatality arising from sulphur dioxide exposure, but on the other hand there may be a risk that is linearly related to concentration; linear risk equations are also presented for hospital days for respiratory conditions and for hospital admissions for respiratory disease. Much of the empirical literature concludes that the sulphur dioxide damage function is linear or is S-shaped with a large portion that is approximately linear.

Suspended particulate matter (PM) generally includes lead and sulphates along with carbon, ash, and other materials. PM is not regarded as carcinogenic, although some compounds that are adsorbed onto airborne particles may be carcinogens. Acute exposures to high levels of PM in conjunction with other pollutants have caused increased mortality and morbidity, but data problems and the conjunction of many pollutants in the air pose difficulties in drawing conclusions regarding the health effects of particulate matter alone (USEPA 1982, 97, 105). Most epidemiological studies have considered SO_2 and PM together; many of these have found an association with health effects. A recent study has found statistically significant increases in mortality from exposure to fine PM (less than 2.5 microns in diameter) even at levels below the current U.S. ambient air quality standards for PM (News Focus 1995, 34).

Lave and Seskin (1977) found that they could not reject the hypothesis of a linear relationship between concentrations of PM and sulphur dioxide and aggregate human health effects, while Ostro (1983) found a logistic relationship with PM and sulphate, which is generally correlated with sulphur dioxide. Cifuentes and Lave (1993) concluded that there was a linear relationship between mortality and the annual average concentration of SO_2 and PM in a city when these concentrations exceed the U.S. primary air quality standards, while air that is on average cleaner than these standards gives rise to a linear relationship that is one-tenth as great. They conclude that the concentration-response function consists of two straight lines, with a break point at the U.S. primary air quality standard. Ontario Hydro (1993, 76) estimated log-linear dose-response models, which means that marginal harm increases with increasing pollution concentrations. Hydro found that 90% of health effects occurred within 100 km of the generating station. Considering the health effects evidence, we conclude that there may be no effects at low concentrations and that marginal health effects increase gradually with concentration until well above the U.S. primary air quality standards. In locations where the electricity generation station contributes less than half of the ambient SO_2 and PM concentration, small variations in the emissions from the generating station would cause little change in marginal harm.

DPA does not list other effects of SO_2 except that there is a threshold for phytotoxicity at 0.02 ppm. The USEPA (1982, 1-35) concludes that for any plant species there is a threshold concentration above which injury will occur, and that this threshold varies with each plant species and with other environmental conditions. There is no suggestion that exposure above the threshold will be fatal; instead there is some visible damage. Plants may recover from short fumigations, so the duration of exposure matters. This suggests a modest environmental damage function that is linear in concentration above a threshold.

Finally, there is the acid rain problem. While soils and lakes may have a buffering capacity, their capacity to absorb acid varies enormously. In any given lake or ecosystem that is beyond this threshold there is a range of precipitation within which more is worse and less is better. Because the emissions from a given source will affect different regions at different times, and because they combine with emissions from other sources, it is likely that some marginal harm should be attributed to all Ontario acid gas emissions. It seems unlikely that the emissions from a given electricity generation source will so influence the total acid rain in the region that marginal damages will change with the emission rate of the source. Therefore, the aggregate damage functions arising from acid rain related to fossil fuel emissions in Ontario should be approximately linear over the relevant range of emissions. Combining human and plant effects should yield an aggregate damage function that may be zero at low concentrations but with a broad range in which marginal harm increases slowly with concentration.

There is some evidence regarding other forms of harm from PM. No direct effect on plants has been confirmed at concentrations common in North America except for sulphates (USEPA 1982, 37). Material damage may occur to the extent that the PM contains sulphates. In addition, PM causes soiling, requiring cleaning or more frequent painting of building surfaces. Reliable damage functions have not been developed (USEPA 1982, 45). PM can affect visibility directly. The extent of impairment depends on particle size and concentration, and on humidity, since suspended particles will attract atmospheric moisture, growing in size. Increased concentration decreases visibility, but not necessarily linearly. The relationship is clearly continuous, and not associated with a threshold (USEPA 1982, 39).

Oxides of nitrogen, collectively referred to as NO_x, include primarily NO and NO_2. The latter is a reddish brown gas that gives photochemical smog its colour. These oxides also interact with VOCs in the presence of sunlight to form ozone. It is not thought that NO_x is harmful to humans at typical urban concentrations, but the ozone that it generates is clearly harmful to humans and to some plants. Ontario Hydro (1993, 77) estimated that ozone arising from Hydro's NO_x emissions caused losses equal to 0.4% of the value of Ontario crop production. Because of the complex atmospheric chemical reactions that form ozone, it is not easy to characterize the benefits of NO_x reduction in general terms. NO_x also participates in the formation of acid rain, discussed above. I conclude that the marginal benefits of NO_x control probably rise slowly with concentration and that they are greater in warm sunny weather, when smog formation is a problem, than at other times.

Volatile organic compounds, VOCs, are organic chemicals that exist as gases. Natural VOCs include methane from swamps and animals and isoprene and turpine hydrocarbons from plants. Burning fossil fuels releases unburned hydrocarbons of many types, some of which are carcinogens. The cancer risk is probably proportional to concentration and population density. Some VOCs contribute to the photochemical smog reaction, producing ozone and other damaging products. The chemical reactions involved in producing photochemical smog are complex, but generally more VOC means more smog. The smog problem is much more serious when the weather is warm and sunny than at other times.

For SO_x, NO_x, VOC, and PM, we conclude that marginal harm in general rises slowly with ambient concentration and that it is higher in areas with a high exposed population density. Marginal harm probably varies little with the emission rate of an individual source except where that source produces a large percentage of the local concentration of that pollutant. The production of ozone from NO_x and VOC emissions in the summer means that the marginal harm from emissions of these two pollutants must be higher in summer than at other times.

Implications for Policy

The conclusions drawn above regarding the shape of damage functions may be merged with the requirements for efficient pollution control discussed in the Introduction to draw some simple implications for the regulation of the air pollutants considered in this paper. Efficiency requires that:

- we equate the marginal cost of abatement for all sources of each of the greenhouse gases;
- for other pollutants we should impose higher marginal costs of abatement in urban areas than in rural areas;
- for NO_x and VOC we should impose considerably higher marginal costs of abatement in high pollution areas in the summer; and
- for SO_x, PM, NO_x, and VOC the marginal cost should be lower in areas where the pollution concentration is low.

IV. EVALUATION OF BIAS IN EXISTING REGULATIONS

We indicated in the Introduction to this paper that we would evaluate the efficiency of environmental regulations at two levels. The first evaluation simply considers whether the regulations differentiate among sources based on the ownership of the source.

Most of the air pollution regulations in Ontario do not depend on ownership or the industry that generates the pollution and therefore do not bias investment in new generation capacity as between Ontario Hydro and NUGs. The Countdown Acid Rain regulations apply only to Ontario Hydro and would discourage the construction by Hydro of new coal- or oil-fired generating capacity. However, if a private NUG proposed to construct a major new coal-fired generating station it seems unlikely that it could proceed without some specific regulatory attention by the MOEE, so it is unclear whether Hydro is favoured over a NUG in such a project. If the Countdown regulation was extended to any new major coal-fired boiler in Ontario, and if the emission limits under the regulation were made marketable, the marginal cost of abatement would be equated among these large sources. The Boiler regulation would apply to a NUG and not to Ontario Hydro and would tend to discourage construction of an oil-fired generating station by a NUG, but this might impose a marginal cost either higher or lower than the cost imposed on Hydro by the Countdown regulation.

Water pollution regulations are either generic, under the OWRA, or they are industry-specific, as under MISA. However, water pollution is likely to be a problem only for large coal-fired plants, so the importance of the apparent bias against NUGs in the existing MISA regulations is not clear.

The second level of evaluation considers whether the regulations vary by location and scale in a way that might favour either Ontario Hydro or a NUG given the type and location of new plant that each might be inclined to construct.

The general air pollution regulations, based on point-of-impingement concentrations, fail to be more strict for those pollutants in high population areas where this is warranted. On the other hand, uniform air quality guidelines are identical in urban and rural areas; if this required greater individual abatement in urban areas because of the contributions of multiple sources, it would impose higher marginal costs, which is appropriate given the greater marginal benefits arising from high population density. It is not clear whether this causes urban marginal costs to be too high or too low relative to marginal benefits; however, only by chance could it be just right. I presume that urban locations will be more attractive to small generators, which would probably be NUGs. Given the predominance of the individual point of impingement standards in regulating the relevant pollutants, small urban generation sources are more favoured under existing point-of-impingement regulations than they should be, and this may favour some NUGs.

SO_x emissions are governed by the Countdown regulation for Ontario Hydro, by the Boiler regulation for NUGs, both discussed above, and by the new combustion turbine regulation and the general point-of-impingement regulations for all sources. It is not obvious what marginal costs the generic regulations impose on new utility sources, and it is therefore unclear whether they will bias new construction toward or away from NUGs in cases where coal or oil is the likely fuel source. Because the concern about SO_x emissions is linked to long-range transport, a policy of marketable SO_x permits that applied to all significant SO_x sources in one or more regions of southern and central Ontario would seem desirable, as it would be less likely to cause the biases that are considered here. These could be combined with point-of-impingement limits to avoid local high concentrations. They should include all significant SO_x sources, not just electricity generators to maximize the number of participants in the market for permits.

NO_x is regulated by the Countdown regulations for Ontario Hydro, by the new combustion turbine regulation, and by the general point-of-impingement regulations for NUGs. As with SO_x emissions, it is not obvious what marginal costs these generic regulations impose on new utility sources, and it is therefore unclear whether they will bias new construction toward or away from NUGs. Because the concern about NO_x emissions is linked to long-range transport, a policy of marketable NO_x permits that applied to all significant NO_x sources in one or more regions of southern and central Ontario would seem desirable as it would be less likely to cause the biases of concern here. Because much of the desire to control NO_x arises from concern about photochemical smog in the summer, the failure to regulate NO_x emissions more strictly in the summer than at other times is clearly inefficient, but it is not clear that it creates a bias as between Hydro and NUGs or as among methods of generation.

The regulation of toxics from coal-burning is relatively modest, but it is not clear whether more stringent regulation is justified by the health benefits. In any event, the point-of-impingement regulations apply equally to all sources, so there is no obvious bias based on ownership. One might argue that these regulations should be more strict in highly populated urban areas; the existing regulations may therefore favour urban generators and thus favour NUGs.

The use of loading-based limits in MISA water pollution may discourage MISA sectors from engaging in cogeneration, to the extent that they would have used technology that gives rise to water pollution, since no increase in the allowable loading would arise from the addition of generating capacity. This is in contrast to the Electric Power MISA regulation which specifies pollutant concentrations and thus would allow increased emissions as electricity generation increases.

The requirement that all Ontario Hydro projects undergo environmental assessment imposes significant costs and delays that are not faced by NUGs in most cases. This seems to create a clear bias in favour of NUGs. Consideration should be given to excluding small Ontario Hydro projects from EA requirements, or allowing a class EA for some types of project that might be undertaken either by Ontario Hydro or by NUGs. In any event, the degree of EA that is required should depend on the size and character of the project and its likelihood to do serious environmental harm, rather than on its ownership.

Ontario has generally not followed the U.S. move toward technology-based standards, which requires the use of Best Available Technology, Lowest Achievable Emission Rate, or some other standard that requires that the source install technology that is equal to the best of some defined set. While technology-based standards are often popular because they appear to require each industry to do its best, they are criticized by economists because they do not consider benefits of abatement at all and because their implementation often discourages cost-reducing measures. Technology-based standards would differ for each mode of generation and fuel. Since Ontario Hydro is likely to burn coal while NUGs may burn gas or waste, there is no reason to expect that equal marginal costs of abatement for each pollutant would occur. We expect that the future adoption of technology-based standards for the pollutants discussed in this paper would create the possibility of greater rather than less bias between Ontario Hydro and NUG sources, and between different generation technologies.

V. PROPOSALS TO IMPROVE EFFICIENCY

The point-of-impingement air emission regulations do not differentiate between areas with a high population density and areas with a low density. While we may reasonably wish to protect air quality in areas of low population density, the benefits of a given unit of abatement must generally be greater in high-density areas. The Ministry should consider adopting point-of-impingement standards with more stringent limits for densely settled areas.

The Countdown SO_x regulations were effective in achieving 50% abatement, but may not serve well in the future if new generation capacity involves a large number of smaller generating stations. I recommend converting the existing regulations into marketable permits and extending them to all large sources of SO_x in the Province or in specified regions of the Province. New sources should have to purchase permits from existing sources, so that the total allowable SO_x emissions in the Province would remain at current levels. The point-of-impingement standards could be retained to protect local air quality.

We should implement MPPs for significant sources of NO_x and VOCs including Ontario Hydro and any NUGs, at least in southern Ontario. The MPP regulation should differentiate summer from other times of year and should include motor vehicles if possible.

If we were not to use MPPs or effluent charges, we should regulate the four main pollutants more strictly in southern Ontario than elsewhere, because the marginal benefits are greater in southern Ontario. The regulations should strive to equalize marginal costs of abatement for a given pollutant for all sources regardless of ownership or industry.

These conclusions are limited because of limited data regarding the marginal costs of abatement by the various technologies that may be used for electricity generation by Ontario Hydro and NUGs in the future. While future marginal costs cannot be known with certainty in any event, it would be useful to conduct further research that assembled existing cost data and analyzed them in accordance with the objectives suggested here.

Appendix A
Selected Text of Countdown Acid Rain Regulations

INCO Sudbury Smelter Complex - 1994
Reg. 660/85
1.(2) INCO Limited shall not emit sulphur dioxide from its Sudbury Smelter Complex in The Regional Municipality of Sudbury on any day in any year after its emissions of sulphur dioxide in that calendar year exceed 265 kilotonnes.

Falconbridge Smelter Complex - 1994
Reg. 661/85
1.(1) Falconbridge Limited shall not emit sulphur dioxide from its Smelter Complex in the Town of Nickel Centre in The Regional Municipality of Sudbury on any day in any year after its emissions of sulphur dioxide in that calendar year exceed 100 kilotonnes.

Algoma Sinter Operation - 1986/94.Reg. 663/85
2.(1)Algoma Steel Corporation Limited shall not emit sulphur dioxide from its Sinter Operation at Wawa on any day after its emissions of sulphur dioxide in any calendar year exceed 215 kilotonnes.

Ontario Hydro
Regulation 355; O. Reg. 281/87
2. Emissions of sulphur dioxide and of nitric oxide from the fossil-fuelled electric generating stations of Ontario Hydro shall not exceed, in the aggregate, 215 kilotonnes in any year after 1993.
4. Emissions of sulphur dioxide from the fossil-fuelled electric generating stations of Ontario Hydro shall not exceed, in the aggregate, 175 kilotonnes in any year after 1993.

Note: In all cases, only the section of the regulation relevant to 1994 and subsequent years is reproduced. All of the limitations reproduced here are in effect on January 1, 1994. Less stringent limitations generally applied in previous years.

NOTES

1 I would like to thank University of Toronto Electric Power Project for support of this paper. I would also like to thank Frank DeLuca and Jeff Brown for able research assistance.
2 See, e.g., damage estimates in Ontario Hydro (1993, 79-81); Ottinger (1990) and much lower estimates in Lave and Cifuentes (1993).
3 R.S.O 1990, c. E19 (Hereinafter EPA.)
4 EPA section 1. (1) (a) through (d).
5 R.R.O 1990, Reg. 346.
6 Algoma Sinter Operation - 1986/94, O.Reg. 663/85.
7 INCO Sudbury Smelter Complex - 1994, O.Reg. 660/85.
8 Falconbridge Smelter Complex - 1994, O.Reg. 661/85.
9 Ontario Hydro, R.R.O 1990, Reg. 355.
10 Ontario Ministry of Environment and Energy, "Guidelines for Emission Limits for Stationary Combustion Turbines" (Toronto: MOEE, March 1994).
11 For natural gas fired turbines, the power output factor, measured in grams per gigajoule, is 600, 240, 140 for turbines less than 3 megawatts, 3-20 megawatts, and over 20 megawatts respectively. The factor for peaking turbines over 3 megawatts is 280. Factors approximately twice as large are applied to liquid fuel turbines.
12 Boilers, R.R.O 1990, Reg. 338.
13 "Management Strategy for Greenhouse Gas Emissions," 16 January, 1995, recommendation by J.R. Burpee, General Manager, Fossil, submitted to the Board of Directors by the President and the Chairman, approved by the Board of Directors of Ontario Hydro in January 1995.
14 R.S.O 1990, c. O.40.
15 Reg. 215/95.
16 Effluent Monitoring and Effluent Limits - Petroleum Sector O. Reg. 537/93.
17 R.S.O 1990, c. O.40.
18 Ibid., c. E.18.;
19 R.R.O 1990, Reg. 334, section 3.
20 R.S.O 1990, c. E.18, section 5(3).
21 Ibid., section 5(1).
22 Ibid., section 7(1).
23 Ibid., section 7(2).
24 Ibid., section 12(2).
25 Ibid., section 11(2).

26 Ibid., section 5.(1).
27 Ibid., section 3.(b).
28 Ibid., section 3.(a).
29 The Environmental Assessment and Review Process Guidelines Order SOR/84-467, June 22, 1984, section 6 provides that the Guidelines apply to any proposal "(b) that may have an environmental effect on an area of federal responsibility; (c) for which the Government of Canada makes a financial commitment; or (d) that is located on lands, including the offshore, that are administered by the Government of Canada."
30 For example, Portney (1990, 57) presents a table derived from benefits estimated by Freeman, in which human health benefits represented almost three-quarters of total benefits arising from the U.S. air quality improvements between 1970 and 1978. Ontario Hydro (1993, 79) estimates that human health effects account for over 80% of the harm caused by emissions from fossil fuel plants in Ontario. Ottinger et al. (1990, 209, 228, 276, 351, 357, 362) find that health effects dominate the harm from SO_2 and NO_X, while visibility is the dominant effect of PM. Health effects represent about three-quarters of the total harm from utility air pollution.

BIBLIOGRAPHY

Ahmad, Y.J. and G.K. Sammy. *Guidelines to Environmental Impact Assessment in Developing Countries.* UNEP Regional Seas Reports and Studies No. 85, Nairobi. London: Hodder and Stoughton, 1987.

Baumol, William J. and Wallace E. Oates. *The Theory of Environmental Policy.* 2d. ed. New York: Cambridge University Press, 1988.

Cifuentes, Luis A. and Lester B. Lave. "Economic Valuation of Air Pollution Abatement: Benefits from Health Effects." *Annual Review of Energy and Environment* 18 (1988):319-42.

Dewees, Donald N. "The Efficiency of Pursuing Environmental Quality Objectives: The Shape of Damage Functions." Manuscript, Department of Economics, University of Toronto, 1995.

DPA. "Estimated Public Benefits of Implementing the Proposed Revisions to Regulation 308: Volume II - Appendices A to C." Toronto: Ontario Ministry of the Environment, July 1990.

Freeman, A. M., III, Dallas Burtraw, Winston Harrington, and Alan Krupnick. "Accounting for Environmental Costs in Electric Utility Resource Supply Planning," Quality of the Environment Division Discussion Paper QE 92-14, Washington, D.C.: Resources for the Future (1992).

Lave, Lester B. and Eugene P. Seskin. *Air Pollution and Human Health.* Baltimore: Johns Hopkins Press, 1997.

Nichols, Albert L. *Targeting Economic Incentives for Environmental Protection.* Cambridge, MA:MIT Press, 1984.

Ontario Hydro. "Task Force on Sustainable Energy Development Survey Team #4: Full-Cost Accounting for Decision Making." Toronto: Ontario Hydro, December, 1993.

Ostro, Bart. "The Effects of Air Pollution on Work Loss and Morbidity." *Journal of Environmental Economics and Management* 10:371-82.

Ottinger, Richard, et al. *Environmental Costs of Electricity*. Oceana: New York, 1990.

Patrick, David R. "Special Sources of Toxic Air Pollutants." In *Toxic Air Pollution Handbook,* ed. David R. Patrick, 461-68. New York: Van Nostrand Reinhold.

Portney, P. R. "Air Pollution Policy." In *Public Policies for Environmental Protection,* ed. P.R. Portnoy. Washington, D.C..: Resources for the Future, 1990.

Shipp, Annette M. and Bruce C. Allen. "Quantitative Methods for Cancer Risk Assessment." In *Toxic Air Pollution Handbook,* ed. David R. Patrick, 79-99. New York: Van Nostrand Reinhold, 1994.

SNC. *Non-Utility Power Generation: Environmental Approvals and Impacts.* Toronto: Ontario Ministry of Energy, 1992.

USEPA United States Environmental Protection Agency. *Air Quality Criteria for Particulate Matter and Sulfur Oxides,* Vol. 1, Washington D.C.: U.S.EPA, EPA-600/8-82/029a, December 1982.

__. "Compilation of Air Pollution Emission Factors." Vol. 1. *Stationary Point and Area Sources.* 4th ed. Publication AP-42, October, 1985, and supplement A, 1986; Supplement B, 1988, Office of Air Quality Planning and Standards, Research Triangle Park, N.C.

__. *Cancer Risk from Outdoor Exposure to Air Toxics, Final Report,* Vol. 1, Research Triangle Park: U.S.EPA, EPA-450/1-90-004a, 1990.

World Bank. *Environmental Assessment Sourcebook.* 3 Vols. Washington, D.C.: World Bank, 1991 .

Comments by

ERIK F. HAITES
ANDREW MULLER
DAVID POCH

ERIK F. HAITES

Professor Dewees discusses environmental regulations that apply to conventional air emissions and water discharges from new generating facilities developed by Ontario Hydro (OH) and non-utility generators (NUGs). In my view the scope of the industry restructuring considered is very limited, so my comments cover two areas: issues raised by other restructuring options and comments that relate to the paper as drafted.

Issues Raised by Other Restructuring Options

The author assumes that the "restructured" Ontario Hydro will be a Crown corporation and hence subject to the Environmental Assessment Act. It also assumes that the corporate limits on acid gas emissions and the corporate target for greenhouse gas emissions remain in effect. This is consistent with a restructuring that retains the fossil business unit as a single publicly owned entity. It is not consistent with restructuring options that envisage splitting the fossil generating units among several independent generating companies, at least some of which may be investor-owned.

The regulations governing conventional pollutants for *new* generating capacity built by investor-owned generating utilities will probably be more like those for NUGs than for OH . In other restructuring options, then, the disparities between the treatment of NUGs and utilities for the construction of new capacity may well be smaller than those identified by Professor Dewees. Given the existing capacity surplus, little generating capacity is expected to be added over the next five to ten years.

The following comments focus primarily on the regulation of conventional pollutants for existing generating facilities. Restructuring options that split the *existing* fossil-fired generating capacity among several entities, some of which may be privately owned, raises several difficult issues for regulation of conventional pollutants.

To prevent an increase in acid gas emissions by the fossil-fired generating stations, the Ontario Hydro corporate cap would need to be allocated in an equitable manner to the new operators of the existing stations. Pollution permit trading would presumably be allowed, since OH currently has this flexibility and reducing the flexibility could raise compliance costs and electricity rates.

Two of the units at the Lambton Generating Station have scrubbers to remove sulphur emissions. That raises the cost of operating those units. If generating units are dispatched on the basis of marginal cost, the scrubbed units would be at a disadvantage relative to unscrubbed units. To prevent deterioration of the environment, the scrubbed units would need to run at least as much as they are run by Ontario Hydro. Conceptually, it should be possible to achieve that result through the allocation of acid gas emissions permits, so that the scrubbed units have surplus permits for sale while other units need to buy permits. To get the allocation right in practice will require very accurate information on the costs of all of the generating units (which is presumably available from Ontario Hydro) and an accurate forecast of the market price of the permits (which is unlikely).

Conceptually, Ontario Hydro's voluntary greenhouse gas emissions target raises similar issues, although that is a voluntary goal rather than an environmental regulation.

Assume that as part of the industry restructuring, existing NUGs lose their guaranteed sales contracts. Then the existing NUGs and the existing OH generating units will be competing to supply the electricity demand. The existing NUGs are generally newer than the OH generating units and hence have been built to more stringent environmental standards. It may be the case that the lower environmental standards to which the OH units were built give those units a cost advantage over the NUG units. A New England proposal for industry restructuring addresses this issue by suggesting that all generating units participating in the competitive market be required to meet current New Source Performance Standards (NSPS). In other words, all existing OH (and NUG) generating units would need to be retrofit as necessary to meet NSPS before being allowed to participate in the competitive market.

Another interesting issue that could arise in a restructured industry is the treatment of externalities in the costs of exports and imports. At present OH is the only exporter of electricity from Ontario. As part of the export licence granted by the National Energy Board, OH estimates the environmental externalities created by generation of the electricity and *includes* these amounts in the price of the exported power. Would NUGs need to obtain licences from the National Energy Board to export electricity and incorporate their externalities into their prices if the industry were deregulated? It is more interesting, perhaps, to ask if suppliers located outside Ontario wishing to sell power in the province would need to incorporate into their price the value of the externalities associated with generation of the electricity? That would mean comparable treatment of imports and exports, but unequal treatment of imports from other countries relative to electricity generated within the province.

Other Issues

The study notes that OH has adopted a voluntary greenhouse gas emissions target and suggests that there is no reason for NUGs to do likewise. In a recent request for proposals to supply power to B.C. Hydro, several NUGs voluntarily undertook to fully offset their CO_2 emissions. They may have done this only because environmental impacts were one of the selection criteria specified by B.C. Hydro, but it does indicate that NUGs may believe that such voluntary commitments could give them a competitive advantage under some circumstances.

Greenhouse gases are not discussed extensively in the paper, presumably because there are at present no regulations governing such emissions. However, regulations governing greenhouse gases are anticipated over the next decade. Any firm planning to build a generating unit with an operating life of 20 to 40 years would be foolish to ignore the possibility of regulations on greenhouse gases. Some speculation as to how regulations governing greenhouse gases would affect NUG and OH generating units is warranted.

Finally on the topic of greenhouse gases, if climate change damages are included in externality calculations for fossil-fired generating units, they are typically one of the larger components of the total. This is not mentioned in note 30.

The discussion of MISA regulations governing NUGs indicates that NUG generating units are not specifically addressed by the regulations. The discussion then proceeds to suggest that the NUG units could be affected by the regulations governing the industry purchasing the steam output. This is unlikely. NUG facilities are typically established as separate entities and so would not be governed by the MISA regulations of the steam purchaser.

The comparison in Table 6 and discussion on page 187 that relates OH emissions to Ontario emissions for a range of pollutants is overly simplistic. The correct comparison is OH emissions against the relevant airshed loadings. In some instances, such as NO_X emissions, the relevant airshed is probably smaller than the province. In other cases, the relevant airshed is larger than the province. The temporal patterns of the loadings may also be important, especially for NO_X and PM.

Professor Dewees finds that the environmental regulations treat *new* generating capacity built by OH and by NUGs differently. He proposes to remedy this situation through the use of marketable pollution permits (MPPs). The reasons for the preference for permit systems rather than emissions taxes (environmental adders) are not given. I concur with his choice for the following reasons. First, Dodds and Lesser show that under reasonable assumptions a tradeable permit system where new sources have to purchase permits equal to their total emissions from existing sources can fully address the environmental externalities associated with the new source, even if the limit on total emissions is not set at the societally optimal level.[1]

Second, estimates of the damages associated with air emissions by fossil-fired generating units vary widely, so it would be very difficult to establish the correct tax level. Although some of the environmental adders are estimated using conceptual approaches that are inconsistent with the valuation of marginal damages, the range of values is often an order of magnitude.[2] Estimates range from roughly 10% of the cost of generation to 100% of the cost of generation. It is simply not possible to reconcile the differences to arrive at the "correct" value for the tax. Setting the tax at the wrong level, especially at the upper end of the range, would have a significant impact on the competitiveness of fossil-fired generating units.

Others might argue that a tax is the preferred mechanism due to the nature of the damage curves. Professor Dewees argues that the aggregate marginal damage curves for most of the air pollutants emitted by fossil-fired generating stations are roughly constant over the relevant range. Economic theory suggests that under such circumstances, a tax is preferred to a permit system as a means of achieving the societally optimal level of emissions. However, the analysis in the paper focuses on the emissions of a new generating unit, not total emissions by all sources. A permit system is effective in addressing the externalities of a new unit even if the total level of emissions allowed is not optimal.

The preferred mechanism – emissions taxes or marketable pollution permits – is an open question if adjustments to existing environmental regulations are being considered for industry restructuring options that involve splitting the existing generating capacity among several entities. The discussion of acid gas emissions above assumes the use of tradeable permits since the existing regulation already takes that form.

NOTES

1 Daniel E. Dodds and Jonathan A. Lesser, "Can Utility Improve on Environmental Regulations?" *Land Economics,* 70, 1 (February 1994): 63-76.

2 Dewees cites two American studies of damage estimates. Studies specific to Ontario are available. Ontario Hydro has developed damage cost estimates, while Chernick, Caverhill, and Brailove have developed control cost estimates.

Takis Plagiannakos (Team Leader), "Full-Cost Accounting for Decision Making," Survey Team 4, Task Force on Sustainable Energy Development, Ontario Hydro, December 1993. Paul Chernick, Emily Caverhill, and Rachel Brailove, "Environmental Externality Values for Use in Ontario Hydro's Resource Planning," Vol. 3, prepared by Resource Insight, Inc. for Coalition of Environmental Groups for a Sustainable Energy Future, undated. Presented to Ontario Environmental Assessment Board, Ontario Hydro Demand/Supply Plan Hearings.

ANDREW MULLER

Professor Dewees asks whether current environmental regulations discriminate against non-utility generators (NUGs) by imposing upon them costs which are not borne by Ontario Hydro. He concludes they do and suggests several approaches, including marketable pollution permits (MPPs) for SO_x, NO_x and VOCs emissions, to reduce the bias. His is an important question and I have little quarrel with the approach or the answer. I will limit my comments to two areas, namely the significance of linear damage functions and the issue of tradeable emission permits (i.e., MPPs).

Professor Dewees stresses that the damage functions from most pollutants are linear over the range that is relevant for policy, and hence that marginal damages are essentially independent of the total emissions of either Ontario Hydro or the NUGs. He also argues that marginal pollution damages tend to be higher in areas of high population density than in areas of low density. In this last he is probably correct, although I admit to some reservations. Over 30 years ago John Dales (1968) pointed out that this reasoning tends to imply that we should pollute relatively pristine areas such as the Muskoka Lakes rather than already heavily polluted areas such as Toronto or Hamilton Harbours. For the argument from population density to hold we must believe that marginal damages per capita are comparable in the high- and low-density areas, and this in turn may blind us to the possibility that differences in original uses may cause marginal damages to vary substantially across geographic areas.

More importantly, Professor Dewees does not fully explain why the shape of the damage function is particularly important and may even give the impression that roughly constant marginal damages are a necessary condition for wishing to equate the marginal abatement cost of all producers. In fact, as Professor Dewees is well aware, this condition is optimal whenever we are dealing with a fully mixed pollutant, regardless of whether the marginal damage is rising, constant, or falling. The real importance of the shape of the damage function lies in its implications for the choice between taxation and tradeable emission permits as the preferred economic instrument for environmental regulation.

In principle, Professor Dewees' goal of equating marginal abatement cost across firms can be achieved either by a uniform tax (preferably equal to the marginal damage cost in equilibrium) or by issuing emission permits in the socially desired quantity and allowing trading to establish the appropriate price. The first is a price instrument, the second a quantity instrument. In a world of certainty one is essentially equivalent to the other. However, in a world of uncertainty, quantity-based instruments provide reasonably precise control over quantities at the expense of a fluctuation in prices, while price-based instruments provide the reverse. If marginal damages are rising rapidly, a small underestimate of the correct tax for a price-based system can lead to large damages and hence a quantity instrument such as permits might be preferred. If marginal damages are relatively constant, however, precise control over quantities is less essential and a fixed price instrument may be preferred.[1] Professor Dewees' finding of relatively constant marginal abatement costs tends to support a policy of environmental charges common across generators rather than the fixed quotas implied by the MPP plan. Of course, an environmental tax may also be desirable from a revenue standpoint.

Nevertheless, Professor Dewees recommends that a MPP system be established for SO_x, NO_x, and VOCs. Several reports have made specific proposals for such a system. (Nichols 1992; Nichols and Harrison 1990a, 1990b.) Much of my recent work and that of my colleagues at the McMaster Experimental Economics Laboratory has investigated the properties of emissions trading under various market rules and structures. I should like to tell you a little about experimental economics and what it has taught us about emissions permit markets.

Experimental economics has evolved rapidly over the past 30 years to fill the gap between abstract theory and policy making on the one hand and field experience on the other.[2] It is based on the proposition that before implementing expensive policies based on unproven theoretical assumptions one should test the theory and prototype the institutions in relatively inexpensive laboratory experiments. In our experiments we create small but genuine markets and reward our subjects with significant amounts of real cash. In our emission trading experiments we have examined the effect of market trading rules, rules concerning banking of permits and trading of shares, and the effect of market power. We have made or confirmed a number of important findings.[3]

First, emissions trading is effective in reducing the cost of achieving pollution control under a wide range of experimental treatments. Cost savings range from 50% to 90% of the potentially achievable gains. This is a higher efficiency gain than found in previous experiments and demonstrates that emissions trading with well-trained and informed subjects can work.

Second, market institutions matter. In particular, double auction markets, in which price information is continuously made available to participants, perform better than markets in which contracts are formed through search and bi-lateral negotiation. This supports remarks made by other commentators about the importance of public price information both in power and emissions trading.

Third, rules about banking unused permits matter. If control over emissions is uncertain, firms may arrive at the end of a compliance period holding too few or too many permits. If these cannot be banked, they will be sold in a thin market at prices which are wildly unstable and sometimes very high. We have generated these price spikes in the laboratory and have demonstrated that banking eliminates them.

Fourth, the choice of trading instrument matters. Canadian proposals for emission trading often distinguish between coupons (the permit to discharge one unit of waste) and shares (the right to a specified fraction of each year's coupons). Trading in shares demonstrably improves the performance of laboratory markets.

Finally, and perhaps most importantly, market power matters. It is common to allege that thin markets for emission permits do not really pose a problem for emissions trading plans because the aggregate cap on emissions will be achieved in any case and any trades that do occur will be pareto improvements.[4] Moreover, many have suggested that appropriate trading institutions, such as double auction markets, can largely suppress market power. Nevertheless, recent economic theorizing has stressed that emissions trading can indeed *reduce* efficiency when firms compete in both a permit market and in a downstream product market. The potential competition between NUGs and Ontario Hydro both in air emissions and power markets provides a perfect example. This theory has been tested in an elegant experiment designed by Jamie Brown Kruse and Steve Elliott and replicated and extended by one of our students, Rob Godby. In this experiment, a dominant firm competes with a fringe of smaller firms in both a permit and a downstream product market. When given the chance, the dominant firm clearly withholds permits in order to drive up competitors' costs.

I read these results first as providing general support for Professor Dewees' proposal to introduce marketable pollution permits into the Ontario power market, but secondly as suggesting that the dangers of a thin market bear careful investigation. Experimental approaches, along with field study of comparable institutions, can help us avoid costly mistakes.

NOTES

1 This argument is made by Oates (1990), who ascribes it to Weitzman (1974).

2 See Davis and Holt (1994) for a textbook survey.

3 See Muller and Mestelman (1994), Godby, Mestelman, Muller, and Welland (1995), Brown Kruse, Elliott, and Godby (1995).

4 A pareto improvement is one which improves the welfare of at least one party without reducing the welfare of any other party.

BIBLIOGRAPHY

Brown Kruse, Jamie, S. Elliott, and R. Godby. "Strategic Manipulation of Pollution Permit Markets: An Experimental Approach." McMaster University, Department of Economics, typescript, 1995.

Dales, John H. *Pollution, Property and Prices.* Toronto: University of Toronto Press, 1968.

Davis, Douglas and Charles Holt. *Experimental Economics.* Princeton, NJ: Princeton University Press, 1993.

Godby, Rob, S. Mestelman, R. Andrew Muller, and Douglas Welland. "Emissions Trading with Shares and Coupons when Control over Discharges is Uncertain." McMaster University, Department of Economics, typescript, 1995.

Muller, R. Andrew and Stuart Mestelman. "Emissions Trading with Shares and Coupons: A Laboratory Experiment." *Energy Journal* 15, 2 (1994):1-27.

Nichols, Albert L. *Emissions Trading Program for Stationary Sources of* NO_x *in Ontario.* Cambridge, MA: National Economic Research Associates, Inc., 1992.

Nichols, Albert L. and David Harrison. *Using Emissions Trading to Reduce Ground-Level Ozone in Canada: A Feasibility Analysis.* Cambridge, MA: National Economic Research Associates, Inc., 1990.

—. *The Impact on Ontario Hydro of Emissions Trading for Nitrogen Oxides: A Preliminary Analysis.* Cambridge, MA: National Economic Research Associates, Inc., 1990.

Oates, Wallace E. "Economics, Economists and Environmental Policy." *Eastern Economic Journal* 16, 4:289-96.

Weitzman, Martin L. "Prices vs. Quantities." *Review of Economic Studies* 41 (1974):477-91.

DAVID POCH

Professor Dewees has provided us with a valuable survey of the current regulatory context including its inconsistencies. His focus is on curing inconsistent treatment as between Ontario Hydro and NUGs and on enhancing economic efficiency, two laudable objectives. However, the reform of regulation that must accompany restructuring demands consideration of some broader questions pertaining to environmental regulation of the sector.

Market Mechanisms Are Good Allocators But Have Serious Inadequacies When It Comes to Setting Limits

Professor Dewees favours Marketable Pollution Permits (MPPs), with regional and seasonal adjustments as appropriate. While the linear damage function he suggests may indeed hold for many pollutants and over a considerable range, at some point we reach a limit or threshold beyond which ecological damage impinges on nature's ability to repair itself or at which point society finds the burden unacceptable. This suggests that MPPs can be employed to increase the fairness and efficiency of the allocation of pollution permits within acceptable caps (appropriately adjusted to take into account regional and seasonal considerations), but we must first have a rational basis for setting the caps.

Caps should respect Sustainable Development dictates (as Brundtland defines it). To be sustainable, caps must respect fairness intra-generationally as well as inter-generationally. Economics has little to say about the ethical implications inherent in this challenge and not a lot of insight to offer us in regard to ecological considerations. Certainly, economic evaluations of damage that have set the value of health and life lower in developing countries (based upon lower GDP/capita) or that discount the value of human life in the future, evidence the need to contain the role of economics in regulation even within the allocation role.

Professor Dewees favours MPPs or taxes over externality valuation (especially based upon implied valuation) because externality monetization has given us order of magnitude variations in the results generated. This seems a red herring. When an official rations MPPs or sets levels for tax-based instruments he or she inevitably considers the trade-off between mitigation costs imposed versus societal damage costs to be avoided. An "economically rational" regulator would set the cap at the level which inspires marginal mitigation measures equal in cost to the societal cost of the next emission. Costing and evaluation of societal willingness to pay is occurring implicitly or explicitly in each case. Implied valuation simply extracts the same official's judgment that would inform tax or permit levels.

The problem is that regulation that attempts to have polluters pay the marginal damage cost (be it by monetizing externalities or MPPs or taxes) is inadequate, both because we usually cannot calculate that cost fully, and because market based damage costs do not reflect the true ecological or societal value of avoiding the impact. (Human and ecosystem health and ethical considerations are not widely traded commodities that the market can appropriately cost.) Damage-costing techniques that do not temper the evaluation with political judgment are the worst offenders in this regard.

Thus "economic rationality" will be a helpful but insufficient qualification for the regulator. This is so not simply because the number will be calculated incorrectly. There are also ethical considerations that economics stumbles over. Permits to pollute premised upon marginal damage cost techniques give polluters the right to expropriate ecological and human health simply by paying "adequate" compensation.

It may be economically rational for a regulator to allow a power generator to cause a random death because cheap power will benefit Falconbridge and Ford to the tune of $110,000, while society is generally willing to spend only $100,000 to save a life by investing in health care or road safety. But there is an ethical distinction to be made between involuntary death being imposed for unevenly shared commercial gain versus societal rationing of expenditure on health care or road safety with costs and risks shared by all.

Regulators must therefore ensure that impacts are acceptable from an ecological and social perspective, as well as compensated for and efficiently allocated. Regulation comes down to the imposition of suffering and death and the loss of ecosystem vitality in the name of economic gain that is unevenly distributed among industry and the public and between generations. This suggests that the cap should be set by an open political process not a narrow technical analysis.

We must also guard against biasing our assessments of acceptability by undue reliance upon economic models that only count impacts that are monetized. For example, the author suggests that impacts are greater in more densely populated regions (based upon studies which find the lion's share of monetized impacts are due to human health impairment). The fact that economists have had a hard time valuing ecosystem health (unless there is a reduction in the lumber yield) should not be taken as evidence that the concern is minor or that the value to be placed upon avoiding ecosystem impairment should be slight. Governments responded to the greenhouse gas problem prior to any rigorous damage costing and they were right to do so.

This is not to suggest that environmental regulation should be devoid of economics and market mechanisms. Economists can help us recognize the hidden costs of pollution and can evaluate how best to phase in tighter caps to achieve ecological and human health goals while minimizing disruption. Market mechanisms can increase economic efficiency in implementation through efficient allocation. But let us not rely on economics as the sole basis to set the goal or cap itself.

So the evil of marketable pollution permits is not so much in the "trading" of the environment as it is in the inevitably accompanying elevation of monetization as the vehicle to determine the level of the cap. Economically justified pollution levels are not always ecologically or ethically justified. The first question should be: on what basis should we set caps? The allocation question is easy by comparison.

When we do select the allocation mechanism we should be cognizant of the political realities. A significant part of the push for restructuring comes from industrial customers looking for lower rates. Any competitive model that goes beyond generation-side-only competition will compromise utility DSM. (If electricity, is simply a commodity, price will be the dominant factor and bundling in of energy services will be shunned by marketers – especially if societal costs are reflected in the valuation of DSM). With utility DSM compromised, the importance of Professor Dewees' observation that externalities must be reflected in the price for economic and environmental sense to prevail is amplified. But will politicians be prepared to see the price rise to include externalities? Will AMPCO suffer it? In contrast, generation-side competition would still allow for utility-based DSM with inclusion of monetized externalities in the DSM screening protocol and a buffering of the price impacts.

If we cannot get the price right, then restructuring beyond generation-only competition will be counterproductive.

Permit Schemes Must Not Give Existing Polluters an Advantage

Professor Dewees does not suggest a mechanism for initial issuance of permits, though he does suggest for SO_x that new sources should have to purchase permits from existing sources (to maintain the existing capped level).

Two questions arise. The first is: Should existing polluters be required to pay for the permits to avoid unfairness and windfalls? Apart from fairness, this is important to consider in so far as we wish to see the societal costs of residual impacts internalized in retail prices and therefore in choices between electricity consumption and energy-efficiency investment. (Alternatively, as discussed above, if we or AMPCO want to keep retail prices down we will continue to need to foster utility DSM investment by regulating least-cost integrated resource planning.)

The second question is more difficult: If parts of Ontario Hydro remain intact, will the financial reward available from the sale of a permit to a NUG be sufficient to overcome the temptation of a market-dominant entity to hold inefficiently or hoard pollution rights to restrict competition? This type of risk requires some degree of open public regulatory supervision.

Tradeable Permits Alone Will Not Level the Environmental Field for Ontario Hydro and NUGS

If one objective is to enhance fairness and efficiency in society's choices as between NUGs and Ontario Hydro generation, then it makes no sense to segregate consideration of "conventional" pollutants from others such as nuclear emissions and risks, as has occurred in these discussions. If Hydro is allowed to maintain limited liability for nuclear accidents, is not required to pay for radiation emission permits, and can ignore risks in the valuation of decommissioning and disposal costs, with two-thirds of its energy being nuclear, it will have an advantage that will swamp the other inconsistencies in the present regulatory regime.

In regard to the uneven coverage and application of the Environmental Assessment Act, the issue most in need of our consideration is not whether a NUG that proposes a major coal plant would be designated as requiring an EA (which I would certainly suggest it should), but rather, whether Hydro or Ontario Nuclear will continue to enjoy the ability to life-extend, modify, or refit its large, aging stations without EA process. Hydro is currently championing an international fusion reactor facility to be located at Darlington or Bruce. It will seek to avoid EA hearings at all costs. (It was successful in that regard when it built its $100 million tritium removal facility on the Darlington site and when it engaged in the $1 billion retubing and life extension of Pickering A.)

While class EA should apply to dispersed generation (be it OH or NUG), Hydro's megaplants do have the potential to cause environmental and social disruption in ways that are not linearly related to electrical output and do warrant full scale EA treatment. Ontario Hydro creates boom and bust local economic cycles, large plant thermal impacts can exceed ecological threshold levels, water patterns upstream and downstream from large dams can be altered dramatically, nuclear accidents or near accidents can cause cities and regions to be evacuated. Megaprojects warrant the time and expense of environmental assessment at a detailed level. However, Hydro has never been required to undergo Environmental Assessment Hearings for generating facilities (apart from the aborted DSP plan hearings).

Environmental Regulations Must Capture Imports and Exports

We have heard how restructuring toward competition will lead to increased international competition due to the GATT and NAFTA. Presumably interprovincial competition would accompany such a development. Regulatory regimes that tax, restrict, or impose costs upon domestic polluters without imposing similar costs upon external polluters will be unacceptable environmentally and politically.

Efficiency in Regulation

Regulation should seek to ensure efficiency in choices between supply and energy efficiency investments, not simply between supply sources. As already noted, restructuring may dramatically change the environment for DSM delivery. Restructurers must include a mechanism to ensure that societally optimal levels of efficiency are approached. This demands either incentive or pricing methods that capture the societal cost of pollution and supply risk, i.e., the full avoided costs. Discussions to date have focused upon centralized utility supply versus decentralized NUG, between public and private ownership, and between monopoly and competitive models. Most environmentalists have concluded that the institutional realities of a monopoly tilt against smaller, cleaner technologies. However, in deciding how to dismantle Ontario Hydro's generation monopoly, we should not lose sight of the need to facilitate energy efficiency and conservation that avoid pollution and risk altogether.

Re-Aligning Human Resource Management and Industrial Relations: Can Hydro Become a Mutual Gains Enterprise?

PETER WARRIAN

The electrical power industry around the world has undergone major changes in its technology, regulatory environment, and fundamental economics in the last twenty years. These changes have accumulated into a qualitative shift away from its traditional natural monopoly status. Ontario Hydro has definitively moved away from its traditional, centralized, top-down organizational model. The question now is: What is the appropriate human resource management and industrial relations model to meet the interests of the corporation and its employee groups?

INTRODUCTION

From the beginning of electrical industry regulation in the early 1900s until recent decades, incremental technological advances in standard turbine, generator, and transmission technologies provided economies of scale and decreasing unit costs. The regulatory regime for the industry was seen by most as the legitimate oversight of a natural monopoly. This was validated by real costs and rates falling continually.[1] Not surprisingly, the labour and management parties at Hydro adopted the same adversarial industrial relations practices as had been adopted in other mass-production industries like auto and steel, which were also characterized by standardized costs and economies of scale in the post-war economy. In the language of industrial relations practitioners it is called Wagnerism or the "Industrial Model."

In the 1990s, however, the regulatory, financial, and operational environment has fundamentally shifted. Ontario Hydro has embarked on a major reorganization into autonomous business units to meet the challenges of this new environment. The Corporation has been restructured into autonomous business units:

Upstream:	Nuclear	Downstream:	Grid
	Fossil		Retail
	Hydro Electric		Energy Services

What has yet to happen is a fundamental change in how labour and management at Hydro do their business. Various initiatives have been taken in the past decade, as discussed below, but it has all been within the existing structure. We will consider here the need to overhaul the Industrial Relations Model at Hydro fundamentally on both sides. Otherwise the system will drift towards a crisis.

BACKGROUND ON EMPLOYEES, BARGAINING UNITS, AND BARGAINING AGENTS

There are two unions at Ontario Hydro operations. The Power Workers' Union (PWU) has 15,000 out of a current workforce of 22,000, and the Society of Ontario Hydro Professional and Administrative Employees ("the Society") has 5,000 members. In addition, there are some 2,000 non-bargaining unit employees. There is a fundamental difference in the relationship and culture of the two unions. The Society relationship is seen as more cooperative while the PWU relationship is seen as more adversarial. As it is the PWU relationship that predominates, this will be the focus of most of the following discussion.

The typical PWU member is a technician. He/she has a Grade 12-13 education and 8-10 years training on the job. The members are very highly skilled but with very firm-specific skills. There is in fact relatively little mobility across business units, e.g., for Nuclear Operators. The skill sets of all of the upstream business units are of this character. Downstream, customer service, and standardized skills are more typical.

The PWU sees employment security as fundamental, even if it is not narrowly defined job security. In management's eyes, the PWU view is that the work does not have to be inherently challenging, it just has to have people employed. The employment relations model is that of the tradesman. Employees come in through a regular apprenticeship or its five-year training equivalent, then they stay in that slot more or less throughout their careers. Some 60% of the PWU in fact are certified tradesmen. This includes "power line maintainers." The other PWU members are clerical and technical. The trades and operators progress through fixed job ladders. The clericals have to apply each time there is a job opening. The techs come in with their diplomas and stay in their occupational group.

By contrast, the Society is mostly comprised of engineering staff; it was given voluntary recognition in 1993. The management views the Society as looking more for careers for people. In contrast to the PWU, members of the Society are interested in challenging careers, not just jobs.

RECENT HISTORY: SEARCHING FOR A DIFFERENT WAY

In recent years the Hydro-PWU relationship has gone through cycles of cooperation and confrontation. In 1991-92 the parties tried to move away from their traditional adversarial relationship and into a more modern interest-based and facilitated model of cooperative negotiations. They achieved some success and produced the Purchased Services Agreement along with initial steps at local labour management change initiatives in the Nuclear units. In 1993-94, under enormous financial pressures and the reorganization into separate business units, there was resort to traditional confrontation with the aggressive managerial objective of a separate collective agreement for Nuclear. At the conclusion there was agreement on an "armed truce" and a temporary employment security agreement. The parties are now exploring new forms of cooperation.

Major conflicts over Hydro restructuring go back over a decade. In 1984 article 11 in the contract was agreed to concerning redeployment. The Surplus Staff Procedure under article 11 takes up 18 full pages of the collective agreement, with 27 official sub-clauses.[2] An indication of the elaborate detail of the provision is section 11.1 which only deals with Definitions, including separate legal definitions for:

Classification
Job Family
Equal Classifications
Seniority
Work Group
LSEOLS
R/D/HO
Site
Occupational Group
Same Classifications
Lower Classifications
Head Office
Exceptions
R/D/B
Location

Each of these contractual terms is invoked for the purposes of defining "bumping" rights and mobility entitlements of employees within each affected job category. The complexities of the system equal or exceed any parallel provisions in large private sector agreements.

MANAGEMENT VIEWS OF THE EXISTING SYSTEM

In management's view, to understand how the system evolved requires going back to 1985-87 when there was, for the first time, significant restructuring but not downsizing. People were moved around within the organization utilizing the contractual work rules. In their view, it worked reasonably well in doing the job of internal redeployment across the incredibly complex system of job classifications and work units that both parties had built up at Hydro.

The article 11 approach worked, in management's view, until the early 1990s when the problem became one of major downsizing and reorganization of the corporation. This required fundamental changes in approach to the collective agreement. Management representatives acknowledge that the existing collective agreement is not able to deal with major downsizing. The sheer scale and nature of the changes means that the process is now bogged down with all sorts of procedural steps. Neither has there been a lot of active cooperation with the voluntary severance packages that have been offered. The PWU agreed with offer #1 but not with #2 or #3.

In part, these issues were addressed by the management proposal for a separate agreement for Nuclear. However, in the final result, the 1994-96 collective agreement did not include a provision to change article 11. A new contract provision on Employment Security was agreed to for the life of the agreement, to take some of the pressure off article 11 through effectively introducing a moratorium on major dislocations. The key provision reads as follows:

Employment Security[3]

For the period April 1, 1994 to March 31, 1996, no regular Power Workers' Union-represented employees will be involuntarily terminated (does not include termination for cause or retirement at normal retirement age). During this time period, staff who are currently surplus or become surplus shall have their search/notice period extended to the earlier of placement in a vacancy/placement opportunity or March 31, 1996. Staff declared surplus on or after December 11, 1995 shall receive a minimum 16-week or 4-week search notice as described in Article 11.4.

To relieve the pressure on payroll costs of the employment security provision, the parties agreed to reduce the Employer's pension fund contribution.[4]

As a tactical move, reliance on a temporary employment security agreement and use of pension monies is understandable, but it is no long-range solution. It can easily be viewed by outsiders as trying to put off the inevitable. And, by definition, the parties can only go back to the pension surplus a limited number of times. In addition, the intricate and cumbersome mechanisms of prolonged bumping rights and entitlements under article 11 do not provide a long-range solution. They at best can buy some time for adjustment and change. The heart of the problem lies with the basic approach to jobs, skills, and work organization. Individual management representatives agree that the Industrial Union model is what has applied in Hydro. And it is even more so for management than for the union. Both the Society and the PWU have sought greater employee involvement but many in management have resisted it. On the other hand, in their view the PWU wants control in the form of veto power. In the management view, the union's model of a "concensus" corporation would be one in which the PWU would hold a veto over all key decisions.

In the view of Hydro management, in 1992-94 the PWU made a major gain through the "Purchase Services Agreement," which established a joint process with the union in respect to contracting out. In management's view, the letter of Understanding on Purchased Services has been used by PWU as a weapon, a broad veto device, to leverage their position on other issues not related to contracting out.

In regard to future human resource management (HRM) strategies, management representatives see that decisions involving fundamental choices over business strategies will have to be made, but the directions have not yet been set. Upstream there is expected to be substantial competition. The corporation itself may shift resources and emphasis downstream to the grid. Most of the employees are upstream and vulnerable to dislocation in such a scenario. The shift would also involve the professionalization of all employee groups and, therefore, movement away from the traditional labour contract work organization and seniority rules.

UNION VIEW OF THE EXISTING SYSTEM

The union's view of the recent history is not dissimilar from management's in its description of the overall directions and current practices of the parties. In the PWU's view, the problem is inconsistency. The 1980s mentality was that Hydro was bigger than the government and money was not an issue. The labour-management relationship was traditional. This changed at the end of the 1980s and was primarily driven by the regulatory side.

In the union's view, the key to change was the AECB's concerns about the nuclear operations including industrial relations hostility, morale, safety, and efficiency. In correspondence, the view was expressed that if certain Hydro nuclear facilities had been operating in the U.S., they would have been shut down by the regulatory authorities because of risks associated with accidents, backlog of maintenance work required, near misses, etc. The implicit threat was that Hydro nuclear operating licences could be lifted unless new behaviour was forthcoming.

In reaction, the management looked to TQM or QWL as a one-off "solution" and introduced an ambitious program of quality improvement in the nuclear sector in 1990. For the Power Workers' Union this represented an opportunity to engage in the change process as equal partners. There would be joint committees operating by consensus, i.e., the union would have a veto. This, in fact, has become the union's standard for assessing involvement across the corporation. It began in the 1990-92 period and by 1993-94, in the union's view, it had begun to get local buy-in by employees and engagement in the workplace.

In this period of cooperation, the parties also dealt with contracting out. The major issue in the 1985 work stoppage was contracting out. It went to arbitration and Kevin Burkett imposed a formula: a 25% rule limiting the extent of contracting out without union agreement. However, critical for the union was the no enforcement provision within the Burkett award. Between 1985 and 1992 the union found it virtually impossible to monitor and police the provision. As a result, a mid-term working group was established to deal with contracting out. A facilitator was involved and the working group eventually agreed to the Purchased Services Agreement. This was just prior to the 1992 official bargaining round.

During the 1992 negotiations, the parties dealt, among other things, with differences and differentiation across what are now the separate business units. There are different sections of the collective bargaining agreement (CBA) to deal with working conditions in the various business units and they have been there a long time. They are usually dealt with by sub-committees, but because there is little incentive to settle, the pattern was erratic. In 1992 they used facilitators and problem-solving techniques and all sub-committees settled for the first time.

The main negotiations went well and both sides went around in public advocating the new way, with interest-based negotiations as the model for labour/management relations in the future.

However, in the 1994 negotiations, in the PWU's view the management completely flipped over into adversarialism and decided to try to bust the union. The corporation announced pre-conditions for negotiations in the form of a $200 million concessions target and a separate collective agreement for Nuclear. The PWU and the company both know that it is the bargaining leverage of the nuclear operators that supports all the other employee groups' demands. Consequently, the Union called a council meeting, changed their constitution, and put in place a strike mandate.

The vote was 98%. As a partial concession, the union allowed that the management could put on the agenda a separate collective agreement as a demand but not as a pre-condition for the talks.

The PWU priority in 1994 was job protection. They would pay the price to keep the jobs, particularly because in their estimation, the wage bill is only 6% of the total operating cost for Hydro. The key trade-off was the employment security commitment for PWU members for the duration of the collective agreement. The mathematics and economics were straight forward: about 500 PWU members were vulnerable so they multiplied this amount of salary for two years and arrived at $200 million, which they took out of the pension fund surplus.

The Union has sought to deal with decentralization of the corporation through its own structure. There are consultative "Boards" for each business unit. At each plant there are also joint Steering Committees of each side. At the most senior level, decision makers on each side are well aware of the issues concerning their relationship and employee involvement. There was a traditional Joint Committee on the Relationship (JCR) and a new Employee Representative Involvement (ERI). At present, they are close to agreement on the principles and procedures for activities in these forums.

The PWU has developed a comprehensive policy position on how they will approach employee involvement and restructuring discussion.

Power Workers' Union: Involvement Policy

It is the PWU's belief that Ontario Hydro should be the leader in the province of Ontario in workplace reforms directed at full employee involvement. To achieve this, a redesigning of the workplace is necessary so that it becomes less authoritarian, more democratic, and safer. Recognizing this, the PWU will enter into negotiations with Ontario Hydro to develop a plan that will lead to full employee involvement in the business.[5]

In the Union's view, this will require:

1. Substantial changes in how work is organized;
2. A significant reduction in bureaucracy and other associated overheads;
3. The creation of opportunity for employees to identify and solve operating problems;
4. The continued upgrading of skills of PWU represented members;
5. The opportunity for full and meaningful involvement in all decisions.[6]

In order to achieve a workplace that continually improves fairness, performance, and safety of operation, for the Power Workers' Union the plan will need to include:

1. Provisions for workers to have greater influences, accountability with responsibility, and control over the day-to-day operations of their workplace, including all decisions at the operating unit level, which would be made by consensus;
2. Development of managers who emphasize coaching and coordinating;
3. A workforce trained in safe work practices to ensure a workplace free of health and safety hazards;
4. Continual improvements in efficiency based on working more effectively, using improved equipment and technology, and producing less waste;
5. A flattening of the organization, resulting in elimination of unnecessary layers of management, administration, and overhead costs.

The union also proposes stringent rules for participation. In its view, the new workplace structure will require the full participation of PWU members to take advantage of their first-hand knowledge and experience. In order to achieve full participation of the members, the plan will require the following:

1. Participation in the plan is premised on their knowledge and any resulting performance improvements will not result in direct or indirect loss of employment.
2. The process for redesigning the workplace and implementation of participation programs will be jointly developed and administered.
3. The structure and content of training programs and the materials used in any course, will be jointly developed.
4. The plan will not be used to bypass the normal Union structure, interfere with or usurp authority of the grievance procedure, or interfere with any collective agreement or process.
5. The plan will not be used to discipline employees.
6. Participation of any member will, at all times, be voluntary.

It is clear from the text that the PWU has proposed an ambitious and aggressive approach to participatory change. No doubt there will be adaptation as the parties discuss an actual protocol for involvement. Such a procedural agreement is currently under discussion. A cynical view might be that the union, in the context of the conditions it seeks to impose on the process, is seeking a no-risk approach to change. On the other hand, the difficult recent history is alive in people's memories. Equally daunting is the search for clarity on what the other side of a new deal might be.

THREE COMPETING MODELS OF PRODUCTIVE EFFICIENCY

One of the reasons that industrial relations strategies have flipped flopped around at Hydro, but also in other major industries in the public and private sectors of the economy, is that there is an ambiguity at the heart of how productive efficiency is perceived. We have in the global economy, three competing models of productive efficiency.[7]

The traditional model of productivity for most of this century has been that of the Mass Production System (MPS). This is in essence the mass production assembly line of Henry Ford and of Alfred Sloan's General Motors. This model stressed low-cost mass production of standardized units based on top-down management, a narrowly skilled work force, but it came at a cost of high energy consumption, low-trust relations with employees and suppliers and low quality.

The Japanese challenged and changed global manufacturing through Toyotaism or the Lean Production System (LPS). This model stressed lower-cost but higher-quality production through work teams, participative management, continuous process improvement, employee involvement focused on evolving but closely standarized procedures, and close relations with suppliers.

A third model, associated with the states of the European Union, such as Germany and Sweden, is the Socio-Tech Model (STSPS). This model stresses high-quality customized products, medium costs, high and fluid employee participation, high and flexible skill sets, with process improvement and organizational democracy as a goal.

The strategic choices for Ontario Hydro, as well as other traditional industries, can be summarized in the following diagram:

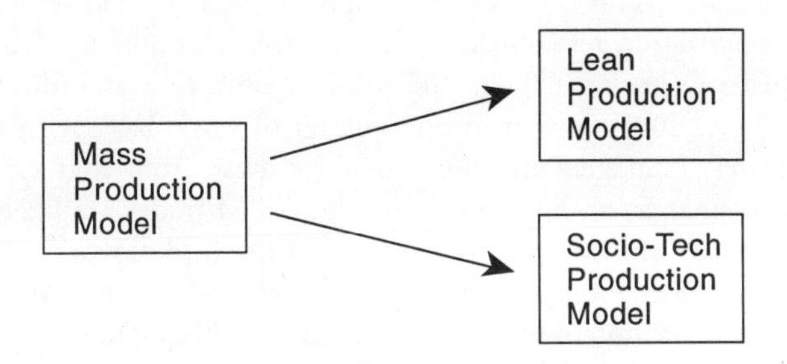

Each of these models of productive efficiency has inherent advantages and limitations. There is an implicit human resource management and industrial relations model associated with each. Therefore the process of strategic choice about the appropriate businesses Hydro is to be in and how it wants to be in those businesses becomes critical to the HRM and IR outcomes and processes.

CRISIS OF THE WAGNER INDUSTRIAL RELATIONS MODEL

At the heart of the issue is the post-World War II Wagner Act industrial union model, and what this means at root is "job control unionism," i.e., the adversarial relationship between labour and management in the workplace as reflected in the system of job classifications, wage rate structure, and seniority-based work rules. To be clear, this is not an attempt to assign blame. The Wagner model was well-suited to the conditions of mass production industries in the 1930s and 1940s. However, the traditional industrial model is not well suited to the economic and organizational environment of the 1990s, in either the public or private sectors. Collective bargaining will continue but the parties will have to find a new employment security-flexibility trade-off that meets their new work force and economic performance needs.[8] Indeed, based on interviews at Hydro, the management identification with the model is every bit as intense, in its management's rights prescriptions, as it is with the Power Workers' Union.

The tensions between job control unionism and the new human resource management model have been analyzed in a recent study by John O'Grady.[9]

O'Grady then analyzes collective agreement data in the private sector from the early 1980s to the early 1990s to see how the above tensions have played themselves out. In summary, his survey data make for the following conclusions and observations. By and large, the majority of blue-collar, unionized workplaces have job grade structures that are moderately to highly articulated. This is consistent with the view that Taylorist approaches to work organization remain the dominant organizational model in Canadian industry. Viewing changes in work organization as indicative of a strategic choice suggests that in one-quarter of workplaces with extended job structures, management chose to move away from that type of job structure. Demarcation lines involving the skilled trades are likely to be a source of operational inflexibility in roughly one-third of workplaces in the manufacturing and resource industries. In roughly 30% of bargaining units, union leverage resulted in management either modifying or setting aside its desire to reduce the number of job classifications. In approximately 70% of cases, management negotiated or consulted over changes in the job grade structure. Most companies do not judge disputes relating to their job classification system to be a major source of conflict with their union. This is consistent with overall stability in the predominant model of work organization. There is some evidence that changes in work organization are being

Table 1
The New HRM Paradigm vs. Job Control Unionism

New Human Resource Management Paradigm	*Job Control Unionism*
Broad Job definitions. Compressed job grade structures.	Narrow job definitions.
Unmediated employee involvement, i.e., involvement of employees is not channelled or filtered by the union.	Emphasis on the grievance system, elected stewards and the union executive as the channels for dealing with management.
Use of semi-autonomous production teams and problem-solving teams.	Scepticism about "employee involvement." Concern that "teams" displace the union as a channel of representation, making peer pressure on workers an instrument of management and weakening the union's ability to defend its members in discipline situations.
Greater flexibility in assignments including extensive adoption of job rotation and multi-skilling.	Emphasis on seniority in the allocation of jobs. Seniority establishes a "property-like" claim to a job. The operation of seniority claims requires that job definitions be treated as demarcation lines.
Gain-sharing element in remuneration.	Preference for standardization of wages across an industry. Disinclination to accept company or plant-specific gain-sharing, which undermines standardization.
Pay-for-knowledge element in remuneration.	Preference for tying pay to jobs. Not attracted to pay-for-knowledge, especially when access to training and design of training is management-controlled.

made independently of new technology and that the need to combine jobs in downsizing situations is a significant factor. The incidence of grievance litigation over classification issues declines when unions enjoy consultation or negotiation rights over the administration of the job evaluation system. To the extent that contract provisions providing for such consultation or negotiations reflect union leverage, there was a modest increase in that leverage during the 1980s. Seniority confers significant protection to long-service workers against lay-off. In the vast majority of workplaces, however, seniority is not the sole factor. Rather seniority claims are counterpoised to varying degrees with skills and ability. The benefit of seniority protection will be eroded by changes in work organization, such as broader job definitions, which increase the relative importance of skills and ability in assigning jobs.[10]

How does the collective agreement regime between Hydro and PWU compare to this scenario?

In total there are approximately 300 job classifications across a grid of some 68 wage grades. Within these grades and classifications, there are dozens of individual salary steps. In management's view, ideally these should be reduced to about 50 job definitions in the PWU bargaining unit. To reach such an objective would require a wholesale restructuring of the collective agreement job classification and work rule systems.

Some significant progress in reforming the job classification system is currently being made in this direction because of Pay Equity regulations. Ironically, while pay and employment equity statutory requirements and regulations are often seen as just part of the regulatory overburden and a cost, the process of meeting the requirements may produce opportunities to rationalize personnel policies and produce efficiency gains. The parties at Hydro are currently engaged in an Internal Relativity Study, which will generate job structure issues for the next round of bargaining. They have produced job documents and new salary grades to meet pay equity policy and legislative objectives. A more coherent system will at least give both parties a more manageable reference point to discuss potential job combinations.

FINDING OPPORTUNITIES FOR MUTUAL GAINS

The Wagner industrial union model is part of the ambient operating environment in both the public and private sectors of the Canadian economy. It has served both parties' interests in the past but in the future it will be necessary to evolve the collective bargaining and workplace relationship in more cooperative, participatory, and productive directions. Most people will now affirm that consultation involving employees is good, and improving working conditions and skills is productive. However, at the end of the day, it is necessary to quantify the impacts of human resource management techniques, both to inform everyone on when and where progress is being made and to send better signals to the bargaining table on potential trade-offs.

Some of the most advanced and creative econometric work on productivity impacts of human resource policies has been done on the steel industry, which along with the auto assembly plant, are the reigning metaphors of the classical industrial age. The study by Casey Ichniowski, Kathryn Shaw, and Giovanni Prennushi quantifies the effects of human resource management practices on productivity and quality.[11] The authors examine the impact of a series of human resource management variables including:

- Incentive Pay
- Recruiting and Selection
- Teamwork and Cooperation
- Employment Security
- Flexible Job Assignment
- Knowledge and Skill Training
- Communications
- Labour Relations

Taken separately the authors find that these individual actions have only minor impacts on productivity. What has much greater impact is when "clusters" of practices reinforce each other. They distinguish four main clusters or "systems" of human resource management practices. These correspond to clear choices in the kind of labour-management regime Hydro, or any other employer, wishes to pursue. The first is the *Traditional*, or adversarial model discussed previously. The second, is a *Communications* model, where the attempt is made to involve the union in consultation and increase the information flow to employee groups but not to change the basic structure of the workplace or the labour agreement. The third is a *Teamwork* model where employee work groups are reorganized around teams and given new training and opportunities to coordinate activities differently than with the traditional top-down management approach. Fourthly, there is a *High Performance* model where the whole approach to employee involvement is combined with changes to the basic compensation system and overall human resource management system.

HRM System	*Designation*
Traditional	HRM1
Communications	HRM2
Teamwork	HRM3
High Performance	HRM4

Each of these HRM models has a measurable impact on productivity, as summarized in the following table.

HRM Model	*Productivity Outcome*	*Quality Outcome*
HRM1	Trend Line	Trend Line
HRM2	2%	4%
HRM3	3.5%	4%
HRM4	7.5%	13%

It is not possible to make immediate generalizations regarding the whole private industrial sector, let alone the whole public sector, from this one specialized study in the steel industry. However, it is certainly suggestive of directions for future research on the potential productivity gains to be made from innovative workplace practices. It also suggests directions for progressive movement forward from within the Wagner industrial model to a more participatory and higher performance model in the future.

The study was done on the steel industry. However, steel mills and power plants share many common attributes: high fixed costs, high capital intensity, and traditionally organized work forces.

An initial estimate is that each 1% productivity improvement in a nuclear station brings the corporation an additional $25 million in revenue. On this basis, the parties have major opportunities to make mutual gains through development of new joint human resource initiatives. If these results are even generally applicable, then they indicate that hundreds of millions of dollars in savings are available to the parties if they could agree to pursue alternative human resource practices and realign their industrial relations practices to match. That is the good news. The bad news is that one-off solutions produce meagre results unless they are associated with general changes or clusters of new policies. This means that quality improvement initiatives by management or employment security proposals by the union, are unlikely, in isolation, to be sustainable successes.

The research cited indicates a direction and potential. What is needed is further analysis and its application to hydro facilities to document the potential for gains and to quantify the trade-offs for the parties.

RESTRUCTURING LESSONS: ELECTRICITY COMMISSION OF NEW SOUTH WALES

The transformation underway at Ontario Hydro is not unique. As indicated at the outset, the process of change in the regulatory, technological, and operating environments is underway in the electric utility industry worldwide. A helpful example may be taken from the New South Wales Electricity Commission in Australia. A five-year long process of redesign and dialogue with bargaining agents has resulted in major progress on workplace reform. The move into autonomous Business Units has been generally similar to that at Ontario Hydro. For management, this has been linked to a focus on results and performance agreements for which local management will be held accountable. On the employee side, there has been a whole change in work organization linked to retraining and entry into work teams and workplace consultative committees.

The whole NSW Electricity Commission has been consolidated into a single salary structure consisting of 40 job grades, within which there are six salary bands.[12] There is a Skills Development and Career Path for each power worker. Over 130 non-trade classifications are being consolidated into four new multi-skilled non-trades groups.

New Structure and Career Path[13]

Probationary Work	Level 4	PW4
Power Worker	Level 3	PW3
Power Worker 2	Level 2	PW2
Power Worker 1	Level 1	PW1
Power Line Worker	Level A	PLW

When the restructuring process is fully implemented, all existing non-trade employees will be reclassified as Power Workers. Training will be provided in accordance with a certified skill development plan (SDP).

CONCLUSION

There are three fundamental issues facing Hydro and the PWU on which the senior decision makers have to make strategic policy decisions:

1. What is the Union-Management relationship that they want?
   The message has constantly changed. It was cooperation in 1990-1991. It was confrontation in 1992-1993. What is the relationship they both want and need for the future? The descriptive and quantitative analysis described above are suggestive of the direction for a different kind of labour-management deal and indicate what some of the tradeoffs and outcomes may be.
2. Business Unit Differentiation
   Hydro has moved operationally and introduced different Business Units but they still have one collective agreement. How different can the units be and what is the relationship between them and to the central agreement? A balance of central and local "establishment" agreements has been negotiated in similar circumstances in Australia. There are lessons to be learned.
3. Employment Security
   Hydro still has some 1200 surplus employees. They still face the problem of organizational renewal. The low-lying fruit has been plucked and both parties now face the tough issues of how to institutionalize what they have been through and to deal with reskilling and outplacement. For example, article 11 presents serious problems in the

implementation of redeployment, e.g., training for which jobs, bumping for jobs that are too specific.

Looking forward to the 1996 negotiations, pragmatic considerations suggest that the parties should negotiate within the context of a new relationship but with limited issues on the table. The parties will have to deal with employment security. The pressures are mounting. The surplus employees are still there. Overtime is higher than it has ever been. The near-term trade-off could contain a short-term plan on how to deal with the union's need for an employment security commitment and for the company's financial needs. The parties could also put in place a longer-term understanding to push the ERI process forward through a participative and decentralized procedure.

Moving around the deckchairs within the existing model will not solve the problems of the future. Moving beyond the Wagner model will be difficult, but the outlines of a new and acceptable trade-off of security-flexibility are in view.

The author would like to thank the representatives of Ontario Hydro and the Power Workers' Union who made available their time and insights to assist in the preparation of this paper.

NOTES

1 Richard Hirsh, "Regulation and Technology in the Electric Utility Industry: A Historical Analysis of Interdependence and Change." In *Regulation: Economic Theory and History*, ed. Jack High (Ann Arbor: University of Michigan Press, 1990), 147.
2 Collective Agreement between Ontario Hydro and Power Workers' Union, Canadian Union of Public Employees – CLC, Local 1000, April 1, 1994 – March 31, 1996.
3 Memorandum of Agreement, Ontario Hydro Corporation and Power Workers' Union, CUPE Local 1000, March 30, 1994, 5.
4 Ibid., 14.
5 PWU, May 6, 1994.
6 Ibid., 2.
7 John Mathews, *Catching the Wave* (Cornell: IRC Press, 1994).
8 Dorothy Sue Cobble, "Making Postindustrial Unionism Possible," Institute of Management Relations, Rutgers University, October 4, 1993.
9 John O'Grady, "Job-Control Unionism vs. the New Human Resource Management Mode: the Impact of Industrial Relations Factors on Changes in Work Organization in Manufacturing and Related Industries," Queen's-University of Ottawa Economic Projects, March 1994.

10 Ibid., 42.
11 Casey Ichniowski, Kathryn Shaw, and Giovanni Prennushi, "The Effects of Human Resource Management Practices on Productivity," Carnegie Mellon University, Mimeo, August 1993.
12 Electricity Commission of New South Wales, "The Future: The Electricity Commissions T.E.A.M. Plan Implementation," Sydney, April 1990.
13 Ibid., 25.

Comments by

JOHN D. MURPHY

A study of the future of the electrical power industry in Ontario would be incomplete without considering the industry's most important asset: the people. I appreciate particularly the opportunity to participate and comment on the discussion of this aspect of the business.

I want to share with you some history and background of the Power Workers' Union (PWU) before I comment on the contribution by Peter Warrian.

Up until a few years ago, the PWU was very much a traditional union. We restricted our attention to collective bargaining, health and safety, etc., and ignored the company business decisions. That approach got us in a lot of trouble. While many of the management decisions of the past were good, others created many of the financial problems we face today.

As an organization, we decided that managing was too important to be left solely to the management. We began to demand a lot more involvement in all aspects of the business. This has been achieved with varying degrees of success throughout the company.

The PWU is determined to have the value system changed throughout Ontario Hydro to ensure meaningful involvement of the workers; the people who best know the business and how to improve it. This is not a quick-fix problem, but a long-term project that will require the commitment and support of senior-level management.

Now to my comments. In general terms, I think this material is good and generally reflects the historical, current, and future options for Ontario Hydro and the PWU. I would, however, like to clarify a few points.

The material leaves us with the impression that the PWU simply wants employment security and the work does not have to be challenging.

A more accurate description would be to describe the PWU's goal as having all its members gainfully employed in challenging positions. Generally speaking, our members are well educated and highly motivated. They have little interest in being locked into a non-challenging position.

The PWU believes that this goal can best be achieved by engaging in a meaningful reskilling and retraining program. This program should be designed to make sure that people are not only trained for the jobs they do today, but for the jobs they will be required to do in the future. Surely, this type of investment makes sense if we truly regard people as our most valuable asset.

The conclusion makes three recommendations that I support. The first recommendation is the most important because it says the parties need to decide what type of relationship they want. I fully agree. For the record, the PWU preference is for a cooperative relationship. Cooperation rather than confrontation is the best way to look after the needs of the customer, the business, and the people we represent.

Comparative Cost of Financing Ontario Hydro as a Crown Corporation and a Private Corporation

MYRON J. GORDON

I. INTRODUCTION AND SUMMARY

The movement towards privatization and deregulation in the electric power industry in the United Kingdom and the United States has encouraged interest in that course of action in Canada. Privatization and reliance on market competition instead of government to regulate the industry has been advocated largely on the grounds that the cost of the service to consumers would thereby be reduced. However, the comparative fuel, labour, and other production costs under the two regimes are measured with great difficulty, so that the case for or against government ownership or control has been made typically on the basis of limited and imperfect cost studies, or on the basis of abstract principles with regards to how the market and government succeed or fail as mechanisms for innovation and control.

One area in which it is possible to develop comparatively accurate and complete cost information under each form of ownership is the cost of financing. For a utility plant with given operating costs, what is the difference between the cost of financing that plant under government ownership or private ownership? That is the primary question to be addressed here. In fact, the movement from government ownership or regulation to private ownership and market regulation has not been completed at one stroke in the United Kingdom or the United States. The financial risks are shared among the various old and new participants in the industry. The limited movement carried out and contemplated in Ontario has involved sharing the costs and risks among

Ontario Hydro's consumers, the new private corporation, and government. It will not be possible here to evaluate all of the mixed arrangements that have evolved in the three countries, but some observations will be made on what has taken place here in Ontario. The conclusions reached with regard to complete privatization as the alternative to crown ownership also holds more or less for partial privatization.

In Sections II and III below, the financing cost of electric power to the consumers of Ontario is estimated for Ontario Hydro as a consumers' cooperative and as a private corporation in a manner that facilitates comparison. Section IV examines the risk and its costs that Ontario Hydro imposes on Ontarians as both consumers and taxpayers under each form of ownership. Section V brings together the above data. The comparative analysis is carried out on the basis of current values of the relevant parameters, such as interest and tax rates, in order to communicate more clearly.

With the cost expressed as a percentage of the capital employed and with 9% the interest rate on Ontario debt, 9.07% is found to be the cost of financing Ontario Hydro as a consumers' cooperative. With private ownership, the cost to Ontario consumers of electric power is estimated to be over 75% higher, or 15.94%. Recognizing that the people of Ontario are taxpayers as well as consumers of power, the financing cost to Ontarians of a consumers' cooperative is raised from 9.07% to 12.42%. It is still below the financing cost of a private corporation by 3.52%. Of this, 1.1% is due to lower risk costs and 2.42% is savings in federal, personal, and corporate taxes. These figures incorporate estimates with regard to the loss associated with future contingencies and the cost of assuming the risks involved, which are somewhat subjective in nature.

The financing cost per kWh under the two forms of ownership depend upon the capital-output ratio of Ontario Hydro. Nuclear and hydro production of power are very capital intensive, and the capital output ratio for Ontario Hydro is about 6.72. With that capital-output ratio and with production costs about $3.00 per 100 kWh, financing raises the cost to Ontarians to $5.51 per 100 kWh for Ontario Hydro as a crown corporation. With private ownership the cost to the people of Ontario is raised to $6.21 per 100 kWh, which is 12.7% higher.

The above figures do not include the cost of the transition from public to private ownership. Investment dealers can be expected to charge at least 5% of the funds provided for the required sales of shares, and investors can be expected to require at least a 5% discount in the issue price as a condition for absorbing such a large initial public offering. Also not included are the costs associated with the loss of sovereignty over the industry associated with its privatization under NAFTA.

The last section considers what discount rate Ontario Hydro should use in evaluating its investment opportunities. Should it be the 15.94% that represents the cost of capital or rate of return that a private corporation would require, or should it be the 12.42% that reflects the risk and tax advantage that the crown corporation enjoys? In addition, the section comments on two relatively popular proposals for enlarging the private sector in the electric power industry, the privatization of Ontario Hydro's hydroelectric generating capacity and the building of gas-fired non-utility generating (NUG) capacity.

With the cost of financing Ontario Hydro to the people of Ontario considerably lower under crown ownership, what are the compensating benefits of privatization? How much will be gained through the more efficient allocation and use of resources under private ownership? The great economist Joseph Schumpeter has argued that the benefits of private ownership and markets under capitalism are not to be found in the more efficient allocation and use of resources under existing technology. They are to be found in innovation – new technologies, new products, and new markets. Accordingly, private ownership has advantages over crown ownership in industries where the capital-output ratio is low and where technological change, product variety, and other attributes make the costs of market control small compared to the benefits. Computer software, toys, fashion clothing, perhaps even telecommunication come to mind as industries that qualify. It is not clear how this is now true of the electric power industry.

II. FINANCING COST OF ONTARIO HYDRO AS A CONSUMER COOPERATIVE

Our objective in this section is to arrive at the capital costs of having Ontario Hydro (OH) provide power to the consumers of Ontario when OH is looked upon as a consumers' cooperative. It is reasonable to look on OH as a consumer cooperative, meaning by this that the corporation is owned collectively by the consumers of electric power in the province. Ontario Hydro's mandate from the province is to sell power to consumers at cost, in which case the ownership equity in the capital structure is intended to be a contingency reserve that absorbs short-run differences between revenues and costs, and changes in consumer rates are used to restore this contingency reserve to a normal level. Furthermore, the funds required to establish the contingency reserve or ownership equity, as well as the funds required to maintain it, come from OH consumers.

The only revenues that the province and the people of Ontario as taxpayers receive from OH are water use charges that are looked upon here as hydro production costs and the debt guarantee fee equal to 0.5% of OH outstanding debt. (Local governments receive payments in lieu of local taxes.) Ontario Hydro pays no income tax, so that the province and federal government do not share in the profits of OH in the way that they share in the profits of private corporations through the income tax. Of course, OH could not function as a consumer cooperative without the provincial guarantee of its debt. That, and the residual obligation the guarantee implies, combined with the authority of the province over OH make the people of Ontario the real owners of OH. We will look at OH from the viewpoint of Ontarians as taxpayers later.

For the purpose of this study, capital costs comprise the interest on debt, profit on ownership equity and all the charges that are related to the financing of OH, such as its debt guarantee charge and income taxes. Capital costs might also be defined to include depreciation charges and property taxes, but they can be ignored here, because our objective is to compare capital costs under crown and private ownerships, and these costs should be the same under both types of ownership. It should be remembered in what follows that we are also ignoring differences in operating costs between crown and private ownership. Some would argue that employment and other production costs are lower under private ownership while others may argue the opposite position. Here we assume operating costs and decisions would be the same.

To arrive at the capital costs under crown ownership, we must arrive at an appropriate capital structure and the cost for each component of capital. The existing capital structures for crown corporations may be looked at for guidance on this question. Table 1 presents the percentage distribution of the long-term financing of OH and two other provincial electric power companies between debt and equity for the years 1989 to 1993. We see that the ownership equity or contingency reserve ranges from a high of 18.8% for OH in 1989 to a low of 2.4% for Manitoba in the same year.

The size of the contingency reserve for a crown corporation that operates as a consumer cooperative should vary with the volatility of its revenues and expenses. Accordingly, Manitoba Hydro, with most of its power obtained from low-cost hydro sources, might require a lower equity than OH, which relies heavily on nuclear power. On the other hand, to the extent that Manitoba Hydro exports power for profit and its output depends on weather, the volatility of its profit is increased with provincial market conditions unchanged.

In my judgment, the common equity for OH should fall between 10% and 15% of its total long-term financing. Our financing cost calculation will be carried out using both of these figures to make clear how variation in capital structure influences financing costs.

The interest rate on OH debt may vary for a large number of reasons. They include its maturity structure, when it was issued, the ability of its financing officers, and the credit quality of OH as a borrower. Since our

Table 1
Capital Structures of Crown Corporations Engaged in the Production of Electric Power, 1989-1993*

	1989	*1990*	*1991*	*1992*	*1993*
ONTARIO HYDRO					
Debt	81.2%	82.4%	83.2%	83.7%	91.4%
Equity	18.8	17.6	16.8	16.3	8.6
MANITOBA HYDRO					
Debt	97.6	97.2	96.4	96.6	96.7
Equity	2.4	2.8	3.6	3.4	3.3
B.C. HYDRO					
Debt	91.1	90.9	90.3	86.2	85.8
Equity	8.9	9.1	9.7	13.8	14.2

*Source: Dominion Bond Rating Service.

objective is to compare OH cost as a crown corporation with its cost as a private corporation, we want the interest rate in both cases to abstract from factors that are irrelevant to this difference. In particular, we want to abstract from the maturity structure of the debt and when it was issued. In fact, the comparison of capital costs can be made by assigning to OH any interest rate within a reasonable range, as long as the relative rate assigned to OH as a privately owned corporation is correct.

To arrive at the interval in which the interest rate assigned to OH debt should fall, Table 2 presents the average yield in each of the years 1989 to 1994 on mid-term bonds for the federal government, provincial governments, and AA-rated private debt. It can be seen that the interest rate on provincial debt ranged from 7.74% to 11.55% over the six years, and we will assign a 9% rate to OH debt for the comparative calculations to be undertaken here. If nominal rates were to fall drastically, say to within 5-6% range, financing costs would become materially less important. The consequences of such a reduction in interest rates for the comparative financing costs may reasonably be deferred until there is some likelihood that such a fall in interest rates will take place.

The contingency reserve for a crown corporation that operates as a consumer cooperative must earn a long-run return equal to the rate of growth in assets. That return enables the ownership equity to grow at the same rate as assets and debt and thereby remain unchanged in relative size. If the utility is growing at a rate of 11% per year the return should be 11%, and if it is growing at a rate of 2% per year the return need only be 2%. For the immediate future OH is not expected to grow, so the long-run return OH must earn on its ownership equity is zero percent. However, OH consumers are in fact

Table 2
Average Annual Yields on Various Classes of Mid-Term Bonds, 1989-1994*

Year	*Canada*	*Provincial*	*Corporate AA*
1989	9.91%	10.35%	10.61%
1990	10.93	11.55	11.77
1991	9.36	10.00	10.18
1992	8.15	8.81	9.01
1993	7.20	7.74	7.74
1994	8.15	8.56	8.57
Average	8.95	9.5	9.65

*Source: Scotia McLeod, *Scotia McLeod's Handbook of Canadian Debt Market Indices, 1947-1994* and *Canada-U.S. 52 Week Yield Spread Matrices* (Toronto, 3 January 1995).

earning the return they require on the money they have invested in OH through lower prices for the power they buy. If consumers required, say, a 15% return on their equity, OH could pay them that amount by raising rates to cover that cost and paying the money out as a dividend. Ignoring risk and taxes for the moment the return that consumers require and earn is the interest rate they could earn on the money. With OH borrowing at 9% its consumers can at most earn 9% by lending money directly. Nonetheless, we will use that figure instead of a lower one as the risk-free interest cost of the ownership equity.

The provincial charge of 0.5% of the outstanding debt to guarantee its payment is a risk cost of financing OH under consumer ownership, and it must be added to the interest cost. Whether or not this charge is adequate to compensate Ontarians as taxpayers for the risk they thereby assume will be considered in Section IV. Is there any additional risk cost to the consumers of OH? The additional risks they face fall under two headings. One is the fluctuations in the prices they pay for power and in the quality of service. These risks, if anything, are lower than the risks they would face under private ownership. Nonetheless, it will be assumed that they are the same, in which case they can be ignored.

The other risk is that the future payoffs on their investment will not be realized by the consumers of OH. The future payoffs are the interest rate of 9% per year plus the principal when the investment is liquidated. Only the date and amount of the loss of principal is open to question, and there will be a loss of principal under two conditions. One is that OH ceases to be able to generate the revenues needed to meet its obligations, and it is liquidated with the consumer equity wiped out. The other is that OH is privatized and the consumer equity is given to the financial community instead of being given to Ontarians as consumers or taxpayers. It will be assumed that both of these risk costs are covered by adding a 0.5% charge to the interest cost of the common equity.

The consumers of OH avoid the personal income tax on the interest they earn through their investment in OH. The amount varies from one person to another depending upon their marginal tax rate. It will be assumed that the tax avoidance is on average 30% of the 9.5% imputed interest charge on the amount invested.

Finally, it may be argued that a cost should be imputed to the financing of OH as a consumers' cooperative, because each consumer's investment does not have the advantages of private wealth. The consumer is not free to determine his or her investment in OH. It is collectively determined. The investment is not liquid. It cannot be sold with the proceeds put to some other use. Hence, 9% is not an adequate return for a consumer with no other risk-free wealth. Finally, on death the investment is passed on to the people of Ontario and not to the consumer's designated beneficiaries. These costs will be ignored on the grounds that there are compensating advantages of compulsory saving and investing via OH.

Table 3 presents the financing cost of OH as a consumers' cooperative under a 10% and a 15% equity component of the capital structure, with 9% the cost rate on debt and 9.5% the cost rate on equity, a debt guarantee cost of 0.5% of debt outstanding, and a tax saving due to equity financing based on the assumption that the marginal tax rate on average is 30%. The total financing cost is 9.215% with a 10% equity, and it is 9.072% with a 15% equity.

Table 3
Financing Cost of Ontario Hydro as a Consumers' Cooperative Under Alternative Capital Structures*

	Equity 10%		*Equity 15%*	
Source	*Fraction*	*Cost*	*Fraction*	*Cost*
Debt @ 9%	.90	8.10%	.85	7.65%
Equity @ 9.5%	.10	0.95	.15	1.425
Debt Guarantee @ 0.5%	.90	0.45	.85	0.425
Pre-Tax Cost		9.50%		9.50%
Personal Income Tax 30% of 9.5%**	.10	-0.285	.15	-0.428
Total Financing Cost		9.215%		9.072%
Capital Output Ratio***		6.72		6.72
Percent Financing Cost of Operating Cost		61.92%		60.96%

*Based on an interest rate of 9% and an equity cost rate of 9.5% as explained in text.

**Personal income tax on equity under crown ownership avoided on the assumption that the average marginal tax rate on personal income is 30%.

***Based on the average of net fixed assets of Ontario Hydro at the start and end of 1993 divided by total operating costs including depreciation.

The impact of the financing cost on the rates charged to consumers depends upon the ratio of capital employed to operating costs. The more capital intensive the method of production, the greater the relative burden of financing cost. The use of hydro or of nuclear energy as the source of electric power creates very high capital-output ratios. For OH the ratio was 6.72 in 1993, so that financing costs raised the cost of power to OH consumers by 61.92% with a 10% equity and by 60.96% with a 15% equity. In other words, for every dollar of operating costs to consumers, the financing of OH adds over sixty cents to the bill.

III. FINANCING COST OF ONTARIO HYDRO AS A PRIVATE CORPORATION

By having much more ownership equity in their capital structures, public utility companies that are privately owned and have no government guarantee for their debt are able to borrow at rates which are little higher than the rates paid by crown corporations The common and preferred equity provide a partial guarantee in that a corporation with a 40% equity must experience a fall in its asset value of more than 40% before the safety of the debt is impaired. When the equity is 40% of assets, the ratio of assets to debt is 1.67, but earnings before interest and taxes are then considerably more than 1.67 times the interest charge. Consequently, the coverage of the debt provided by the common equity is higher than 1.67, since earnings before interest and taxes can fall by more than 40% before falling below the interest charge.

The capital structures employed by private electric power companies vary over some range with their objectives and their circumstances. A regulated company that is experiencing no difficulty in earning its allowed rate of return and is not being allowed to earn more than a fair rate of return will move to the most conservative capital structure that it can obtain. The reason is that the cost of a reduced debt ratio falls on consumers while the benefit accrues to the company's management and stockholders. A regulated company that is having difficulty in earning its allowed rate of return, perhaps because of regulatory lag, will allow its debt-equity ratio to rise. The alternative is to sell shares in a hostile market, and that is too costly. The policy of an unregulated company with regard to capital structure will depend on the relative importance of security and profitability for the management. It also will depend on whether losses fall on the company or can be shifted to other parties. For instance, NUGs have high debt ratios since losses are shifted to captive customers.

Table 4 presents the percentage use of debt, preferred stock, and common equity in the financing of investor-owned electric power companies in the United States and selected companies in Canada. It can be

Table 4
Capital Structures of Bell Canada, Private Electric Power Companies in Canada, and a Group of Companies in the United States, 1989 to 1993

	Percentage		
Year	*Debt*	*Preferred*	*Common*
BELL CANADA			
1989	40.2%	6.6%	53.2%
1990	40.3	6.5	53.2
1991	40.2	7.2	52.6
1992	42.2	6.5	51.4
1993	42.4	6.3	51.3
TRANSALTA			
1989	41.5%	20.7%	37.8%
1990	45.0	20.2	34.9
1991	40.1	22.7	37.2
1992	40.6	20.4	39.0
1993	46.7	14.7	39.5
CANADIAN UTILITIES			
1989	42.0%	21.4%	36.6%
1990	46.2	20.1	33.6
1991	42.4	25.7	32.0
1992	44.3	23.3	32.3
1993	46.0	21.7	32.3
MOODY'S ELECTRIC UTILITIES GROUP			
1989	50.6%	7.3%	42.1%
1990	52.1	6.9	41.0
1991	51.9	6.6	41.6
1992	49.8	6.6	43.6
1993	53.3	7.4	39.2

Sources: Dominion Bond Rating Service and *Moody's Public Utilities Manual* (1994).

seen from this data that a capital structure of 45% debt, 15% preferred stock, and 40% common equity does not depart materially from practice, and the financing cost of Ontario Hydro as a privately owned corporation (OHP) will be calculated on the basis of this capital structure. However, preferred stock financing is a peculiar source of funds, and it will be helpful to calculate also the financing cost of OHP with a capital structure of 50% debt and 50% common equity.

To arrive at the financing cost of OHP we must assign a cost rate to each source of funds and then calculate a weighted average cost of capital. For debt we used 9.25%, which is 25 basis points higher than the 9% used for OH as a crown corporation. Looking back at Table 2, we see that the yield on provincial bonds were from zero to 26 basis points lower than the yields on AA corporate bonds. The provincial yields are an average that includes Alberta and British Columbia on the one hand, and Quebec and the Maritime Provinces on the other hand. Ontario is somewhat below the average, so that a 25-basis point spread is quite reasonable.

Preferred stocks have a claim to assets that is subordinate to debt, and they ordinarily pay at most the stated return. Hence, their risk characteristics suggest that their yields should be materially higher than the yields on AA corporate debt. Instead, their yields are much lower. In late February 1995, when Ontario debt with a coupon of about 9.5% and a ten-year maturity was selling at a yield of about 9.5%, high-grade preferred stocks were selling at yields of between 6.5% and 8.5%. The large spread is due to various reasons including call provisions and unreliable data due to thin trading. The reason these yields are on average substantially below bond yields is the tax advantage of dividend income over interest income to individuals and to corporations. Dividend income is practically tax free to corporations and taxed at preferential rates to individuals, while interest is taxable at ordinary rates. A yield of 7.5% is assigned to the preferred component of the capital structure.

To arrive at the cost of common equity capital, the spread between the return actually earned on common and the yield on AA-rated corporate bonds was calculated for the years 1989 to 1993 for three Canadian utility companies. The data appear in Table 5, where it can be seen that the average spread ranged from 0.7% to 4.40%. Regulatory lag prevents the rate adjustments needed to make the return on common equity vary so as to keep the spread constant. The overall average was 2.71% over the five years. This seems to be a reasonable figure for the premium over the yield on AA corporate bonds. It reflects both the tax advantage of the dividend and capital gain return on common stocks over the interest return on bonds, and the favourable treatment of utilities by their regulators. Notice that 2.71% is not the risk premium that investors require on public utility stocks. In the absence of taxes, the risk premium would be considerably more than 2.71%. On the other hand, the tax advantage of dividends and capital gains over interest income makes the premium lower. My research leads to the conclusion that the premium is about 1.5%.[1] Utilities commonly are allowed to earn more than their cost of capital, as evidenced by the fact that their market-to-book value ratios are typically well above one. A premium of 2.70% over the interest rate was used to arrive at the rate of return assigned to the common equity, to reflect the tendency of regulators to allow a rate of return that is greater than the cost of capital.

Finally, to arrive at the financing cost to consumers of having OHP provide electric power to them, we must take account of the corporate income tax and the tax on corporate capital. It may be argued that taxes on

Table 5
Rates of Return on Common Equity Earned by Canadian Utility Companies and Interest Rates, 1989 to 1993

	Year				
Company	*1989*	*1990*	*1991*	*1992*	*1993*
Bell Canada*	13.2%	13.5%	12.9%	12.5%	10.3%
Canadian Utilities*	12.7	11.8	12.5	13.3	13.4
Trans Alta*	11.1	12.2	13.1	13.5	12.6
Interest Rate**	10.6	11.8	10.2	9.0	7.7
Average Spread	1.73	0.70	2.63	4.10	4.40

*Based on earnings on common equity for the year divided by the average of the common equity at start and end of the year.

**Scotia McLeod yield on mid-term AA corporate bonds for the year.

private corporations that do not fall on crown corporations should not be considered an advantage of crown ownership, and that question will be considered further in Section IV, where we take account of the taxpayers of the province. However, now the comparison is with OH owned by its provincial consumers. For them, there is no question that the cost of electric power would be higher by the corporate income and capital taxes as well as any other difference in financing cost under private ownership.

Under the 1995 federal budget the corporate income tax on large corporations not engaged in activities for which a reduced rate is granted is 44.62%, with 29.12% to Canada and 15.5% to Ontario. In addition, there is a tax on the capital employed that works out to be about 0.525% with 0.3% to the province and 0.225% to Canada, the latter in the current budget.

Table 6 presents the financing cost of OHP under the two capital structures arrived at earlier. With preferred stock in the capital structure, the financing cost is 15.35%, and without preferred stock the financing cost is somewhat higher at 15.94%. The financing cost is over 100% of operating costs under both capital structures, meaning by this that for every dollar of operating costs financing raises the cost to consumers by more than one dollar.

IV. COST TO THE TAXPAYERS OF ONTARIO OF ONTARIO HYDRO AS A CROWN CORPORATION

Although the people of Ontario have decided that it is in their best interest to have OH operate as a consumer cooperative, they are not concerned solely with its consequences for them as consumers of electric power. They are also concerned with its consequences for them as

Table 6
Financing Cost of Ontario Hydro as a Private Corporation Under Alternative Capital Structures*

	Preferred Stock		*No Preferred Stock*	
Source	*Fraction*	*Cost*	*Fraction*	*Cost*
Debt @ 9.25%	.45	4.163%	.50	4.625%
Preferred @ 7.5%	.15	1.125	—	—
Common @ 11.95%	.40	4.780	.50	5.975
Pre-tax Cost		10.068%		10.600%
Capital Tax @ 0.525%	1.00	0.525		0.525
Corporate Income Tax @ 44.62%**		4.758		4.814
Total Financing Cost		15.351%		15.939%
Capital-Output Ratio***		6.72		6.72
Percent Financing Cost of Operating Cost		103.16 %		107.11 %

*For choice of capital structures and cost rates, see text.

**44.62% is the combined tax rate in Ontario based on the 1995 budget. To arrive at the income tax as a percentage of the capital employed, the income to preferred and common (5.905% or 5.975%) is divided by one minus the tax rate to establish the pre-tax income. It is then multiplied by the tax rate to get the tax of 4.758% or 4.814%.

***Based on average of net fixed assets of Ontario Hydro at the start and end of 1993 divided by total operating costs including depreciation.

taxpayers. The next section will examine the difference in tax revenues to Ontario between OH and OHP. Here, we will consider the costs and benefits of the provincial guarantee of OH debt to the taxpayers of Ontario. The guarantee poses two questions. (1) Does the guarantee raise the interest rate the province pays on its debt other than OH debt? Consumers, not taxpayers, pay the interest on OH debt. (2) Does the 0.5% provincial charge per dollar of debt for the guarantee cover the cost to the taxpayers of providing the guarantee?

There are two conditions under which the guarantee of OH debt would make the interest rate on the province's other debt higher than it would be without the guarantee. (1) The guarantee of Hydro debt may increase the risk of default on the total debt and thereby push up the interest rate on all debt. (2) The OH debt may be increasing so fast that a premium must be paid on all debt to get the market to absorb the growing provincial debt direct and guaranteed. The evidence that we can obtain on these questions is examined below, and it supports the conclusion that OH debt is not making the interest rate on provincial debt materially if at all higher than it would be without the guarantee.

To discover whether the provincial guarantee of OH debt increases the interest rate on the province's debt, we may consider the data in Table 7. This table presents estimates of the premiums in the yields on new bond issues by various provinces over the yields on Canada government bonds on March 16, 1995 by Nesbitt Burns, a leading Canadian bond dealer. The actual spreads in yields for specific bond issues will vary for various reasons such as coupon, maturity, and size that are extraneous to the quality of the issuer. The spreads in Table 7 abstract from these extraneous factors, other than maturity.

We see in Table 7 that Alberta and British Columbia carry the lowest premiums, about 20 basis points on a ten-year bond. Their premiums are so small that they can be attributed entirely to the greater liquidity of federal over provincial debt. The market considers the likelihood that these provinces will default on their debt to be practically zero. The greater premium on the debt of other provinces represents some combination of default risk, liquidity premium, and absorption costs.

The premium on a ten-year Ontario bond is 40 basis points, about 20 basis points higher than the premium on Alberta and British Columbia debt. Some part of this 20 basis point difference may be due to growth in provincial debt. Hence, default risk adds at most 20 basis points to the cost of Ontario debt. Furthermore, it is quite possible that this premium would be larger in the absence of provincial ownership of OH. Ontario Hydro's revenues are greater than its costs including interest on its debt, so that OH debt is being retired. By contrast, provincial tax revenues fall short of expenditures so that provincial debt is growing at a rapid rate that some consider to be alarming. Hence, provincial ownership of Hydro is

Table 7
Nesbitt Burns Estimates of the Yield Spreads in Basis Points on New Issues of Provincial Bonds Relative to Canada Government Bonds, March 16, 1995

	Term in Years			
	3 yr	*5 yr*	*10 yr*	*30 yr*
Ontario	25	28	40	46
Québec	32	61	79	85
British Columbia	10	10	21	24
Alberta	8	7	18	N/A
Saskatchewan	25	35	45	50
Manitoba	19	23	33	40
New Brunswick	19	23	33	40

Source: Extracted from Nesbitt Burns Bond Market Closes (March 16, 1995).

now both reducing the growth rate in total provincial debt, and it is increasing the province's ability to service its total debt.

Is the fee that the province charges to guarantee OH debt, which is 0.5% of the amount outstanding, adequate compensation to the province for (1) the present value of the expected value of the future loss, plus (2) compensation for the risk involved?[2] In considering this question it should be realized that the province has a contingent liability that kicks in if and when OH is unable to raise the revenues needed to meet its obligations through rates charged to the users of power. OH has a monopoly over the sale of electric power, and the demand for power is still highly inelastic with respect to price. In addition, when relieved of the obligation to contribute to employment in Ontario and to the further development of the nuclear power industry in Canada, OH is a relatively low-cost producer of electric power in North America.

Hence, the question posed at the start of the previous paragraph involves the following questions. What is the probability that OH will not be able to generate the excess of revenues over operating costs that is needed to service its debt one, two, three,...or one hundred years in the future? What would be the loss to the province if it became necessary to liquidate OH in that year? What is the discounted present value of that loss? In addition, what rate of profit do Ontarians require as compensation for being burdened with the uncertainty involved in the possible loss?

Before considering this question further, it should be noted that we should exclude some events which might generate the loss. There is a very small but non-zero probability that a terrible nuclear accident will create an enormous liability or that continued operation of the nuclear plant will be found to be environmentally unacceptable. The possible loss from such catastrophes can be ignored, I believe, because it is very likely that a condition for private ownership is that the liability for such events be transferred to the government. A cost or loss that falls on government regardless of ownership does not enter into our comparative analysis.

Regardless of what contingencies are included, there is no point in attempting to make the above calculation, because it cannot be done with any confidence in its accuracy. It may be thought that the alternative is to take, as the cost of bearing the risk, what the market charges for doing so in the case of privately owned electric power companies. The holders of the debt and common shares in privately owned companies put a price on the cost to them of bearing the risk involved, regardless of how difficult it is to measure the risk and its cost. Why not take as the cost to Ontario taxpayers what investors would charge if it were owned privately?

To arrive at what the debt guarantee charge should be under the above reasoning, we could make use of the data established previously, as follows. Recall from Table 3 that the charge to consumers for the return on capital employed by OH, including the debt guarantee charge, was 9.5% on the basis of a 15% common equity, while from Table 6 we see that with no preferred in the capital structure, the OHP charge to consumers would be 10.6% of the capital employed, with provincial and federal income taxes excluded. The difference – 10.6% - 9.5% = 1.1% of the capital employed – is the amount by which the province should raise the debt guarantee charge, if the province is to recover the cost of guaranteeing Hydro's debt, under the assumption that the risk costs of financing Ontario Hydro should be independent of ownership.

Closer examination reveals important differences between a 100% guarantee of 85% of the capital employed through a provincial guarantee and the guarantee of 50% of the capital employed by having the other 50% subordinated to the debt. The provincial guarantee is superior, since it covers 100% of the debt while the guarantee provided by the subordinated equity under investor ownership falls below 100% to the extent that the liquidation of the assets would recover less than 50% of their cost. It may therefore be argued that the debt guarantee charge should be raised by more than 1.1% of the capital employed.

Let us examine the premiums in the debt and equity of OHP in Table 6 more closely. The 9.25% rate on the debt is the nominal and not the true interest rate. It includes the expected value of the default loss as well as the liquidity and risk premiums over the Canada government rate. If the Canada government rate is 8.80%, and if the expected value of the loss on default is 30 basis points, the true rate of return or interest is 8.95%, and the risk premium is 15 basis points. On the other hand, the 11.95% on the common equity is a true return, since it is an average of the possible future returns. They may vary over a wide range from one year to the next. The excess of the 11.95% over the government bond rate is due primarily to (1) the return investors require for bearing risk, and (2) the ability of the company to persuade regulators to allow a return in excess of its cost of capital, the return investors require on its stock.

We are now in a position to examine the differences between a debt guarantee and a subordinated claim to earnings, and we can consider whether the cost of the debt guarantee may be smaller than the risk premium in the return investors require on their equity in a privately owned corporation. With a provincial debt guarantee the ownership equity belongs to consumers. Revenues above (below) costs by the amount needed to maintain the 15% ownership equity accrue to the owners. As long as default and liquidation do not take place, rates are raised (lowered) to restore the 15% ownership equity. Hence, the province's taxpayers do not gain or lose when revenues are above (below) what they should be.

The same is not true when the debt guarantee is provided by a subordinated equity ownership. The regulator, either an agency of government or the market, is supposed to adjust rates to provide a fair return on the capital employed. However, government regulation is subject to regulatory lag, and the lag may be even greater when the regulator is the market. Furthermore, there is no legal or economic obligation on the government or the market to have lean years followed by fat years or vice versa. That is, profits above or below a fair return on the capital employed take place due to regulatory lag, and they need not be compensated for by subsequent adjustments in revenue.

Consequently, the risk premium in the rate of return investors require on utility shares over the risk-free interest rate is not motivated by the need to guarantee the debt. It is motivated by the uncertainty of the future return on capital. That uncertainty generates large fluctuations in the market value of their investment from one period to the next. In so far as regulatory lag is greater under market regulation than it is under government regulation, the cost of capital is thereby increased. The cost of equity capital increases as the debt-equity ratio is raised because the uncertainty of the return on the equity is thereby increased, but it is also true that the quality of the guarantee declines as the debt-equity ratio is raised.

However, regulatory lag on the part of a government agency has compensating advantages just like regulatory lag under the regulation through the market. Government tends to allow corporations to keep abnormal profits to the degree that they are achieved without raising prices, and they tend to allow losses to continue if a rise in prices is needed to eliminate them. Hence, regulatory lag under government regulation tends to function like the market in encouraging firms to discover profitable innovations and to cut costs when profits fall, regardless of the reason.

A complaint about government regulation is that the regulator is captured by the industry, and it is persuaded to allow a higher rate of return than the cost of capital. This takes place to the degree that there is a long-run tendency for the ratio of market-to-book value of the corporation's stock to be above one. In Table 6 only 1.5% of the 2.7% excess of the return on common over the interest rate is a risk cost. The balance is due to regulator generosity, in my judgment. As the allowed rate of return rises above the cost of capital, the cost of capital also rises, since uncertainty about what the utility will earn increases as the allowed return is raised.

It should now be clear that the premium investors earn on a subordinated equity is far different from the cost of the contingent liability arising from a debt guarantee. In one case 50% of the total capital is invested, and the market value of that capital fluctuates with its expected future earnings. In the other case we have the expected present value of the loss to consumers and taxpayers on default and liquidation of a crown corporation plus the risk costs to both with regard to the uncertainty of that event.

What is the risk cost to the consumers of OH? Their investment is not capital in the ordinary sense of the term, since their position as investors in OH is subordinate to their position as consumers. The risk they face is uncertainty about future prices and quality of service. That risk is quite unlikely to be any greater than the risk they would face as consumers in OHP, in which case it can be ignored. What is the risk cost to taxpayers of the contingent liability on the debt equal to 85% of the capital employed? What monetary compensation do they require for the psychic cost involved in the fluctuations in their net worth due to the guarantee? I do not believe that the risk premiums involved in the private ownership of OH carry over to the consumers and taxpayers of a crown corporation. We are left with the expected present value of the loss on the collapse and liquidation of OH. As indicated earlier, we cannot say whether the cost of the guarantee is more or less than the 0.5% charge for providing it, but in the absence of evidence to the contrary my judgment is that the charge is about equal to the cost.

The evidence in support of the above conclusions is not so compelling as to preclude doubt. This may lead some to turn back to the proposition suggested earlier that apart from taxes, the cost of financing Ontario Hydro is independent of ownership. Those who do so may be comforted by the fact that under a widely accepted body of economic theory called neoclassical economic theory, that proposition is true. However, I should point out that under this body of theory there is no such thing as involuntary unemployment. The unemployed simply refuse to take work at the prevailing wage rate. Under this body of theory it is also true that the cost of financing a private corporation is independent of how it is financed. In other words, the value of a private corporation is independent of its dividend policy, of its capital structure, and of other components of financing policy.

Some neoclassical economists maintain that in reality there is no involuntary unemployment and there is no dependence of corporate value on financial policy. On the latter subject, neoclassical financial economists maintain that anyone who believes that a rise in a corporation's dividend per se raises the price of its stock is simply confused. Even the presence of various taxes and other "market imperfections" do not invalidate the empirical truth of the above employment and financing propositions, according to some neoclassical economists. Others maintain only that full employment and value independence are true in a world of perfectly competitive labour and capital markets, and they concede that various "market imperfections" impair the truth of these propositions more or less in the real world. Little has been done to establish the nature, significance, and direction of the market imperfections that might impair the perfect substitutability of private and public ownership with respect to financing costs within the framework of neoclassical theory.

V. COMPARISON OF TWO FINANCING ARRANGEMENTS

To compare the costs to consumers of electric power in Ontario of having Ontario Hydro a crown corporation with what the cost would be if it were a privately owned corporation, we need only bring together the data in Tables 3 and 6. With two capital structures in each case that would involve four comparisons. Since the cost does not vary materially with capital structure in each case, it will simplify matters to base the comparison on just one capital structure for each form of ownership. Table 8 brings together the costs with a 15% equity for OH and with a 50% debt and 50% equity for OHP, that is Ontario Hydro under private ownership.

The total cost expressed as a percentage of the capital employed is 9.07% for the crown corporation and over 75% higher at 15.94% for the private corporation. The cost to the consumer of this difference in financing costs in terms of price per kWh, depends upon how capital-intensive the operation is. In manufacturing, capital-output ratios of 0.2 to 0.5 are common, in which case differences in financing cost would have a small impact on product cost. However, with a capital output ratio of 6.72, the impact of this difference in financing costs on consumer rates is very large.

To see how the difference in financing costs affects consumers, assume that the operating costs are $3.00 per 100 kWh. With crown ownership, the charge to consumers to cover financing costs increases by 60.96%, for a total cost of $4.83 per 100 kWh. Private ownership would add 107.11% to the $3.00 production cost, for a total cost of $6.21 per 100 kWh. Private ownership raises the cost to consumers by 29%.

The people of Ontario are taxpayers as well as consumers of electric power. Hence, the true tax advantage of crown ownership is the difference in taxes attributable to it. Hence, with 9.50% the financing cost of OH with taxes ignored, the additional tax cost imposed by private ownership is only

Table 8
Comparative Cost of Power to Consumers with Ontario Hydro a Crown Corporation and Ontario Hydro Privately Owned*

	Crown			*Private*		
Source	*Cost Rate*	*Fraction*	*Cost*	*Cost Rate*	*Fraction*	*Cost*
Debt	9.0%	0.85	7.650%	9.25%	0.5	4.625%
Equity	9.5	0.15	1.425	11.95	0.5	5.975
Guarantee	0.5	0.85	0.425			
Pre-Tax Cost			9.500%			10.600%
Personal Income Tax on Equity Avoided	-30.0	0.15	-0.428			
Capital Tax				1.0		0.525
Income Tax				44.62		4.814
Total Cost			9.072%			15.939%
Capital-Output Ratio			6.72			6.72
Percent Financing Cost of Operating Cost			60.960%			107.110%

*Source: Tables 3 and 6.

the federal personal and corporate taxes that are avoided through crown ownership. However, for the comparative analysis of the cost of capital that should be used in investment decisions, to be carried out later, 15.94% should be the figure for the private corporation, and the cost to OH should be raised by the provincial and federal taxes that are not avoided. In other words, it is six of one or half a dozen of the other, whether we add the provincial taxes to the pre-tax financing cost of the crown corporation or add the federal taxes to the pre-tax financing cost of the private corporation in comparing the two alternatives.

Table 9 reproduces Table 8 except for taxes. The provincial income and capital taxes plus one-third of the federal taxes on the private corporation are added to the cost of the crown corporation.[3] In addition, the personal income tax avoided is reduced to two-thirds of the federal component of the 30% tax rate. The financing cost of OH is raised from 9.07% to 12.42%, but it is still 3.52% lower than the financing cost under private ownership. The cost to consumers of power with operating costs $3.00 per 100 kWh under crown ownership is now raised by 83.49% to $5.51 per 100 kWh. Private ownership results in a cost per kWh that is now 12.7% higher than the cost under crown ownership, with the latter remaining at $6.21 per 100 kWh.

Table 9
Comparative Cost of Power to Consumers and Taxpayers with Ontario Hydro a Crown Corporation and Ontario Hydro Privately Owned*

	Crown			*Private*		
Source	*Cost Rate*	*Fraction*	*Cost*	*Cost Rate*	*Fraction*	*Cost*
Debt	9.0%	0.85	7.650%	9.25%	0.5	4.625%
Equity	9.5	0.15	1.425	11.95	0.5	5.975
Guarantee	0.5	0.85	0.425			
Pre-Tax Cost			9.500%			10.600%
Personal Income Tax on Equity Avoided**	-12.0	0.15	-0.171			
Capital Tax	0.375	1.0	0.375	0.525	1.0	0.525
Income Tax	25.21		2.720	44.62		4.814
Total Cost			12.424%			15.939%
Capital-Output Ratio			6.72			6.72
Percent Financing Cost of Operating Cost			83.490%			107.110%

*Source: Tables 3 and 6. The corporate tax rates with Ontario Hydro a crown corporation are the provincial tax rates plus one-third of the federal rates applied to the equity of the private corporation, as in Table 6.

**With the combined federal and provincial tax rates in Table 8 assumed to be 30%, the federal rate here is assumed to be 18%, and two-thirds of it is avoided.

The initial reaction of some economists to the treatment of taxes presented here will be negative, on the grounds that taxes "distort" economic decisions and they should be ignored. That is, when a tax leads an individual or firm to make a different decision than would be made in the absence of the tax, the tax-induced decision is wrong, and the decision-making process should be reframed somehow so as to induce the individual or firm to make the correct decision.

For instance, a paper by Jack Mintz[4] argued that in deciding whether or not to undertake an investment, Ontario Hydro should use a discount rate or cost of capital in arriving at the present value of the project's future revenues and costs that reflects the higher tax and risk costs that are faced by a privately owned corporation. Ontario Hydro should not use its own much lower cost of capital for the following reasons. First, the low nominal cost of capital for OH is more apparent than real. Its true cost of capital is the higher figure that private investors face. Second, by investing on the basis of its lower cost of capital,

OH is taking funds away from and displacing a private sector investment that could earn a higher rate of return and thereby be of greater benefit to society. The discount rate that OH should use in deciding whether or not to make an investment will be considered further in the next section. Here we are concerned with the cost of power to the people of Ontario.

As in this paper, Mintz estimated what the risk and the tax costs of financing Ontario Hydro would be if it were a privately owned corporation. His approach and numbers were somewhat different than mine, but that is not the fundamental difference between us, so that these differences do not warrant a detailed analysis here. The fundamental question is whether the risk and the tax costs that fall on a private power corporation should be imposed on a crown corporation.

With regard to risk, we argued in the previous section that there are good reasons to question the presumption that the costs are the same for the two forms of ownership. For consumers there is no reason to believe that private ownership reduces the risks of fluctuating prices and uncertain supply by comparison with consumer ownership. Also, there is no evidence that 0.5% of the outstanding debt fails to compensate the people of Ontario as taxpayers for the present value of the expected loss on their contingent liability for OH debt.

With regard to the federal income and capital tax that would be paid by a private corporation, it is clear that these taxes do not fall on a crown corporation. We are not concerned with the *social costs* of providing power to the people of Ontario, meaning by social costs the monetary costs to the *people of the world*.[5] Our concern here is with the comparative monetary cost to the people of Ontario of the power they consume. Hence, we do not see why the cost to Ontarians under crown ownership in Table 9 should be increased by more than the provincial corporate income and capital taxes that they would pay under private ownership. The federal taxes they avoid through crown ownership is a cost advantage of crown ownership.

VI. COSTS OF LIMITED PRIVATIZATION

The previous pages have established the comparative financing costs of Ontario Hydro as a crown corporation and as a private corporation, without considering the cost of the transition from crown to private ownership. That transition would involve a cost of at least 10% of the equity funds raised. To my knowledge, the complete transfer from crown to private ownership is not now under serious consideration. I believe that the privatization of some facilities, in particular the hydro plant, has been proposed, and the construction of private gas-fired capacity has and continues to take place. Both of these avenues to privatization impose costs on the people of Ontario that are in addition to the costs discussed in the previous pages.

It requires no great imagination to see how the privatization of the hydro facilities of Ontario Hydro would develop. The hydro plant is a very low-cost producer of power, since it has been depreciated to a fraction of its historical cost and the latter does not include the imputed value of the natural resource. Hence, the plant could be purchased at a price well in excess of its book value, so that the sale would show a profit to Ontario Hydro as well as reduce its debt by the amount of the sale.

Furthermore, the new company, call it "Private Hydro," could sell power to industrial and municipal users at a price well below the price charged by OH and still cover its substantially higher financing costs. Raising the billions of dollars needed to carry out the privatization would require a return well in excess of the cost of capital. The investment dealers who arrange the initial public offering can be expected to charge at least 5% to arrange such a large stock issue, while the investors who buy the issue can be expected to require a considerable discount off the price that provides the going rate of return.[6] The equity cost rate in Table 9 should be raised by at least 10% to cover the cost of moving to private ownership.

After giving up its low-cost hydro plant, the average cost of power produced by Ontario Hydro would be raised materially. The total cost incurred by Private Hydro would be greater than the cost when the plant was a part of OH, so that the total cost of crown and private power to the people of Ontario would be increased. With OH mandated to sell power at cost, the excess of its price over the price charged by Private Hydro would be increased, and that would be cited as further "evidence" in support of the greater efficiency of private power production. In the interest of a sound evaluation of the privatization of the hydroelectric plant, OH should follow good accounting practice and both write this plant up and write the nuclear plant down to their fair market values.

The privatization that is now taking place is non-utility generation (NUG), that is the building of relatively small gas-fired plants that sell part of their output to industrial or municipal users and the balance to Ontario Hydro. Regardless of the comparative merits of gas and other sources of energy, building these plants with the current excess capacity in the province is a waste of money that is being charged to the people of Ontario. In Ontario Hydro's 1992 Annual Report (page 10), Allan Kupcis stated that with the Darlington facility coming into production, "We're adding 3,600 megawatts of capacity with no immediate prospects to sell it." In a memo of February 16, 1993 to the Ontario Hydro Task Force, Felix Chee stated that the marginal cost of generating power with Hydro's capacity is one to two cents per kWh, which is 25-50% less than the cost of purchased NUG power.

Nonetheless, NUG plants are being built under the following terms. All the power that is not being taken by the municipal or industrial customer is sold to Ontario Hydro under a long-term contract at a price that is somewhat below Ontario Hydro's selling price but well above Ontario Hydro's marginal cost of production. When the NUG plant is down for maintenance or for some unexpected reason, power is purchased from Ontario Hydro for the NUG customer at a reasonable price, so that the NUG has no responsibility for security of supply. Finally, each NUG is financed with a very high debt ratio, often about 80%, because most of the risk associated with its operation is transferred to OH through the sale contract. Each new NUG, therefore, raises costs and price to the consumers of Ontario.

Ontario Hydro is not considering whether or not to build new capacity in competition with NUG generation at the present time. However, OH is faced with decisions whether or not to undertake investments that repair and maintain existing capacity, and some day it will be faced with the decision whether or not to add to its capacity. In either case, an alternative is NUG capacity. What discount rate or cost of capital should OH use?

Should OH use the 15.94% discount rate that would be appropriate for a private corporation under the condition that produced this figure in Table 9? Or should OH use 12.42% as its discount rate, to reflect its lower risk and tax costs? The reasoning in support of the higher figure with regard to its risk component is that the best estimate of the risk and its cost on a public investment is what these figures would be on a private investment.[7] The University of Chicago economist Ronald Coase received a Nobel Prize for pointing out that economic activity and its coordination are carried out more economically through markets under some conditions, and under other conditions it is carried out more economically within firms by bureaucratic methods. By analogy, risk costs to the consumers and taxpayers of OH under the present financing arrangement need not be the same as the price investors would charge for bearing the risks they incur when they are inserted in the financing of OH investments. Without evidence to the contrary, a 0.5% charge on the capital employed may be taken as an adequate charge for the present value of the expected future loss to the taxpayers and consumers of Ontario due to the possible financial collapse of OH.

Turning to taxes, the argument here is that the taxes imposed on a private investor must be added to the cost of capital on a tax-free public investment, because taxes are not a true social cost. The private corporation includes taxes in its cost of capital, and if the public corporation does not make the imputation, investment is biased in favour of the public corporation. The more economical or profitable private investment is crowded out. We are here concerned only with effective tax rates, not stated rates, so that in so far as the private corporation avoids or evades taxation there is no tax bias in favour of public investment. We are also only concerned with taxes that fall on the people of Ontario. In so far as a private corporation is subject to federal and foreign taxes that are avoided by OH, the taxes that are a cost of capital for the private corporation are not a cost of capital for OH.

The above cost of capital figures should be used when Ontario Hydro is faced with two mutually exclusive investment opportunities. That is, if the choice is between having OH build an addition to generating capacity or authorizing the establishment of a NUG to do it, OH should use 12.42% for its own investment proposal and 15.94% in evaluating the NUG's proposal. However, the other terms under which NUGs are allowed onto the system are frequently more important than the discount rate in deciding how a demand for generating capacity should be satisfied.

More often, OH is not faced with two or more mutually exclusive means for doing the same thing. The investment it has under consideration is more or less complementary to other investment opportunities in the province. If the province has full employment of its labour force, the discount rate it should use is the 12.42% in Table 9. In so far as there is unemployment, the discount rate should be reduced to the non-tax 9.5%, and the employment benefits may be factored into the cash flows.

NOTES

The author benefited from comments on an earlier draft by Larry Gould, Trevor Chamberlain and Don Brean.

1 See, for example, M.J. Gordon and L.I. Gould, "Memorandum of Evidence: Before the Canadian Radio-Television and Telecommunications Commission, In the Matter of the AGT Limited Review of Revenue Requirement and General Rate Proceeding 1992" (February 1992).

2 It is now generally recognized that risk is a cost, in the sense that $1,000,000 with certainty is preferred to nothing or $2,000,000 each with a probability of one-half. Hence, if a risk-free investment earns, say, 9%, the expected value of the return or profit on a risky investment must exceed 9%.

3 The assumption is that the smaller tax revenues to the federal government through the tax-free status of Ontario Hydro raises other federal taxes by a

corresponding amount. The relative size of Ontario in the Canadian economy is such that about one-third of the increase in other taxes falls on Ontario.

4 J.M. Mintz, "Risk, Taxation, and the Social Cost of Electricity," *Financial and Technical Bulletin of the Independent Power Producers Association* 1, 7 (1993):4b.

5 There are environmental costs of producing power that fall on people in Ontario and elsewhere. These costs should enter into the investment and other decisions of OH and OHP in exactly the same way, so that they need not concern us here.

6 In August 1992, Nova Scotia Power was privatized through a $650 million share issue on which the issue costs were $34.3 million, according to the 1992 Annual Report of the company. For the price discount on share issues that are not enormous initial public offerings, see P. Asquith and D. Mullins, "Equity Issues and Offering Dilution," *Journal of Financial Economics* 15 (1986):61. For discounts on initial offerings, see M.J. Gordon and J. Jin, "Risk, Asymmetric Payoffs, and the Underpricing of Initial Public Offerings," *Research in Finance* 11 (1993):133.

7 See J.M. Mintz, and the references cited there for more on the reasoning in support of this conclusion.

Comments by

WILLIAM A. FARLINGER
A. STEPHEN PROBYN
STEPHEN F. SHERWIN

WILLIAM A. FARLINGER

Privatization

The arguments around privatization of Ontario Hydro are often aimed at trying to determine the net numerical cost advantage of public ownership, which is typically related to freedom from income taxes and lower costs of government capital. However, the most compelling arguments are in fact the obverse of this approach. That is, why should any competitive service remain in government hands? Outside of electric power, there are few examples of competitive services that are successfully offered by government-owned enterprises. And those that have been in the past (e.g., Air Canada, CN Rail) have been or are being privatized. (See Appendix for a list of all major privatizations in the G-7.)

The reasons for this are simple. In a competitive environment, the marketplace works. Allocation of capital, human resources, and entrepreneurial effort are driven by the demand for the service and competitive supplies available. Though government organizations can strive to emulate or recreate market sensitive structures, there is no substitute for actually being directly driven by these forces.

No doubt, much of the U.S. $60 billion-worth of companies that were privatized in 1994 and again expected in 1995 is driven by governments' budgetary considerations including the need for improved cash flows and debt reduction. But this activity is also driven by other economic forces. Businesses in competitive fields (like transportation and energy) cannot offer services as effectively and efficiently while operating under the structures and restrictions of government control.

Some observers oppose the idea of privatization based on arithmetical calculations which show that the added cost of private capital and taxes is such a large number that it is inconceivable that efficiencies and productivity and marketing innovation could overcome these costs and still produce marketable products and services at reasonable costs. The increased cost of private sector capital and the impact of income taxes are indeed significant. Based on the evidence from the U.K. and previous Canadian privatizations, so are the improvements in the operations of a company freed from the negative impacts of government ownership. Even in the four U.K. regulated industries that were privatized, the scope for efficiency gains was consistently underestimated by their regulators. This is even more important in a situation like Ontario Hydro, where the company has had a monopoly to protect it from competition but is now faced with the threat of full-scale competition. And fundamentally, one has to ask whether future capital decisions should be made without being subject to private sector disciplines.

Until recently, the tax argument in Canada was not really powerful in utilities because of "PUITTA" legislation that required the federal government to rebate any federal income taxes paid by a privatized utility to the province, who in turn could rebate it to the corporation. The province could also rebate its portion of income taxes to the newly privatized utility, leaving it more or less tax free. However, the most recent federal budget has eliminated PUITTA. While the province could still rebate its share of income taxes to a privatized company, one could argue that in the absence of federal rebates it is unlikely that it would do so for long given the state of public finances.

In pure economic terms, one can argue that the impact of taxes is overstated. About one-third of the total taxes payable would flow directly to the Provincial treasury and, in the long run, the citizens of Ontario, who make up 35% of Canada, are indirectly beneficiaries of a major portion of the federal taxes payable.

Some observers also express concerns over environmental and safety issues if privatization of generation were to occur. One of Ontario Hydro's objectives has been to be a leader in promoting sustainable development and energy efficiency and there is no reason to believe that this would not continue on a commercial basis after privatization. In fact, private-sector owned companies such as TransAlta Utilities are also active in this area, as is evidenced by its recent agreement with IPL Energy to work as its agent to improve IPL's electrical efficiency. With regard to environmental and safety issues, there should be no change in regulatory oversight on these fronts, which will remain exclusively government regulatory functions.

Government Equity Is Not Free and Government Guarantees on Debt Are a Potential Liability

Although often ignored in government accounting, the accumulated equity in a government-owned entity is in fact costing the taxpayers the foregone return on that investment. In a "special purpose" entity such as Ontario Hydro, where most of the equity has been contributed by ratepayers, this argument becomes somewhat clouded. But for the most part, the ratepayers are also taxpayers and citizens and therefore the equity is in fact segregated for future use and is equally not "free." As well in Ontario, if anything goes wrong, the taxpayers will have to face up to the guarantee on the debt. Both of these issues motivated the Province of Nova Scotia to privatize Nova Scotia Power in 1992.

Privatization will also result in transferring competitive and operational risks from the government to the private sector. In an increasingly competitive industry, these risks could become more acute and would likely be better managed by private sector owners.

Improved Efficiency and Profitability

In most jurisdictions, the evidence shows that privatization results in more efficient operations, lowered costs, and improved operations.

Management in Ontario Hydro has been given accolades over the past two years for reducing their labour force by nearly one-third. This praise is well deserved. On the other hand, one needs to ask how Ontario Hydro came to the position where such reductions were necessary? Government monopolies are less efficient and relatively slower to react to changing market conditions. In changing times, where competition is increasing, the inherent rigidness of such structures becomes more evident. A government-owned business operates under a number of very substantial difficulties:

Political and policy handcuffs. When any crown corporation takes on new initiatives that would be followed by any private sector corporation, it is criticized by the media, its regulators, and the government. As an example, Ontario Hydro has been criticized by private sector companies for bringing in innovative new products which the private sector says are being financed by a government-owned organization. As well, its international investments have attracted criticism. Yet, if Ontario Hydro is going to compete in the private sector, it is going to have to introduce new products and new ways of making money and capitalize on its international expertise.

Just as Canadian power companies were instrumental in the electrification of countries such as Mexico, Brazil, and Venezuela in the first half of this century, so today Canada should be benefiting from its electric utilities using their expertise to pursue the many opportunities that now exist around the world.

Looking over the shoulder. Given government ownership in a high profile industry, the effect on management of any crown-owned agency or corporation of having "government and politicians" looking over their shoulder is all pervasive. Management will often presuppose what actions or decisions the government might want and act accordingly, even if no explicit direction is given by the government.

Revenue enhancement. Revenue enhancement is not likely to be as aggressively pursued by management driven by weak incentives and under certain handcuffs, as discussed. Innovative products and methods of improving (or retaining) revenues would undoubtedly be enhanced by a private sector management who would be rewarded for their new initiatives. This factor, in all likelihood, will be very significant in the competitive and changing power industry.

Cost reduction and efficiency. Cost reduction is difficult in all organizations. Private sector managements continue to perform cost reduction in waves as they are driven to improve their bottom lines. It is not pleasant, no one likes doing it, and it requires a great deal of motivation to get management at all levels to cooperate in the process. The drivers for this effort are much stronger in a private sector environment than they are in a government-owned corporation.

There is much room yet for cost reduction in the Ontario electricity industry, but it will take privatization to fully harness it.

Management and employee incentives. Senior management of government organizations, including Ontario Hydro, have traditionally been paid considerably less than their private sector counterparts. While Ontario Hydro has provided some results-oriented management incentives in recent years, it is difficult or impossible to match private sector management and employee motivation schemes in a government-owned corporation. The introduction of incentive pay schemes and employee and share purchase programs as part of other privatizations have been well received by all levels of employees and are an important factor in improving corporate performance over time.

Labour relationships. Many observers have remarked that the Ontario Hydro labour contracts are among the most favourable to labour in the country. This is a situation that has developed over many years and many governments. It is a simple reality that a government-owned company faces different pressures in dealing with a union than would a private sector operator. Management and the board of directors are, in the final analysis, responsible to the government of the day and few governments encourage high-profile disputes with labour unions.

Over the past few years, Ontario Hydro has tried to emulate a competitive environment by dividing the company into operating units with transfer pricing processes among the various units. This has improved Ontario Hydro's skills and internal decision making, even though it is somewhat artificial. The full benefits of this move can only be harnessed through the pressures of truly competitive processes available only in private sector ownership structures.

International Experience with Privatization

These cultural effects of government ownership are negative to efficiency and profitability, and this has been substantiated through a number of studies of the impact of privatizations on corporate performance, including a major paper published in *The Journal of Finance* in 1994.[1] This study compared the pre- and post-privatization performance of 61 companies around the world from 1961 to 1990 and documented strong performance improvements after privatization, even in non-competitive industries. This was especially the case when 100% of the equity had been disposed of and when there had been significant changes in the board of directors due to privatization. This study did not include the major privatization of the British electricity industry, which took place in 1991. Similar conclusions were found in other studies of the British electricity industry.

Professor Stephen Littlechild was recently in Canada addressing a meeting of academics and industry observers on the privatization of the electric industry in Britain. Stephen Littlechild is the regulator of the industry and has held that position since privatization up to the present time.

During these sessions he was asked on several occasions whether or not privatization was necessary in order to achieve efficiency or whether it could be achieved simply by introducing competition. Professor Littlechild said that it was clear to him that both factors should be implemented where possible. He cited the Regional Electricity Companies as an example where competition was not introduced initially but where privatization was resulting in significant cost reductions in this sector. He also indicated that the significant reductions in coal prices would not have been achievable without privatization of the electricity industry. Nuclear Electric, which was not privatized, also achieved major cost reductions, but this was driven by both competition and the goal to be privatized.

NOTE

1 "The Financial and Operating Performance of Newly Privatized Firms: An International Empirical Analysis," *The Journal of Finance* vol. XLIX, 2 (June 1994).

APPENDIX

Selected Privatizations in the G-7

Canada	*France*
Air Canada	Société Generale
Petro-Canada	Rhone-Poulenc
Cameco	Paribas
Canadair	UAP
DeHavilland	Banque Nationale de Paris
Skydome	Total
Urban transit	Alcatel
Telus	Renault
Nova Scotia Power	St. Gobain
Saskoil	Crédit Local de France
CN	*Germany*
	Lufthansa
Britain	Deutsche Telecom (upcoming)
British Telecom	*Italy*
British Gas	INA
Britoil	BCI
British Airways	Credito Italiano
BAA	IMI
British Steel	ENEL (pending)
Jaguar	STET
Water companies	*United States*
Regional electricity companies	Conrail
National Power/PowerGen	
Nuclear Electric (pending)	*Japan*
Scottish Nuclear (pending)	NTT

The Financial and Operating Performance of Newly Privatized Firms
An International Empirical Study
% Average Improvement after Privatization[1]

	Total	*Type of Industry*[3]		*Degree of Divestiture*[4]		*Degree of Director Turnover*[5]	
	Sample[2]	*Competitive*	*Non-Competitive*	*Full*	*Partial*	*Higher*	*Lower*
Profitability							
Return on sales	45.2	57.8	16.3	42.5	46.9	56.7	20.7
Return on assets	24.7						
Return on equity	-1.3						
Efficiency (per employee)							
Sales	111.1	14.6	3.1	11.5	10.9	11.3	12.1
Net income	31.8						
Investment							
Capex. to sales	44.6	52.9	6.3	128.8	12.1	117.3	10.4
Capex. to assets	16.5						
Output							
Real sales	26.8	27.7	25.21	35.6	21.2	36.9	22.7
Employment							
Total	5.7	5.6	5.9	13.9	0.0	14.1	6.4

[1]Percentages are the "Mean Change as a % of the "Mean Before" of all firms in the sample, using the mean of three years before and three years after privatization for each firm.
[2]Calculated from Table III.
[3]Calculated from Table IV.
[4]Calculated from Table V.
[5]Calculated from Table VIII.

Source: Prepared by the Financial Restructuring Group.

A. STEPHEN PROBYN

As someone involved in project finance for Canada's independent power industry and the president of the Independent Power Producers' Society of Ontario (IPPSO), the industry's trade association here in Ontario, I have some views which may or may not accord with the official word from our organization. What all of us believe is that we are already embarked on a very exciting, if not a little perilous journey – which, of course reminds me of a story. A project financier died the other day and (this may lead you to doubt what I'm about to tell you) he went to heaven. As St. Peter was taking him around, he observed his surroundings: The sky was blue, the clouds white and fluffy, the grass was green, the air was fresh and clean. Being a cautious man and wishing to make an informed choice before he decided where to spend eternity, he asked St. Peter if he might take a look at hell just to make sure. Certainly, said the Saint and in an instant the two were whisked off to hell. Our man looked around. The sky was blue, the clouds white and fluffy, the grass was green, the air was fresh and clean. "But, Saint Peter." he exclaimed, "there doesn't seem to be any difference at all!" St. Peter paused for a moment, looking reflective. "That's true," he said. "Heaven and hell do look very much alike. But down here the deals never close."

My first employer, the Rt. Hon. Margaret Thatcher, states in her autobiography that, through the power of competitive markets – particularly the kind of privatization which leads to the widest possible share ownership by members of the public – "the state's power is reduced and the power of the people enhanced." While that seems like a truism today, it was deemed dangerously radical when I worked as the energy policy analyst for the British Conservative Party, of which Mrs. Thatcher became the leader in the mid-70s.

As anyone reading the newspapers knows, the final years of this century have, in a very real sense, been about the emergence of competitive markets throughout the world and the shift of power from producers to consumers. Now, competitive forces and the power of the marketplace to create economic efficiency – and hence, wealth – are recognized from Beijing to Bratislava, while the ideology of statist command economies retreats into insignificance. Now too in the Western economies, the regulated monopolies with their closed networks and systems are vanishing. "Open Access" is the *leitmotiv* of the end of the 20th century, just as "mass production" was to its early years and "trade follows the flag" was to the mid-19th century.

"Open access" to computer networks in the '70s broke the near-monopoly on information technology enjoyed by IBM in the '60s. In the '80s, the telephone network enjoyed the creation of competitive markets in a sector dominated by giant monopolies such as AT&T. The result has been an explosion in information technology. The "infobahn" is a public expressway, open to all.

In the energy sector itself, the monopolies, beginning with natural gas, saw the emergence of competitive market forces within their previously regulated systems in the 1980s. In fact, I was part of the federal government's team responsible for natural gas deregulation. On one occasion, I was dispatched to Toronto to brief a highly apprehensive – and pretty vocal – Ontario Minister of Energy and his Deputy. In the Minister's opinion, Ontario had been sacrificed to the West. Had he been able to look into a crystal ball to see the billions of dollars of benefit to Ontario which have since flowed from that decision to insert competition into the gas industry, he might have scolded me for not moving quickly enough!

The Benefits of Experience

What is true of other economic sectors is not a revolutionary concept for the electrical system either. Electricity privatization and deregulation has already been proven to work.

The British economy is now reaping real, quantifiable benefits from deregulating the U.K. power grid.

In the United States, the birth of the independent power industry dates from the Public Utilities Regulatory Policy Act of 1978. From near zero, the industry now provides roughly half of all new capacity in the U.S. today, with a total operating capacity of more than 51 gigawatts. And this amount is likely to double in the next 10 years – creating a $100 billion asset base.

Now, the U.S. market has taken another stride with the implementation of the Energy Policy Act of 1992. This legislation removes most of the barriers to inter-regional trade in electrical energy and effectively makes utilities "common carriers" of electricity. Under the increased flexibility allowed by the Act, a new class of Exempt Wholesale Generators (known by the initials EWG) has begun to emerge.

EWGs are potentially the "merchant" power plants of the '90s, selling across utility and state boundaries to the highest bidder – whether that be a utility, industrial customer, or even (in some states, including California by the turn of the century) the retail consumer. Power brokers now connect customer to generator, negotiating wheeling deals with the utilities that separate the two. These power brokers – often marriages between investment banks and power companies – are moving to use financial instruments, such as derivative products to shape power to meet increasingly sophisticated customer demand. There are now well over 100 firms in this market that simply did not exist 18 months ago.

We in Canada live in the world, and more particularly in an integrated North American economy. Ours is one of the most open economies anywhere. We have the single most energy-intensive industrial base on the face of the planet. If the world has figured out a new way of organizing the business of generating and transmitting electricity to customers, we had better make sure that we know that the Canadian way is the right one.

The Canadian Scene

For these reasons, I am going to focus on a single example: Ontario. It is fair to say that other provincial utilities, such as B.C Hydro, have moved further and faster than we have in Ontario. B.C Hydro has already posted wheeling rates and through its subsidiary, Powerex, is poised to become a significant player in west coast energy markets.

In Alberta, we have two of the most successful investor-owned utilities in North America. As of January 1, 1996, Alberta will be moving to a deregulated electrical system in which generators will bid daily into a power pool, which will buy from them on a least-cost basis. Financial hedging contracts will enable customers to work with generators to shape power costs.

Here in Ontario, I think it is fair to say that the senior management of Ontario Hydro recognizes the sea change in their business environment and is taking steps to change the way the corporation does business. A new transfer pricing system, for example, is designed to act as a pseudo-market in allocating production costs to the various components of Hydro. Al Kupcis, the President of Hydro, has stated that deregulation is coming – for sure – and that we have, at a maximum, three years to get ready for it.

Yet aside from a few true believers in the marketplace, Ontario's political system is largely somnolent. Wherever the economics may be, the politics do not seem to have caught up.

Transfer of Ownership

The irony is that major players in the energy marketplace are far ahead of the politicians. Major energy consumers, many smaller customers, environmental groups such as Energy Probe, and even the utility itself have recognized the need to reform and to transform the energy marketplace. Of course, there are many different routes to this objective. Let me give you mine.

To achieve the end of deregulated energy markets requires far more than the privatization of the utility. As I see it, there are four essential steps:

1. Unbundling;
2. Regulating an "Open Access" system;
3. Dealing with stranded assets and debt;
4. Privatizing.

Unbundling the System

Taking them one at a time, the first step is to "unbundle" the system itself, separating out the distinct businesses it contains. In fact, the first element is already unbundled here in Ontario – namely, the distribution network of the municipal utilities.

The major problem with the municipal utility structure is its fragmented nature, ranging from essentially micro-operations to Toronto Hydro. This can be rationalized over time.

Next are the business lines which Ontario Hydro contains. Here again, we are not starting from scratch. The process of unbundling has already begun, in the sense that Hydro has been reorganized into functional companies. But it will be important to segregate one function from another completely, to deal with the competitive issues raised by each. Different Hydro business lines range from natural monopolies to entrepreneurial producers. But unless the unbundled companies have the absolute right and ability to function as stand-alone entities, changes cannot be more than cosmetic.

The Need for a Common Carrier

Take the transmission grid. This is a vital infrastructure asset for the Ontario economy – and the ultimate product of Sir Adam's vision. It is also a natural monopoly for the simple reason that we do not want competitors stringing a variety of wires around our natural environment.

Like the natural gas transmission industry, Hydro's grid represents a common infrastructure paid for and shared by all users. Unlike TransCanada Pipelines Ltd., however, Hydro is not a common carrier. Under the Power Corporation Act, not only is Hydro not required to transmit power between third parties, but also third parties are prohibited by law from buying electricity from or selling to anyone except Hydro. Hydro is not even required to wheel energy between two facilities belonging to the same owner.

Common carrier or power pool? This is the question at the heart of establishing the new market place in electrical energy. Common carrier proposals, such as the California model, provide direct bilateral access between generator and customer. This may permit the by-pass of system costs, especially where generators are able to connect directly without the benefit of the transmission grid.

Many proposals, including the Alberta reforms I mentioned earlier, do not contemplate going all the way to the common carrier model of the transmission system. In these models, the transmission system would retain its position as the monopoly buyer and seller of power, with no direct contact between buyers and sellers in terms of contracting for the physical delivery of power. Pricing would be accomplished on an hourly basis, with deregulated generators competing through daily auctions. Hedging contracts will enable generators and customers to shape power financially. An example could be this: a customer wanting long-term fixed rates would sign a contract with a generator (possibly through the intermediation of a market) who will, in effect, transfer revenue received above an agreed hedging price to a customer who, in turn, will make payments to the generator when the pool price is less than the hedging price.

The power pool moves us part of the way towards deregulation, but it does leave us somewhat short of the full deployment of market forces. Generators, for example, are not allowed to contract directly with customers for the sale of power. Embedded costs of existing generation are entrenched and no bypass of the existing system is permitted.

While the pool concept does give us the tools to deregulate generation, it also ensures that we continue to have the ability to impose the costs of the existing system on consumers. For this reason, it is popular with government and may be less favoured by consumers. This gives rise to one of the central facts of the structuring of deregulation. Behind every technical solution, there are billions of dollars on the table to be divided up between producers and consumers.

Dealing with the Debtosaurs

In the politics of deregulation, it is the stranded assets which are the really big news.

"Stranded assets" refer to the investments in generating capacity which have not stood the test of time. Here in Ontario, as elsewhere, these "debtosaurs" are largely nuclear-powered – although some of the fossils are also fossil-fueled. Under the old ways of monopoly economics administered by compliant regulators, they could graze peacefully in the swamp of rate base regulation. Competitive market forces dry up the feeding grounds. Rather than being able to hide their enormous costs in a blended average rate, the unbundled system is transparent to users. It exposes the "debtosaurs" as the inefficient monsters they are – and direct burdens on the ratepayers.

If users are then able to find other cheaper power sources from neighbouring utilities, it is highly likely that they will seek ways to acquire it. Unbundling means that the pressure to deal with the "stranded assets" will become irresistible. That is the *real* reason why we are told by members of the power élite that unbundling is unthinkable.

In Britain, the solution was the creation of a publicly-owned Jurassic Park – a state-owned nuclear company armed with legislation forcing the consumer to buy a proportion of his electricity from nuclear sources. In the U.S., the home of economic Darwinist thought, it appears likely that any good from these giant carcasses will be devoured by fitter creatures, leaving the remains to be digested by utility shareholders. Admittedly, neither prospect is attractive.

But why should reluctant users be chained to these stranded assets? Windsor and Kingston attempted to develop their own energy resources – in both cases, independent power plants – which would partially mitigate the impact of Darlington on their system. In both cases, Ontario Hydro looked at marriage with the municipals as being "in sickness or in health" and took strong administrative exception to threats to the integrity of the power pool. No divorces here in Ontario.

The problem with Darlington is not inefficient production, but the awful weight of debt incurred to build it. The sunk costs may only be partially recoverable. But, so what? The fact that the revenues attributable to Darlington's energy production do not cover debt service costs should be of no concern to the consumers.

"Power at Cost" does *not* mean "power at whatever it costs Hydro." Power should be supplied at its cost in a competitive marketplace, the only real measure of economic efficiency. The fact that the utility made some unfortunate investment decisions in the past is really a matter for the shareholder of Hydro – namely, the Government and the citizens of Ontario.

This does not imply that all the stranded debt needs to be written off. Judging by the rapidly improving financial situation at Hydro, it strongly appears that, given a phased implementation of open access, Hydro can manage its way into a competitive market within a minimum of three years. To the extent that there is excess debt in the system, I believe that it is imperative to ensure that the write-off of debt is the responsibility of the shareholder/taxpayer rather than the ratepayer. I say this because in the deregulated world, costs will tend to be reallocated from the customers with options (mainly large industrial users) to those that do not (the residential/small commercial sector). To the extent that residential customers – also known as voters – feel victimized by deregulation, the viability of such a reallocation must be in doubt.

In my view this implies a definite timetable and a clear legislative framework towards the achievement of open access on the electricity grid within a defined period of time – let us say three years. Taking an idea from an old professor of mine, Bill Hogan, I believe that this may involve a transmission access fee to be paid by all consumers. This approach allows an immediate move to open access, with the proviso that excess costs of stranded debt will be paid by all customers. However, I believe that this fee must sharply decline over the period of the phase-in of deregulation to zero by the end of the third year.

I think that the beauty of this approach, as opposed to the British system, is that it immediately puts the entire system into a competitive framework without any implicit technological bias. Rather than providing signals to utility planners that distort the market, the transmission access fee simply provides a buffer to enable the utility time, but not a great deal of it, to manage the system to a low-cost, market-responsive base.

Privatizing the Grid

So now we have tackled the three most difficult steps to a competitive energy economy. First, we unbundled the system. Second, we created a transparent and competitive market place for electricity. Third, we dealt with the accumulation of debt and inefficiency in the old system. We are ready to privatize.

More than that, because we have unbundled the system, we are able to maximize value by selling the constituent parts for the highest price.

First, the transmission grid assets. Under our privatization strategy, the Hydro grid would be divested of generation capacity, as well as the non-regulated businesses carried on by the utility. Effectively, Hydro would become a regulated common carrier of electricity which also buys from suppliers and sells to consumers. Independent power producers would remain as suppliers of system capacity under existing long-term contracts, and new ones would be negotiated to take advantage of efficiencies due to cogeneration and high debt ratio financing. Other energy would be acquired through contracts with other utilities, short-term deals for economic energy. But the principal source of supply would be the newly unbundled generating assets. (More on that in a moment).

The most likely method of privatization of Ontario Hydro (now reduced to the transmission company) would simply be through the sale of shares to the public. Although this would be one of the largest share flotations in Canadian history, there are precedents – particularly Air Canada and Petro-Canada. I believe that the sale of the grid would stimulate enormous investor interest, both at home and abroad.

Incidentally, should the government wish to preserve ownership rights, the mechanism of the '"golden share"– a share owned by the government giving it certain rights to block takeovers – has proven market acceptance in a number of U.K. privatizations.

My preferred structure for the privatization of the generating assets is to restructure them into a number of generating companies – probably two or three, to facilitate competition and preserve economies of scale. Ideally, each of these operating units would hold a balanced portfolio of nuclear, thermal, and hydro capacity. Once established, each would participate as unregulated suppliers of energy to the grid through the auction mechanism.

But before coming to market with the generating companies, we have three more rivers to cross:

- First, debt write downs to create realistic balance sheets;
- Second, for the nuclear assets, long-term fuel disposal contracts and environmental indemnities negotiated with federal and provincial government agencies;
- Third, the provincially guaranteed debt could be dealt with either by the type of defeasance mechanism used in Nova Scotia – more likely – simply a continuance of existing guarantees. Then, as the debt is retired, new issues could be marketed without provincial support.

Rationalizing Distribution

What about our distribution system? As we have already seen, with over 300 municipal utilities in Ontario, we have a structure ripe for rationalization. Further, the economic environment is right.

Many municipalities will want to exchange public ownership of their utilities for cash to meet other infrastructure needs no longer funded by senior levels of government. And as privatization transactions flourish, we are likely to see the acquisition of smaller utilities by bigger ones, much as has occurred in the natural gas distribution industry. Given the right regulatory structure, common facilities and economies of scale should provide efficiency gains that translate into benefits for both shareholder and consumer.

Ultimately, I believe that we will end up with a hybrid of privately-owned *and* publicly owned distribution utilities in Ontario. As long as regulation is consistent among all players, I do not see why this should be a problem. As is the case in Alberta, legislation would be required to allocate federal and provincial tax rebates under the Public Utilities Income Tax Transfer Act, equalizing the consumers of public and private sector utilities.

What about Independent Power?

I strongly believe that we in the independent power sector must adapt to the new realities. In the future, all generation will be independent power. But people in my industry must realize that the rules of the game have already changed. "No money down" power projects have probably run their course. The era of the highly leveraged, project-financed gas generation project, which is essentially financed on a long-term fixed price contract with known escalation has come to an end. Future independent power projects will be increasingly exposed to market risk and will adjust their capital structures accordingly. Debt-to-equity ratios will be more characteristic of those found in conventional businesses. The rationalization and consolidation of the industry into larger units, already well underway, will accelerate.

A critical consideration in the discussion of the future of power must be the realization of the context of the times. In our time, our deliberations are taking place at a time of large capacity surpluses. In an administered price or regulated monopoly framework, this implies that a great deal of economic rent, which would in a competitive market end up in the hands of consumers, is being retained by the producers. The more competitive the new framework, the greater will be the transfer of wealth. This is why producers tend to favour the poolco model over the direct bilateral access paradigm.

But the development of new production represents the other side of the coin. In our industry, one of the most capital intensive in a modern economy, new capacity has been financed – one way or another – on the rate base. The logic of the market system is that risk is shifted from the producer to the consumer.

Financing new projects will become vastly more complex in the new market environment. If we begin at the most extreme end of the spectrum, we would envision projects that are entirely financed from equity. Since that equity is taking market risk, its return will be correspondingly higher. Compared to today's financing costs, this may represent a greater than doubling of the cost of capital for a typical project. In other words, as we come to the end of the surplus situation around the turn of the century, we could expect drastic increases in power costs as new equity-financed, merchant plants come on to provide marginal supply.

This is obviously not what the proponents of the brave new world of competitive markets had in mind. What can we do to help them out.

As a first step, lenders will not completely turn off the tap. However, rather than being able to rely on strong contractual relationships with utilities, they will be compelled to finance against projections of market prices.

As a matter of public policy, it will likely be considered desirable to provide for the development of renewable resources as part of the integrated energy system. Renewables often pose unique solutions to environmental problems posed by energy production and represent an investment in long-term supply diversity. For example, while a gas cogen plant may deliver very low emission standards, it is an undeniable truth that a wind plant delivers no greenhouse gases. Biomass plants actually reduce the environmental impact of wood waste residues as they generate power. And while every form of energy has an environmental price, the cost of renewables over the span of time measured in decades is low.

The preferred vehicle for the encouragement of renewables should be the so-called "renewable set aside." These solicitations, which are being undertaken in the U.S., the U.K., and now B.C and Ontario, provide long-term fixed contracts to renewable producers. I believe that these long-term contracts should not exceed the sum of regional long-run avoided costs plus any externalities (such as cleaning up a pollution problem) so that renewables are not seen as costly drains on the ratepayer.

Conclusion

By the turn of the millennium, the wheel of public ownership, which began moving in the opening years of the century, could come full circle. But the development of a competitive market-driven environment for the electrical energy sector is *less* of an option as each day passes.

I believe the "why" in the case for competitive markets is clear and compelling. The "what" is the benefit of competition in encouraging innovation and reducing cost. The "how" is increasingly well-defined in the experience of markets throughout the world. The "when" for Ontario is *now.*

STEPHEN F. SHERWIN

Dr. Gordon expresses doubts that privatization of Ontario Hydro would reduce the cost of electricity to the customer or improve the efficient allocation and use of resources, essentially because power generation is a capital-intensive industry, and the cost of capital of a private utility significantly exceeds that of a crown corporation. He concludes that privatization would raise the cost of capital by about 29%, and the price of electricity to consumers by 12.7%.

He suggests that privatization and reliance on market competition has been advocated largely on the grounds that the cost of service to consumers would thereby be reduced. The case for privatization of Hydro rests on a much broader base than reducing the cost of electricity. There is no factual, as distinguished from anecdotal, evidence that Hydro suffers from productive inefficiencies, but there is ample evidence that past investment decisions have resulted in allocative inefficiencies, as indicated by the fact that Hydro's generating capacity now exceeds its peak demand by about 57%, compared to 20% for the typical privately owned U.S. electric utility. The case for privatization of Hydro's generating units – not its transmission grid – rests on creating level playing fields with other energy sources, on avoidance of future over-investments, on giving customers greater choices through unbundling of rates, on creating effective competition (the most powerful force for technological innovation), on capturing new markets, and last, but not least, on the need to reduce Ontario's debt burden, of which Hydro now accounts for close to 30%.

Dr. Gordon suggests that Hydro should finance its investments with a capital structure that contains only 10-15% equity, but that a private utility, with a high proportion of nuclear generation, might need a 50% equity ratio. The principle governing capital structures of utilities – accepted throughout North America – is that capital structures should be consistent with the utility's fundamental business risks. The operational risks of nuclear facilities, and the competitive demand risks encountered in the markets served, are not affected by whether ownership is in private or public hands. The reason a crown corporation can finance its assets with a significantly lower equity than is possible for a private utility is either an implicit, or explicit, provincial government guarantee. Dr. Gordon suggests that a 0.5% fee for this guarantee amply compensates for the risk. In an era when excessive government debt, at both the federal and provincial levels, has adversely impacted the Canadian dollar and has impeded, through higher interest rates, economic growth and employment, a 0.5% fee understates the true economic cost of an overleveraged capital structure of a crown corporation.

Turning to the cost of equity, Dr. Gordon places the cost at 9.5% for Hydro, compared to 11.95% for a private utility. However, the 9.5% is reduced to about 6.7% by taking account of avoided personal income taxes. Traditional economics holds that the cost of equity is determined by the risk and productivity of the asset in which it is invested, and is not dependent on the happenstance of the ownership. The suggested differential between 9.5% and 11.95% is far too large. The reduction of the 9.5% estimate for avoided personal income taxes is unwarranted; moreover, to tie such a reduction to the level of personal income taxes fails to recognize that many institutional investors – which provide a high proportion of capital to utilities – pay no income taxes. The cost of equity may be marginally lower than the cost of debt in periods of high prospective inflation, but the premise that the cost of equity for a crown corporation lies 44% below the cost of equity for a private company is not only counter-intuitive but would invite uneconomic investments.

Dr. Gordon's concept of the cost of capital lies outside traditional economic criteria; he suggests that the choice between Hydro and a non-utility generator building additional generating capacity should be dependent on the Province's employment level; at full employment the appropriate discount rate for evaluating the economics of the investment is placed at 12.42%; if there is unemployment, the discount rate is reduced to 9.5% and "the employment benefits may be factored into the cash flows." Such differential rates may be justified as experiments in social engineering, but not in a competitive environment. Employment benefits cannot be factored into the cash flow, not even if one assumes that a crown corporation were to mandate that incremental investments were to be built exclusively by Ontario labour.

While Dr. Gordon's estimate of the cost of capital for Hydro is understated, he accepts income tax normalization for the private sector. However, it is quite uncertain that it would be acceptable to an Ontario regulator; rates for the entire Canadian natural gas pipeline and distribution industry – with one exception – are set on the basis of "flow-through" taxes, which typically result in an effective corporate income tax rate of 30% rather than the normalized rate of 44.6%. The impact on Dr. Gordon's model (Table 8) for comparing public vs. private ownership is significant; substituting a 30% effective income tax rate for the 44.6% rate would reduce the differential between private and public ownership from 29% to 14%. Moreover, the assumption that the capital/output ratio would not be affected by privatization may be unwarranted. It is highly likely that – with the large excess capacity available to Hydro – privatization would successfully expand exports of electricity; an 8% rise in total electric sales (at constant rates), together with the reduction in the assumed income tax rate, would totally eliminate Dr. Gordon's cost differentials.

Disagreement with crediting income taxes to the cost of equity for a crown corporation does not warrant ignoring – in the context of privatization – the impact of income taxes. Indeed, it is a critical factor. Virtually every privately owned utility incurs income taxes, which are generally passed on to the customers. How significant that tax would be if Hydro's generating assets were privatized is, in large measure, dependent on whether Revenue Canada will permit the newly privatized companies to take capital cost allowances on the original cost of Hydro's assets – which has been the rule in other privatizations – or whether Revenue Canada would limit capital cost allowances to the purchase price of the assets. The former would be more advantageous than the latter, but in either event the newly privatized companies would be subject to income taxes after a few years. If regulation took the form of price cap regulation, then prospective income taxes would adversely impact on the valuation of Hydro's assets; under cost-of-service regulation income taxes could probably be passed on to the customer. Income taxes are simply a cost of doing business; the incurrence of income taxes should not be a factor in the choice between private and public ownership.

Dr. Gordon also touches on a number of broader issues relating to a potential restructuring of Hydro, endorsing a revaluation of asset values, for both nuclear and other plants, suggesting that the cost of environmental protection impacts similarly on public and private investments; and that in the event of privatization the risk of closing nuclear plants, on environmental grounds, should not be transferred to the private sector, but be left with the government. I share these views; broad agreement on these critical issues should help to move the restructuring process forward.

With respect to the implementation of privatization, it should be apolitical. That condition can be met only if all stakeholders will derive net benefits. The government must be made whole, in the sense that privatization should permit Hydro to sell its assets – after revaluation – at no less than their present book value; the customers must be protected against further rate increases, in terms of real (not nominal) cost, preferably through price cap regulation; the customers must be afforded greater choices through unbundling of rates; medium and large industrial as well as commercial customers should be given the opportunity to buy electricity through an exchange pool, creating a spot market, as well as to negotiate their own contracts for electric power. Hydro's operational expertise must be preserved; divestiture of generating assets must ensure effective, not just potential competition. The precondition for effective competition is the creation of companies with approximately equal generating capacity, reasonably close average and marginal cost. This can be achieved through disparate distribution of generators. Privatization of some, or all, generating capacity is essential. If public policy precludes the privatization of nuclear facilities, then the capacity of nuclear plants should be leased to the privatized companies, permitting each of them to share in providing the base load. Privatization of the transmission grid is not essential; privatization of the municipal electric distributors may ultimately be desirable, but that objective is secondary to the need to ensure that these distributors would become effective participants in a competitive market for power; retail competition coupled with a regulatory surveillance that would establish performance benchmarks – for rate-making purposes – would provide incentives for "rationalization" of the distribution industry through mergers.

Nuclear Environmental Consequences

DONALD N. DEWEES

I. INTRODUCTION AND ANALYTICAL FRAMEWORK[1]

Is the choice between new fossil-fuelled generation and new nuclear generation fully informed by the likely environmental consequences of each? Only if the answer to this question is yes can we be confident that the correct choice will be made between these energy sources when increased demand requires capacity expansion in the future. We will review the literature on the environmental consequences of Canadian nuclear generation, including mining, operation and decommissioning of nuclear plants, and on the extent to which nuclear generation is not bearing its full social costs. We will look briefly at the corresponding literature on fossil-fuelled generation, specifically coal-fired generation, which has historically been the principal non-renewable competitor for nuclear power. We will then consider whether nuclear is bearing a lesser share of its full costs than is coal-fired generation, and if so what is the magnitude and significance of that shortfall.

We begin from the theoretical principle that economic efficiency requires a polluter to pay the full cost of all externalities generated by its emissions. (Baumol and Oates,1988, ch. 4.) If the polluter pays for this external harm, then it has an incentive to control its emissions to the point where the marginal cost of one unit of abatement equals the marginal harm caused by one unit of pollution discharge, the welfare-maximizing or optimal degree of abatement. Even when the optimal degree of abatement has been achieved, the polluter should pay the remaining marginal external social cost of its emissions in order to ensure that the price of the product, in this case electricity, fully reflects the social costs of the product and, therefore, that the product is not over-produced. We will therefore look for evidence regarding the external costs associated with existing generation technology and with technology that would be installed if expansion occurred today.

Note that theory requires that we consider only external costs, that is, costs that are not already paid for by the polluter. If electricity generation causes health risks for residents of the area around the plant, compensation for these risks is not generally borne by the generator and the risks represent external costs. On the other hand, if electricity generation causes risks of injury or disease for the workers at the plant, and if the workers are fully informed about these risks and demand a wage rate that fully compensates for them, the generator has paid for these risks and they do not represent external costs.

The methodology of this study is to review the existing literature on the external costs arising from electricity production by nuclear power and by coal. We will review and discuss empirical studies of the harm or risks caused by nuclear generation and by coal generation. Because the principal concerns arising from both generation technologies are local and regional, and because the CANDU nuclear plants differ in important ways from nuclear plants produced in the United States and elsewhere, the best evidence would be from studies concerning Ontario power plants. Because much of the concern about external effects is related to human health, any studies of non-Ontario plants will be more relevant if they involve power plants whose location is such that the population density of the surrounding area is similar to that of Ontario power plants.

Estimating the external costs of electricity generation from fossil fuels has been a matter of considerable interest for several decades, and a significant amount of work has gone into such estimates in Canada, in the United States, and elsewhere. The difficulties in producing reliable estimates are daunting, however. Medical science has not estimated accurate dose-response functions from which one could accurately predict morbidity or mortality arising from exposure to a specific concentration of a given pollutant for a specific period of time. The health effects of the combinations of pollutants emitted from coal-fired power plants are even more uncertain. Furthermore the variability of pollution dispersion around an air pollution source means that we can estimate only approximately the exposure of the human population to pollutants arising from a given power plant. Even if we knew with great precision the health effects of operating a power plant, placing an economic value on those effects remains an uncertain and controversial process. With respect to nuclear generation, similar problems arise in estimating the health effects of routine emissions, and their value. In addition nuclear plants present the risk of accidents that cause increased radiation exposure, and estimating the probability and severity of such accidents is highly uncertain.

These pervasive information deficiencies and uncertainties mean that any estimates of the risks arising from power generation will be subject to considerable uncertainty. The literature does not offer precise numbers that are generally accepted. Even widely varying estimates are hotly contested. This review cannot resolve these uncertainties. Our ambition is only to present the data that are available, to discuss them, and to try to assess their relevance for the particular conditions that exist in Ontario. We will then identify what seems to be known and agreed upon and the issues on which further information will be required before firm conclusions can be reached.

The principal externality arising from nuclear power generation to be considered here is the increased risk of premature mortality. Evaluating this risk requires us to place a value on the loss associated with an increase in the risk of premature mortality. Some argue that it is immoral to place an economic value on human life. But in fact individuals make choices every day that involve risks to life, and thereby imply that they are prepared to trade off some increase in such risks for reduced costs or activities or goods that they value. Collectively we make such choices when we decide not to spend more on highway safety, health care, or a dozen public programs that have the proven capability of reducing risks of premature mortality. If we recognize such behaviour, then it is a small step to infer the value of such increased risks from the choices that are made. The main source of such data is studies of workplace risks and the wage premium that is associated with increased risks. Good surveys of this literature are contained in Moore and Viscusi (1990, ch. 2) and Freeman (1993, chs. 10 and 12). While the studies produce a range of estimates, $5 million per statistical life is in the central portion of that range, and we will use that figure. Some of the studies to be cited here use $4 million, and we will report their figures without adjustment, since the accuracy of these estimates does not allow us to reject either estimate.

Some will suggest that we might look to court awards in civil lawsuits to determine the "value of life" as perceived by the legal system. Such a look reveals that the courts determine awards for wrongful death based essentially on the loss to the survivors. The central element in this calculation is the lost wages of the deceased. The result of such a "discounted future earnings" approach is often one-tenth the amount estimated using wage risk studies (Dewees 1986). Because discounted future earnings looks to the financial loss to survivors and not at all to the loss to the *victim* economists regard it as an improper means of measuring the loss associated with an increased risk of premature mortality (Freeman 1993, 323).

An alternative means of determining our collective value of reductions in the risk of mortality is to examine cost per life saved of public programs that promote public safety. If the government is wise and rational and reflects the wishes of its constituents then the marginal cost per life saved of government programs for highway safety, occupational health and safety, and environmental protection should reveal what we think a life is worth. Unfortunately we find vastly differing values from such studies. (See, e.g., Morrall 1986, Table 4.) Indeed, if we examined the safety precautions taken at nuclear reactors we might well find very high costs per life saved. One could argue that this represents the public's fear of the type of risk posed by nuclear power, perhaps the risk of a catastrophic accident, and that this is precisely the value of life to use in evaluating nuclear power. However, the inconsistency of the various values that this produces combined with the implausibility of the full set of assumptions necessary to interpret it lead us to prefer to use the values derived from the wage-risk studies.

There is a further caveat regarding our implied value of a statistical life. Because health care costs are paid for by OHIP, any health care costs associated with occupational health problems will not be paid for by the workers or the employer and are, therefore, external to Ontario Hydro. We believe that health care costs would be a small fraction of $5 million and that there is little error in ignoring these costs here.

Some nuclear wastes maintain their radioactivity for hundreds or thousands of years. How should we treat any risks caused by radioactive emissions far in the future? Generally cost-benefit analyses will discount future costs when calculating the net present value. But the choice of a discount rate for such calculations is controversial and the appropriateness of discounting itself has been called into question when today's decisions affect future generations (Freeman 1993, ch. 7). This issue is important because discounting even at a rate as low as 2% will cause effects more than two centuries in the future to have a negligible weight,[2] while adding up thousands of years of small effects without discounting could yield large aggregate costs. This study will note situations in which differences in estimates of future harm arise from differences regarding discounting.

Section II discusses the environmental costs of nuclear power generation. Section III discusses the environmental costs of coal-fired power generation. Section IV compares these estimates and draws conclusions regarding biases that may exist regarding new investment arising from the external costs identified in Sections II and III.

II. ENVIRONMENTAL COSTS OF NUCLEAR POWER GENERATION

The environmental costs of nuclear power generation (NPG) can be classified according to the three main stages of the nuclear fuel cycle. The first stage, the production of nuclear fuel, encompasses the mining and milling of uranium ore, the processing of the uranium concentrate into a form fit for use in a nuclear reactor, and the delivery of the final product (nuclear fuel) to nuclear power plants. The second stage consists of the fission of the nuclear fuel inside a nuclear reactor, where heat generated by the splitting of uranium atoms is used to produce steam, which drives turbogenerators. A nuclear reactor, like any other piece of equipment, has a finite operating life, at the end of which it must be decommissioned. Decommissioning implies the permanent disposal of spent fuel and other contaminated or inactive material. This forms the third stage of the nuclear fuel cycle. Spent fuel is not reprocessed in Canada, therefore issues relating to this activity will be ignored in this paper.

Each stage of the nuclear fuel cycle carries its own environmental costs. It is useful to classify these costs further according to where or on whom they fall. We can thus distinguish between occupational health costs, incurred by workers involved in the nuclear fuel cycle, and public health costs, borne by the population at large. Property damage costs, consisting of harm to public and private physical assets, and natural environment damage costs, consisting of harm to fauna and flora, also represent cost categories identifiable at every stage of the nuclear fuel cycle.

A still finer level of classification is justified, in order to distinguish between costs arising from routine operations and costs arising from accidents. This distinction is relevant: the former costs will necessarily be incurred, while the latter might not. Their actual value is an expected one.

This section presents a summary of the existing evidence on the environmental costs of NPG, classified along the lines suggested in the previous paragraphs.

Stage One: Production of Nuclear Fuel

Virtually all of the uranium used by Ontario Hydro is mined in Northern Ontario. The first stage of the nuclear fuel cycle must therefore be considered when estimating the external costs of NPG in Ontario.

OCCUPATIONAL HEALTH COSTS

The first line of workers involved in the nuclear fuel cycle are the uranium miners. Uranium is not a stable element: it naturally decays, emitting radiation in the process. One of the elements in the uranium decay chain is radon, a gas that itself decays and produces radiation. In a uranium mine, exposed rock faces allow radon to escape in the air, where it can be inhaled by miners and cause an increase in the risk of lung cancer.

Gofman (1981, 447-51) reported a lung-cancer rate for American uranium miners that was 5.1 times as great as would be expected for an equivalent group of non-miners from the general population. Incidence of death due to other respiratory diseases was 3.2 times as great.

Because of dramatic improvements in mining safety standards and practices over the past four decades, occupational health costs computed from past data may overestimate current costs. Roberts and coworkers (1990, 110) observed that "annual radon doses in Canadian (uranium) mines have fallen from about 300 mSv in 1956 to less than 2 mSv in 1980." The reduction reported is drastic, yet risk is not completely eliminated.

According to Roberts and coworkers (1990, 98), an individual receiving a dose of 2 mSv a year for 40 years would increase his chance of eventually dying of cancer from about 0.200 (the average U.K. figure) to between 0.201 and 0.204. This represents an increase of 0.5% to 2% in the risk of developing cancer.

Radon is present in most metal mines, not just uranium mines. Roberts and coworkers (1990, 99) reported that in 1988, the average annual dose to non-coal miners in the U.K. (uranium is not mined in the U.K.) was 14 mSv. The annual dose to Canadian uranium miners is very low compared to this figure.

Mining is a high-risk occupation by its very nature; however the types of risks of death and injury from cave-ins, rock bursts, etc. are not different in kind for uranium mines compared to other mines. Workers are likely to be well informed about these risks, as they are well known in the industry. Furthermore workers today must be well informed about any radiation risks associated with uranium mining, considering the extensive publicity given to radiation risks in recent decades. We would therefore expect these risks to be compensated by higher wages. (Moore and Viscusi 1990.)

Wages in the mining industry are significantly higher than wages in other similar activities. In 1994 the average seasonally adjusted weekly wage in Ontario mines was $970. This compared to $697 in construction, $797 in logging and forestry. Miners therefore enjoyed a premium of 39% vis-à-vis construction workers, 22% vis-à-vis forestry workers. The risks inherent to mining seem to be compensated for by wage differentials, and since uranium miners do not appear to face higher risks than other metal miners, it does not appear that there are external occupational health costs at this stage of the nuclear fuel cycle. Ontario Hydro (1993) did not include external occupational health costs in its estimates of the upstream external costs of NPG in Ontario.

No studies have been found on the occupational health costs for workers involved in the fabrication and transportation of nuclear fuel.

PUBLIC HEALTH COSTS

Uranium ore, subsequent to its extraction, must be crushed, ground, and chemically treated (the whole operation is referred to as "milling") in order to recover its uranium content. Milling produces "yellowcake," a high-grade uranium concentrate from which nuclear fuel is manufactured, and "tailings," the reject material from the milling process. Since uranium makes up only a small fraction of uranium ore (much less than 10%), a proportionally large volume of tailings is associated with the recovery of a given quantity of uranium. Moreover, the milling process being not perfectly efficient, some proportion of the uranium contained in the ore is necessarily lost in the tailings.

Because of their uranium content, as well as the presence of elements from the uranium decaying chain, uranium mill tailings represent a source of long-lived radioactivity. For example, the thorium isotope Th-230 (a radon emitter) present in a tailings pile will lose only half of its radioactivity after 80,000 years. This radioactivity would have been present anyway in the mineral deposit had it not been mined, but a surface uranium tailings pile represents a much greater potential public health hazard than an underground ore body. Radon released from an uncovered tailings pile disperses in the atmosphere, and other radioactive substances (radium and thorium, for example) can dissolve in water and migrate, or be carried by the wind as dust.

According to Roberts and coworkers (1990, 108), covering a tailings pile with a few metres of clay and topsoil will reduce the exhalation of radon to a level low enough to make it indistinguishable from variations in natural levels (radon is naturally released into the atmosphere from the earth). As an example, the authors offer the following hypothetical scenario: for a mine/mill complex operating for twenty years, processing 600,000 tons per year of 0.2 per cent uranium ore (enough to produce 5 GW of electricity per year), located in a region with population density of 25 per km^2, the collective dose to the population out to 2000 km from the mill would be 54 person-Sv, compared to 600,000 person-Sv from natural background. One could argue that such a low dose increase (0.009%) is insignificant and cannot be translated into a meaningful health impact figure, particularly when we are not certain that there are actual risks from such low exposure levels. Alternatively, if one calculated a risk based on such a low dose and if one added up the risks over 80,000 years, the risk would be quite significant (Resnikoff 1992, Tables 2, 3). Because of the uncertain effects at these low doses and because of the long time horizon, I am inclined to treat these risks as insignificant. If one took the alternative view, one would have to calculate similar types of risks from radiation from coal ash piles to make a balanced comparison of the energy sources.

Gofman (1981, 462) estimated that the lung-cancer deaths per year due to radon release from an *uncovered* tailings pile would be 1/1000 of the natural rate, i.e., 3.9 lung cancer deaths per year for a population of 250,000,000. But since the radioactive life of the tailings is very long, the total number of deaths will be considerable in absolute numbers. The model used to compute this estimate assumed 1,000 nuclear power plants each producing 1,000 MW a year. Ontario's nuclear capacity in 1993 corresponded to less than 15 of these plants. And since tailings piles in Ontario are not left uncovered, Gofman's estimates cannot be applied to gauge the radiological impacts of uranium mill tailings in Ontario.

The Atomic Energy Control Board (AECB) requires that tailings piles be stabilized once a mill closes. According to Ontario Hydro (1993, 82), radiological impacts of stabilized tailings piles are nil; it estimated that the total external cost from uranium mining, fuel fabrication, and transportation in Ontario was between 0.002165 and 0.006015 cents/kWh.

Because of the long-lived nature of the radioactivity of uranium tailings, long-term monitoring is necessary. Beak Consultants Ltd (1986, i) estimated the cost to decommission (i.e., to stabilize) a generic 40-hectare above-ground tailings disposal site at between $2.5 to $3.75 million, depending on site remoteness. An "all inclusive" environmental monitoring program would cost $34,000 on average per year for the first five years, and $11,000 per year subsequently.

No estimates of the health impacts of seepage from tailings piles have been found. Pohl (1982, 130) reported that the rate of migration of radium and thorium has been observed to be not more than 100 metres in approximately 50 years. Tailings piles in Ontario are located in areas of very low population density in the northern part of the province, hundreds of kilometres away from population centres. Seepage in that context appears to be a negligible source of public health concern. Clarke and coworkers (1991) found that the incidence of child leukaemia deaths in Ontario in the vicinity of uranium mining, milling, and refining facilities was slightly above normal, but that the difference was not statistically significant. This study only considered proximity to the emission source, and no measures of radon in the air or radium in the water were used. It is, therefore, not possible to disentangle the effects of radon emissions and radium seepage from this study.

From a purely technical point of view, the safest solution to the tailings disposal problem would be backfilling at the mine, virtually eliminating all up-front decommissioning costs and subsequent monitoring costs. Backfilling is a proven method of tailings disposal and is currently being used in Canada at the present time, for example at the Kiena gold mine in Abitibi. The risks from a well-managed, above-ground disposal site, however, are sufficiently small that there may be no justification for any added cost of backfilling.

PROPERTY AND NATURAL ENVIRONMENT DAMAGE COSTS

No studies on these types of costs were found.

Stage Two: Fission of Nuclear Fuel

All of Ontario Hydro's nuclear reactors are cooled and moderated by heavy water (deuterium), and fuelled by natural uranium, unlike most other reactors, which require enriched uranium. The unique technology of these CANDU (Canadian Deuterium Uranium) reactors must be taken into account when assessing the environmental costs arising from their operation.

The fission of uranium in a nuclear reactor generates hundreds of radioactive fission products inside the fuel rods. Additionally, some neutrons released during fission are captured in materials inside and nearby the reactor, creating radioactive neutron-activation products. Both types of reactor by-products, fission products and neutron-activation products, are released into the environment during the normal course of reactor operation, either as liquid or gaseous emissions. Accidents can be defined as releases that exceed usual levels. Torrie (1986, C-14) described the inventory of radioactive material created inside a CANDU reactor as "low" compared to a light water reactor, but did not draw conclusions for emissions levels.

Radioactive emissions from nuclear power plants into water bodies or the atmosphere can result in radiation doses to the public, through direct irradiation, inhalation, or ingestion. The large number of radioactive products involved, combined with uncertainties concerning their transfer coefficients and the dose-response relationship for humans, complicate the evaluation of the harm caused by these emissions, just as the variety of pollutants emitted from coal-fired power plants combined with uncertainty regarding dispersion and dose-response coefficients complicate the evaluation of the harm caused by these emissions.

Heavy water circulated in a CANDU reactor gradually becomes contaminated with tritium through neutron absorption. Tritium is a radioactive hydrogen isotope with a half-life of 12.5 years. Periodically Ontario Hydro drains heavy water from the cooling system and decontaminates it. According to Charlton (1988), spills are possible at that stage; no data on this seems available however.

Kim (1981, 106) reported that tritium is the major pollutant emitted by a CANDU-type reactor. Routine emissions from such a reactor represent only 0.22% of the legally permitted level in Canada. For other emissions (noble gases, iodine, and particulates), levels are even lower as a proportion of allowable levels.

Annual dose limits to the public are set in Canada by the Atomic Energy Control Board (AECB). These limits are based on recommendations made by the International Commission on Radiological Protection (ICRP). The ICRP establishes a dose limit for workers involved in the nuclear fuel cycle that translates into a health risk no greater than that observed on average in other industries. Maximum doses to individual members of the public are then computed as 10% of the acceptable dose for workers. An AECB official in Ottawa stated that the statutory dose limits to the public indicate exposure levels that do not generate any measurable public health impacts.[3] In theory, even such small exposures may generate small risks that could be quantified, however we have not been able to provide this information here.

Nuclear generating stations in Ontario have set a policy of not exceeding 1% of the statutory dose limits. According to the AECB (1990, 2), for the most part this operating target is achieved. Given the considerable degree of prudence built into the statutory dose limits, it should be expected that any risk arising from such emissions by Ontario nuclear power plants should be inconsequential.

OCCUPATIONAL HEALTH COSTS

Estimates of the radiation exposure of site personnel vary. Kim (1981, 106-7) reported two estimates of the accumulated group exposure resulting from the operation of one 1,000 MW reactor for one year (an electricity output of 6.57 GWh): a low figure, 350 whole body person-rem, and a high one, 2,000 whole body person-rem. Assuming that 6,000 person-days are lost per cancer case at a cost of $50 per person-day lost, the total occupational health cost is 0.0025 mills per kWh using the low exposure estimate, and 0.02 mills per kWh for the high exposure estimate (1975 dollars). If we used recent estimates of the value of a lost person-year of life, these costs would increase by about an order of magnitude.

Hill (1977, iii) suggested an occupational mortality risk for nuclear power plants of 0.1 fatality per 1,000 MW year, or less. At about $5 million per statistical life this would cost $500,000 per year. The author did not specify whether this risk was exclusively related to radiation exposure.

Torrie (1986, C-18) reported that in April 1979 two Ontario Hydro workers "received severe overdoses of radiation while cleaning up after a refuelling accident... The exact level of overdose received by the workers...is uncertain because they had not been equipped with all the required dosimeters." According to Torrie, by 1986 there had been over sixty cases of radiation overdoses to Ontario Hydro workers. Unfortunately, Torrie presented no data regarding these alleged accidental exposures and we have found no estimates of their health impacts.

Ottinger and coworkers (1990, 372-88) distinguished between immediate occupational mortality and latent occupational mortality. They assumed a value of a statistical life of $4 million 1989 U.S. dollars and reviewed several studies of occupational accidents and radiation exposure of workers, using the high exposure estimate from one study. They estimated the occupational health cost from routine operations at 0.007 cents/kWh for accidents and 0.09 cents/kWh for radiation exposure, for a total of 0.097 cents/kWh. See Table 1. The authors did not mention the possibility that compensatory wages would effectively internalize these costs.

Roberts and coworkers (1990, 98) reported that several of the original studies purporting to find statistically significant links between occupational exposure and cancer incidence, such as the Hanford (U.S.) study, were refuted by more recent studies looking at the same data. The authors suggested that there was no solid evidence of such links. Ontario Hydro (1993) did not quantify any occupational health costs on the grounds that the workers are compensated directly or indirectly by Ontario Hydro for those risks.

We believe that workers in nuclear plants are well aware of their exposure to radiation and of the health risks that these exposures may cause. These workers are well trained and unionized. The risks have been discussed publicly for years. We assume that the wages paid to these workers include a premium for any perceived health risk, which therefore is not an external cost that needs to be added to Ontario Hydro's expenditures.

PUBLIC HEALTH COSTS

There seems to be agreement that the radiation doses to the population at large resulting from normal emissions from nuclear power plants are low compared to doses attributed to natural background radiation, which originates from cosmic rays and the Earth's own radioactivity. There also seems to be agreement that the impact of such low exposure levels on the incidence of cancers is very low, practically negligible (Ottinger 1990, 375).

According to Christy (1981, 138-9), the normal dose from natural background sources is 80/1000 of a rem per year. To that can be added approximately 80/1000 of a rem per year from person-made sources, largely diagnostic X-rays. The combined dose from the *entire* nuclear fuel cycle to sustain a 1,000 MW plant for one year is 1,000 person-rem to the entire U.S. population. The background radiation is more than 10,000 times greater than the exposure from such a plant. The dose for workers at the same plant is 1,500 person-rem. These two risks combined translate into about a one-half chance of a cancer fatality occurring sometime to someone in the forty years following the operation of that plant for one year.

Kim (1981, 106) reported a public exposure of 4 to 25 person-rem from the operation of a 1,000 MW plant for one year (electricity output of 6.57 GWh). This is very low compared to the dose received by site personnel (see previous section). Since occupational health costs were estimated to be quite small (see previous section), public health costs can be deemed negligible; Kim did not quantify them.

Roberts and coworkers (1990, 93) estimated the average annual dose to the U.K. population from *all stages* of the nuclear fuel cycle at less than 1/1000 of 1 mSv a year. For people living within a few kilometres of nuclear power stations, the dose was about 1/100 of 1 mSv a year. The maximum dose observed, 1/2 of 1 mSv a year, was for a small group of people eating unusually large amounts of seafood caught in the vicinity of the Sellafield reprocessing plant. The dose received on average from natural background radiation is 2 mSv a year in the U.K.

Roberts and coworkers (1990, 92-3) reported that statistical studies have failed to identify any positive link between radiation dose increases at low exposure levels and the incidence of cancers. On the contrary, evidence from epidemiological studies carried out in the U.S., China, and India suggest that cancer rates are *inversely* related to natural background radiation levels (which vary with altitude and location), suggesting some potential health benefits from radiation at low doses. While such results may be consistent with the supposed health benefits of bathing in radium springs, a health benefit cannot be explained with our current knowledge of cell biology. In any event, these epidemiological data fail to demonstrate that actual radiation exposures from nuclear power plants cause significant harm.

Clarke and coworkers (1991) found that the occurrence of leukaemia in Ontario children living in proximity of nuclear power plants was greater than expected in the general population, although they could not conclude that the difference was statistically significant. The authors did indicate that more research was necessary to settle the issue.

Ottinger and coworkers (1990, 390) estimated the external cost from routine operations at 0.11 cents/kWh, but this erroneously included 0.097 cents/kWh in occupational health costs, as noted above. Only 0.001 cents/kWh was attributed specifically to public exposure to radiation from routine operations. Ontario Hydro (1993, 89) reported external costs from routine emissions that varied between 0.000031 cents/kWh (Bruce B, "low" estimate scenario) and 0.012755 cents/kWh (Pickering A, "high" estimate scenario), with the Darlington estimate of 0.00125 cents/kWh in the middle, a result that is consistent with the Ottinger estimate. See Table 1.

Ontario Hydro's external costs estimates vary from plant to plant because of differences in such factors as plant age, performance, design features, operating characteristics, and, most importantly, surrounding population densities. In all three scenarios studied for each plant (low, nominal, high), Bruce always comes out the least costly, and Pickering the most costly. Bruce is located on the shores of Lake Huron, far from significant population centres, while Pickering is within the Greater Toronto Area. All estimates find very low risks.

We have found no studies that demonstrate significant public risks from the routine operation of nuclear power plants in Ontario. See Table 1.

While there seems to be agreement that the external costs of routine nuclear power plant emissions are quite small, it is natural to suspect that massive releases of radioactive products following an accident could cause extensive harm. The external costs of nuclear accidents are "expected costs" by their very nature; although the possibility of accidents is undeniable, accidents are exceptional events by definition. In order to compute an estimate of the expected cost of an accident, one must estimate both the probability of an accident occurring and the damage resulting from an actual accident.

There are two ways of assessing the probability of accidents occurring: one can study the historical record, or one can rely on estimates derived from simulations on computer models of nuclear power plants. As of the end of 1991, 6038 reactor-years of operating experience had been accumulated worldwide according to the Uranium Institute of London, U.K. Two accidents resulting in damage to the reactor core and the release of radioactivity to the environment were on record at that time: Three Mile Island (1979) and the vastly more damaging accident at Chernobyl (1986). The empirical frequency of accidents per reactor year was thus $3.3*10^{-4}$ as of the end of 1991. Since then hundreds of reactor-years of operating experience have been accumulated, and no new accidents of this magnitude observed, so the empirical frequency of accidents has dropped, while remaining in the same order of magnitude. It could be argued that improved standards and practices, especially following the two aforementioned accidents, have reduced the probability of future accidents. Ottinger and coworkers (1990, 379) suggested a probability of $3.0*10^{-4}$, which might overstate the risks, given this experience.

It is instructive to consider the actual damages inflicted by the two very different major nuclear accidents on record. Roberts and coworkers (1990, 128) reported that the total quantity of radioactivity released at the Three Mile Island accident should result in about five cancer cases in the thirty following years, compared to the 540,000 cases that will be developed in the affected population during the same period. Five cases, valued at $5 million each, would entail a total cost of $25 million. The cost of the clean-up was estimated at one billion dollars.

Emissions levels at Chernobyl were higher, perhaps one thousand times greater. Two hundred and three fire fighters and site personnel

Table 1
Comparison of Nuclear Externality Cost Estimates — (¢/kWh)

Parameter	*Ottinger (1990)*	*Hohmeyer (1988)*	*Shuman & Cavanagh (1982)*	*Chernick (1992) (Darlington)*	*Ontario Hydro (Darlington)*	*Energy Research Group*		
						Unnamed PWR	*CT Yankee & Millstone 1, 2 & 3*	*Unnamed BWR*
Upstream Fuel			0.25 to 16.54			0.40[3]		
Routine Operations								
• Fossil fuels					0.00297	0.000156	≤.000101	0.00027 to 0.0022
• Radiological emissions	0.0011				0.00125	0.000122	≤.000172	0.0045 to 0.0113
• Occupational	0.097							
• Wildlife	0.01							
Low Level Waste					0.0004		0.000285 to 0.00304	0.00340 to 0.00400
High Level Waste					0.000036	0.0044	≤.000076	0 to 0.00006
Accidents	2.3	0.000012 to 2.288	0.0 to 2.1	2.6 to 15	0.00086	0.00353		
Decommissioning	0.5			0.0001	0.00005	0.0119	0.000522 to 0.011	0.00165 to 0.0102
Other		0.0129 to 0.0283[1]	0.0 to 11.26[2]		0.00057			
Total	2.91	0.0129 to 2.31	0.25 to 26.90	2.6 to 15	0.0061	0.418	0.00012 to 0.014	0.00982 to 0.0278

Source: Ontario Hydro (1993, 63, 89).
[1]Public subsidies and uranium depletion.
[2]Weapons proliferation.
[3]The fuel use characteristics and level of enrichment cause this value to be a factor of 2 to 10 higher than for other LWRs.

Notes to Table 1: Comparison of Nuclear Externality Cost Estimates

	Comments by Ontario Hydro (1993, pp. 61, 64)
Study	*Comment*
Ottinger (1990)	• Estimate for wildlife impacts based on arbitrary estimate. • Assumes Chernobyl-type accident risk in calculation of nuclear external cost estimates. • Assumes there does not exist a fund for decommissioning – therefore overestimates decommissioning external cost.
Hohmeyer (1988)	• Inappropriate conversion of Chernobyl accident risk. • Considers use of public funds an externality. However, it should be considered an internal cost of governing that is received by taxation.
Shuman and Cavanagh (1982)	• Nuclear accident estimate based on arbitrary manipulation of a Probability Risk Analysis study of a nuclear plant known as WASH-1400 (1975). • Based on arbitrary assumption regarding radon exposure and nuclear weapons proliferation. • Unreasonable and excessively high estimates of the externalities associated with the upstream fuel cycle. Assumes that radon releases from mill taillings occur infinitely. Ignores U.S. federal toxics legislation passed in 1978.
Chernick (1992)	• Applies arbitrary accident probabilities and consequence values to Darlington station. Ignores up-to-date probability risk assessment information that exists on that Station which gives externality costs orders of magnitude lower. Relies on "Chernobyl-type" accident – unrealistic for CANDU reactors given published Station Probability Risk Assessments and AECB safety requirements. • Decommissioning estimate based on unreasonably high exposure during decommissioning.

were diagnosed as suffering from acute radiation syndrome, and thirty had died by 1991. Impacts away from the reactor were, of course, much less intense. In the first year following the accident, doses to the population of Eastern Europe amounted to 32% of background, 11% of background for the whole USSR, and 2-3% of background in the U.K. Over the following thirty years (during which time most of the dose commitment will be released), doses to populations of Western Europe will amount to an additional three weeks of background, and six months for Eastern Europe. Fall-out from atmospheric nuclear weapons tests up to 1980 was larger.

According to Roberts and coworkers (1990, 131-2), the Chernobyl accident will result in an additional 5,000 to 10,000 fatalities in the former Soviet republics over the next thirty years. This is an enormous toll, and these 10,000 fatalities would be valued at $50 billion in Canada using the estimate of $5 million per statistical life. If there was a significant risk of such an accident at an Ontario reactor it could give rise to significant expected costs, as shown below. However, most of the estimated Chernobyl-related fatalities will occur in an area with a population of 75 million, in which nine million cancer deaths due to other causes will be observed during the same period. Surprisingly, given the large number of estimated fatalities, the radiological impact of the accident is unlikely to be detectable in mortality statistics, and the extra doses received elsewhere are comparable to variations in background doses between different areas.

Why was the Chernobyl accident more than a thousand times more damaging than the Three Mile Island accident, when both involved a loss of control so serious that part of the reactor core actually melted? A major factor is the vastly greater safety and containment systems built into the Three Mile Island reactor (and all western reactors including CANDU reactors) that were not present in the Soviet design. Western designs include more redundant systems to provide extra coolant in case of overheating and to control the reaction if it should begin to run away, and they include containment systems to trap any radiation released from a damaged reactor and to evacuate radioactive emissions to a special vessel to prevent releases to the atmosphere. These systems worked to prevent all but a very small amount of radiation from escaping to the environment at Three Mile Island.

It is common practice in the nuclear power industry to quantify the risks of various types of accidents at a nuclear power plant with the help of computer simulation models. Such studies are referred to as "probabilistic risk assessments" (PRAs). A complete model of the plant is needed in order to carry out a PRA, and the computational requirements necessary to simulate random events and their possible consequences are considerable. Torrie (1986) questioned the logic of considering expected costs, because of the serious consequences of nuclear accidents for the very survival of living organisms in the biosphere. This line of argument leads to a complete ban of nuclear power. The accident experience so far does not support Torrie's dismal view. We believe that probabilistic risk assessment is a valid method for evaluating these risks.

According to Atomic Energy of Canada Limited (1987, 3) "probabilistic safety and risk evaluations show that the CANDU offers a level of safety at least as good as, and in most cases better than, other commercially available reactor designs." The probability of a severe accident occurring per year of reactor operation, as derived from PRAS, is estimated to be in the order of 10^{-7} to 10^{-6}. And even a severe accident may result in minor radiation releases to the environment because of the containment systems. Such low probabilities result in very modest expected costs. Ginoza (1982, 180) reviewed the various generic Environmental Impact Statements prepared by the U.S.. Federal government for the nuclear fuel cycle, and found that probabilities assigned to accidents were less than one in a million per year.

Hill (1977, ii) reported a public mortality risk of 0.1 fatality per 1000 MW reactor-year, including both routine and accidental releases. Christy (1981, 139) estimated the expectation of fatal cancers resulting from accidents per 1000 MW reactor-year at 0.02.

Kim (1981, 109) reported an estimate of $14,400 (1975 U.S. dollars) for the expected annual cost of accidents for one 1,000 MW reactor, equivalent to 0.002 mills per kWh. This estimate considered not only deaths (valued at $300,000) and illnesses (valued at $3,000), but also property damages, the expected cost of which ($13,480) dwarfed all public health costs. If we multiplied Kim's value of life and illness by 10 to approach current values, we would approximately double Kim's total damage estimates.

Ottinger and coworkers (1990, 378-83) estimated the external cost from accidents to be enormously greater, at 2.0 cents/kWh for human health effects plus 0.3 cents/kWh for crop damage, assuming an accident and release of radioactivity similar to that at Chernobyl and using probabilities from the U.S. Nuclear Regulatory Commission. The sum of these two values, 2.3 cents/kWh, is almost 1000 times the total external cost estimate of any of the Ontario Hydro estimates. See Table 1. Ontario Hydro (1993, 84) rejected as totally inappropriate the assumption of a Chernobyl-type radiation release, because of the differences in design between the CANDU and the Soviet reactor. The Chernobyl reactor had very modest containment with no separate external containment vessel, while all CANDU reactors have a containment system that should minimize releases even in the event of a meltdown. The

Chernobyl reactor employs graphite, which is combustible, to moderate the nuclear reaction, while the CANDU reactor uses heavy water as a moderator. In a CANDU reactor the moderator would cool the reactor while in the Soviet reactor the moderator feeds the fire. The Three Mile Island accident involved a core meltdown that approached the Chernobyl meltdown in severity, but at TMI the containment system prevented the escape of all but a very small amount of radiation, leading to estimated health risks that were orders of magnitude less than at Chernobyl. Ontario Hydro therefore argues that even a meltdown accident at a CANDU reactor would have consequences no worse than those at Three Mile Island rather than like those at Chernobyl. Its estimate of the external cost of accidents, based on plant-specific PRAs, ranged from 0.000011 cents/kWh (Bruce B, "low" estimate) to 0.095909 cents/kWh (Pickering A, "high" estimate), with Darlington in between at 0.00086. See Table 1. The Chernick (1992) study, assuming a Chernobyl-type radiation release and high accident probabilities, estimated accident risks even greater than those of Ottinger. Here, then, is a major source of divergence among the accident risk estimates, centred on the probability of a Chernobyl-like radiation release at a CANDU reactor. We believe that the existence and effectiveness of the containment system should be reflected in the risk estimates, and therefore find it plausible that even if a meltdown accident is part of the risk assessment, the radiation releases should be more like those at Three Mile Island than those at Chernobyl, giving some credence to Hydro's risk estimates.

Oxman and coworkers (1987) observed that all quantitative studies of nuclear reactor safety in Ontario were produced by either Ontario Hydro or Atomic Energy of Canada Limited. From their review of a number of studies, they concluded that only the Darlington Probabilistic Safety Evaluation was scientifically sound. They also pointed out that critics of nuclear power cannot be expected to carry out their own PRAs since they do not have access to the information and resources necessary for such a complex endeavour. The authors suggested however that reviews by critics of nuclear power often fell far short of the professional level that they expected of the nuclear industry. This would tend to support the Ontario Hydro estimates in Table 1 over the much higher accident cost estimates reported by other authors in that table.

Mackenzie (1982), Senior Staff Scientist at the Union of Concerned Scientists, Washington, D.C., argued that the Price Anderson Act, limiting the liability of nuclear power plant owners and their contractors in the U.S. at $560 millions, has artificially removed the normal market barriers to a potentially hazardous activity. Hebert (1987) described the similar legislation adopted by Canada in 1970: the Nuclear Liability Act,[4] which transfers to the operator liability that might accrue to suppliers to the reactor and in return limits the operator's liability related to any one nuclear power plant to $75 million. Mackenzie (1982) observed that between 1957 and 1982, the amount of private insurance available to the U.S. nuclear power industry had declined, suggesting that insurers, whose business is to assess risks, were becoming increasingly wary of the potential risks of nuclear power. If each nuclear plant owner was required to insure fully for all accident risks then we might argue that the cost of these risks was internalized, to the extent that liability would fully cover all losses. However, I believe that tort damages would probably under-compensate victims, at least to the extent that the discounted future earnings measure of loss falls short of the true individual loss from a risk of premature mortality. Furthermore, the limits on liability mean that the public, either individually or represented by the government if there were public compensation of individual victims, bears most of the loss. For the present in Canada it appears that a large fraction of the expected accident costs discussed above are external to the utility. Still, the most plausible of those estimates suggest expected accident costs that are small fractions of a cent per kilowatt hour. The omission of these costs should not significantly affect the choice of power sources in the future.

PROPERTY AND NATURAL ENVIRONMENT DAMAGE COSTS

Ottinger and coworkers (1990, 376-7) report that these types of damages from routine emissions have not attracted the attention of researchers. They estimated the expected cost of damages to agriculture from accidents at 0.3 cents per kWh, based on estimated crop losses after the Chernobyl accident.

Stage Three: Decommissioning of Reactors, Waste Disposal

Two main activities can be associated with the last stage of the nuclear fuel cycle: the decommissioning of plants that have reached the end of their operating life, and the permanent disposal of wastes (radioactive as well as inactive). This section reviews the external costs associated with each activity in turn.

DECOMMISSIONING OF REACTORS

Once a nuclear reactor reaches the end of its operating life, a decision has to be made as to what to do with the plant. There are numerous options available, ranging from simply forbidding access to the site until all radioactivity has naturally decayed to background levels (which could take thousands of years), to immediately restoring the site to a "greenfield" condition. This last option entails complete dismantlement of the plant, and shipment of radioactive waste (spent fuel, contaminated equipment, and material) and inactive waste (mainly rubble and steel) to appropriate disposal facilities.

Radioactivity released during decommissioning clearly constitute an externality that translates into occupational and public health impacts. According to Ontario Hydro (1993, 85), the releases will not exceed normal routine operating levels.

Do the costs expected for decommissioning constitute external costs? The answer is not obvious. Should the owner of the plant be incapable or unwilling to absorb decommissioning costs when due, public funds would need to be used, and decommissioning would in that case constitute an external cost. In the case of a private investor contemplating the acquisition of a publicly owned nuclear power plant, future decommissioning costs would be considered as a liability.

Atomic Energy Control Board (AECB) regulations seek to have plant owners in Canada internalize decommissioning costs. The AECB requires plant owners to submit a decommissioning plan when applying for a licence to construct a new facility, or applying to renew an operating licence for an existing plant. To be effective, such regulations necessitate accurate decommissioning cost estimates.

In 1982, Ontario Hydro (1982, 3) estimated the present value cost of decommissioning the Pickering A and Bruce A plants by immediate dismantlement at respectively $201 million and $228 million, in 1980 dollars. At each site, 3,300 person-rem of radiation would be released during the nine years required to carry out the dismantling operation. Deferring dismantlement by thirty years at Pickering A would save $39 million in 1980 dollars, because of the natural decrease in radioactivity levels with time. Radiation released would also drop, to 1,824 person-rem. Six years later, Ontario Hydro (1988, iii) estimated the present value cost of decommissioning Darlington by piece-by-piece dismantling (following 30 years of shutdown storage) at $821 million in 1988 dollars. Then in 1991, Jayawardene and Stevens-Guille (1991, 186) estimated the present value costs in 1989 dollars of decommissioning Ontario nuclear power plants by piece-by-piece dismantling following 30 years of shutdown storage. See Table 2.

It is instructive to compare the 1991 estimates with the ones computed in 1982 for Pickering A and Bruce A, after converting both sets of estimates to a common basis (1986 dollars). Table 3 shows that decommissioning cost estimates have grown substantially in real terms over the 1982-1991 decade, despite the fact that the decommissioning procedure assumed in 1991 (dismantlement following 30 years of shutdown storage) was intrinsically more economical than the one assumed

Table 2
Estimated Decommissioning Costs (millions, 1989$)

Plant	*Cost*
Pickering A	$499
Bruce A	$576
Pickering B	$499
Bruce B	$576
Darlington	$860

Source: Jayawardene and Stevens-Guille (1991, 186).

in 1982 (immediate dismantlement). In fact the delayed dismantlement option has been shown in 1982 to be over 19% cheaper for Pickering A. This raises serious doubts as to the quality of the cost estimates, and means that any funding scheme initially devised on the basis of the 1982 figures would have turned out to be grossly inadequate.

Such cost escalations are not particular to Ontario Hydro. Biewald and Bernow (1991, 235) found that during the period 1976-1991, the average estimated cost for decommissioning a large nuclear plant in the U.S. had increased at an average annual real rate of about 16%, resulting in a cumulative real cost increase over the 11-year period of more than 400%. The authors also determined that for the 75 U.S. nuclear plants with construction starts between 1966 and 1977, actual construction costs were about three times higher than estimated at the start of construction. An argument could be made that decommissioning could be subject to cost increases similar to those observed at the construction stage. This would place further upward pressure on decommissioning cost estimates.

MacKerron (1991, 16) reports estimates of the average undiscounted decommissioning cost for a British Magnox station computed by the Central Electricity Generating Board in the U.K., in 1989 pounds. The

Table 3
Estimated Decommissioning Costs (millions, 1986$)

Plant	*Year of Estimate* 1982	1991	*% increase 1982-1991*
Pickering A	$275	$434	57.8
Bruce A	$312	$501	60.6

Source: Based on Ontario Hydro (1982, 3); Jayawardene and Stevens-Guille (1991, 186).

estimates go from 211 million in 1982, to 312 in 1988, to 599 in 1989. MacKerron (1991, 27) concluded:

From having been a subject of mild academic interest until 1988, the decommissioning of nuclear reactors became, in Britain, an issue of real political significance... The prospect of privatization compelled a proper and public appraisal of the real costs of nuclear power. This appraisal revealed...that the liabilities at the back end of the nuclear fuel cycle were much larger than had been previously disclosed... The British experience seems to show that once real plans for decommissioning are made, rather than the generic studies that have generally predominated, the costs of decommissioning are likely to be much higher than expected...

Plans initiated by the Thatcher government to privatize nuclear power plants failed in Britain, arguably because investors were unwilling to take on decommissioning liabilities that seemingly kept growing with time.

Ontario Hydro has never decommissioned a reactor; the first one scheduled for decommissioning is Pickering A in 2012. This obviously complicates the evaluation of the quality of cost estimates. Nonetheless, actual decommissioning cost data are available for Gentilly 1, a modified CANDU reactor that was partially decommissioned in 1984-1986 in Quebec, at a cost of about $25 million. This decommissioning did not dismantle the reactor. Instead it was secured and placed into an indefinite "static state" requiring ongoing monitoring and security. Because the site will present risks for thousands of years, it is anticipated that at some time when the radiation levels within the reactor have declined substantially the reactor will be dismantled and the hazardous waste disposed of. The timing of this next step is uncertain, but it may occur by the middle of the next century. The Nuclear Energy Agency (1991, 95) estimated the full cost of this decommissioning at $202 million in 1986 dollars, but the magnitude of the remaining costs are uncertain.

Let us compare the decommissioning cost estimates per MW of plant capacity for the five Ontario nuclear power plants, to the actual costs incurred at Gentilly 1, using the 1991 estimates reported earlier for the Ontario plants, and expressing all figures in 1986 dollars. See Table 4. This simple comparison assumes that costs are proportional to plant size, which may or may not be true; it is easy to believe that there are fixed costs associated with decommissioning any plant regardless of size. Unfortunately, we do not have an estimate of a cost function, so we will assume linearity. For all reactors, decommissioning cost estimates per MW of plant capacity are greater than the actual costs incurred at Gentilly 1 but less than the estimated cost of complete decommissioning at Gentilly 1. For example, the cost estimated for Bruce B is less than 20%

Table 4
Estimated Decommissioning Costs per MW Capacity (millions, 1986$)

Plant	*Cost*	*Capacity*	*Cost per MW*
Pickering A	$434	2060 MW	$0.21
Bruce A	$501	3056 MW	$0.16
Pickering B	$434	2064 MW	$0.21
Bruce B	$501	3345 MW	$0.15
Darlington	$748	3524 MW	$0.21
Gentilly 1	$25[1]	250 MW	$0.10
	$202[2]		$0.82

Source: Based on Nuclear Energy Agency (1991, 95), Table 2 and 3; Jayawardene and Stevens-Guille (1991, 186)

[1]Actual cost for placing reactor in static state. First stage of decommissioning.
[2]Estimated cost for complete decommissioning including first stage already done.

of the estimated cost for Gentilly. Economies of scale in decommissioning could theoretically justify some discrepancies; however, the fact that the unit cost at Darlington is 42% higher than at Bruce B, despite a 5% larger capacity, suggests that other factors may be more important.

Internalization of decommissioning costs requires not only accurate cost estimates, but also proper funding procedures to ensure that sufficient monies are available when needed. According to Jayawardene and Stevens-Guille (1991, 186),

> Ontario Hydro calculates an annual amount under the annuity method for the costs of decommissioning and applies it to the rates charged to customers... Total decommissioning provisions ($28.8 million) represents approximately 0.4 percent of the average electricity cost (some 0.8 percent of the costs of nuclear generated electricity).

The $28.8 million annual charge is an annuity that should provide sufficient funds for decommissioning the existing nuclear reactors, using the 1991 decommissioning cost estimates. Hydro re-estimates the decommissioning costs every five years.

While customers are charged for decommissioning costs, Ontario Hydro has not established a separate fund to pay for decommissioning. Instead,

> the monies collected from customers have been utilized in the normal course of business to displace borrowing which would otherwise have been required to finance fixed assets. There is no separate fund in existence (Ontario Hydro 1993, 94).

The fact that there is no actual fund set aside to cover decommissioning costs would affect the value of nuclear assets from the point of view of private investors, if such assets were offered for sale. If the funds collected for decommissioning are attributed to the nuclear reactors, then the debt associated with those reactors should be reduced by the attributable amount of the fund collected. The buyer would receive less debt, but would be responsible for all decommissioning costs. Because it seems likely that actual decommissioning costs will exceed the amounts on which the 0.8% is based, we anticipate that even the implied debt reduction will be insufficient to fund the actual decommissioning costs. We note, however, that the estimated costs are a fraction of 1% of the cost of electricity, so that even a five-fold increase in the estimated cost would not greatly affect the current operating economics of the reactors.

We conclude that decommissioning costs may have been seriously underestimated but that they will remain uncertain until some operating reactors have been actually decommissioned. There is no special fund to pay for these costs, so they will be a liability for the owner of the reactor at the time of decommissioning, although the debt associated with each reactor should be reduced by the amount of the funds that have been collected through the decommissioning charge. Therefore the actual costs are external costs to the extent that this charge turns out to be insufficient.

WASTE DISPOSAL

Nuclear fission inside a reactor creates byproducts which gradually impede the nuclear reaction. Fuel rods must, therefore, be replaced periodically. Spent fuel is placed in temporary on-site storage pools, where it is allowed to cool down. At the end of the plant operating life, a permanent disposal site will have to be found for this so-called high-level waste. Up to the present time there has been extensive study and planning for such disposal, but no such facility has been developed in Canada. Instead temporary disposal sites at each reactor have had to be enlarged.

Routine operations also generate so-called low-level wastes. Ontario Hydro (1993, 85) reported that 40% of this type of waste produced in its plants is incinerated, while the rest is interred in temporary facilities.

Decommissioning will also produce waste in the form of contaminated material and equipment that will require disposal in an appropriate repository.

Few estimates of the external costs of nuclear waste disposal are available. Ontario Hydro (1993, 63) reported estimates computed by Energy Research Group (a consultancy) for three U.S. utility clients. Ontario Hydro (1993, 89) has produced its own estimates, which do not consider wastes generated by the dismantlement of decommissioned plants.

See Table 1. Extensive plans have been made for such disposal, including the evaluation of alternative types of sites. (AECL 1994.) It has been argued that while the politics of such disposal are difficult the technology is not, so that the actual costs should not be unreasonable. The Energy Research Group has estimated fuel disposal costs to be in the range of one mill per kWh.

According to Ottinger and coworkers (1990, 387-8), external costs for waste disposal cannot yet be reasonably calculated because of the uncertainty of site selection, disposal method, and security. The estimates produced by Energy Research Group (1993) and Ontario Hydro (1993) should therefore be viewed as indicative at the very best. This is a potentially serious cost problem. The experience in Ontario with the Ontario Waste Management Corporation spending over $100 million over more than a decade without finding a means of disposing of hazardous waste in the province does not auger well for the possibility that the problem of disposing of nuclear waste will be solved quickly or cheaply, whether the problems are technical or political.

III. ENVIRONMENTAL COSTS OF FOSSIL FUEL POWER GENERATION

The external costs of NPG in Ontario can be better understood if they are compared with the costs of a plausible alternative. Coal-fired power generation (CFPG) is such an alternative for Ontario, which already boasts almost 9,000 MW of CFPG. Most sites with hydroelectric potential have already been developed in Ontario, and natural gas and oil do not seem to be fuels of choice in the province (only 1200 MW and 2200 MW of capacity respectively), although natural gas generation is expanding rapidly for cogeneration.

Like uranium, coal has a cycle, at every step of which environmental costs arise. Coal is not mined in Ontario, therefore the first stage of its cycle will be ignored here.

Environmental costs arise from the burning of coal, which releases various pollutants into the atmosphere. These pollutants are known to affect human health and to damage physical structures and crops. In addition, coal plants must dispose of a large volume of ash: a typical 2000 MW plant (such as Lambton) produces over 2000 tons of ash daily. This residue from coal burning is rich in pollutants, such as heavy metals that can leach from an ash pile and contaminate groundwaters.

Coal-fired power plants give rise to concerns regarding the following air pollutants: sulphur oxides (sulphur dioxide and sulphate: SO_2 and SO_4, jointly known as SO_x); particulate matter (PM); nitrogen oxides (NO_x); volatile organic compounds (VOCs); carbon dioxide (CO_2), a greenhouse gas; and some trace toxic contaminants associated with coal and waste combustion: arsenic, cadmium, lead, and mercury. These pollutants may cause health effects on a local or regional basis except for CO_2, which is a cause of global warming. SO_X and NO_X are also responsible for acid rain over a substantial region.

Sulphur dioxide is not a carcinogen, but chronic or acute inhalation may give rise to health effects in humans. While the evidence on these effects is not satisfactory, some tentative conclusions have been drawn. The USEPA (1982, 1-105) concludes that respiratory symptoms and pulmonary function decrements may occur in children for exposures above perhaps 0.88 ppm (with some concurrent particulate exposure), but that the risks at annual average ambient sulphur dioxide concentrations prevalent in the United States are likely nonexistent or minimal.

Suspended particulate matter (PM) generally includes lead and sulphates along with carbon, ash, and other materials. PM is not regarded as carcinogenic, although some compounds that are adsorbed onto airborne particles may be carcinogens. There have been studies of physiological responses to PM, but the early studies could not draw conclusions regarding the health effects of non-sulphate particles (USEPA 1982, 82).

Epidemiological studies of pollution levels and health have tended to show an association between them, but the presence of multiple pollutants renders difficult the estimation of a quantitative effect of PM alone (USEPA 1982, 97, 105). Most epidemiological studies have considered SO_2 and PM together. A recent study has found statistically significant increases in mortality from exposure to fine PM (less than 2.5 microns in diameter) even at levels below the current U.S. ambient air quality standards for PM.[5]

Lave and Seskin (1977) found that they could not reject the hypothesis of a linear relationship between concentrations of PM and sulphur dioxide and aggregate human health effects, while Ostro (1983) found a logistic relationship with PM and sulphate, which is generally correlated with sulphur dioxide. Cifuentes and Lave (1993) concluded that there was a linear relationship between mortality and the annual average concentration of SO_2 and PM in a city when these concentrations exceed the U.S. primary air quality standards, while air that is on average cleaner than these standards gives rise to a linear relationship that is one-tenth as great.

There is some evidence regarding other forms of harm from PM. Material damage may occur to the extent that the PM contains sulphates. In addition, PM causes soiling, requiring cleaning or more frequent painting of building surfaces (USEPA, 1982, 45). PM can affect visibility directly; the extent of impairment depends on particle size and concentration, and on humidity (USEPA 1982, 39).

Oxides of nitrogen, collectively referred to as NO_x, include primarily NO and NO_2. The latter is a reddish brown gas that gives photochemical smog its colour. These oxides also interact with VOCs in the presence of sunlight to form ozone. It is not thought that NO_x is harmful to humans at typical urban concentrations, but the ozone that it generates is clearly harmful to humans and to some plants. Ontario Hydro (1993, 77) estimated that ozone arising from Hydro's NO_x emissions caused losses equal to 0.4% of the value of Ontario crop production. Because of the complex atmospheric chemical reactions that form ozone, it is not easy to characterize the benefits of NO_x reduction in general terms. NO_x also participates in the formation of acid rain.

Volatile organic compounds, VOCs, are organic chemicals that exist as gases. Natural VOCs include methane from swamps and animals and isoprene and turpine hydrocarbons from plants. Burning fossil fuels releases unburned hydrocarbons of many types, some of which are carcinogens. The cancer risk is probably proportional to concentration and population density. Some VOCs contribute to the photochemical smog reaction, producing ozone and other damaging products. The chemical reactions involved in producing photochemical smog are complex, but generally more VOC means more smog. The smog problem is much more serious when the weather is warm and sunny than at other times.

Carbon dioxide is a greenhouse gas responsible for a substantial fraction of the global warming predicted for the next century and beyond. While there is agreement that atmospheric CO_2 concentrations have increased and will increase if current emission rates continue, there continues to be considerable debate over the increase in global temperature that will result and of the environmental consequences of this warming. There is even greater debate as to the effect of increased future emissions on the global climate in the future. Because CO_2 is a global pollutant, its effects do not depend at all on the location of emission; there are no local effects.

The emissions of the other coal pollutants are quite small and their effects uncertain. At high concentrations, they can cause serious health injury or death. Low concentrations arising from coal combustion may cause little damage, depending on the shape of the damage function. There is also concern about their effects on ecosystems generally, but again there is little evidence regarding the harm from the emissions from coal-fired power stations.

Because the principal harm caused by all of these pollutants, except for CO_2, arises from human exposure, and because their effect is local or regional, the harm caused by a given quantum of pollution depends greatly on the location of the source. For this reason, the best studies would be those which examined the emissions from the source in question, taking into account local pollution dispersion patterns and concentrations and population and density. Estimates of average pollution damages for emissions in the U.S. are of little use for Ontario. Unfortunately the majority of existing studies provide U.S. average damage estimates.

The external costs from coal-fired power plant operations are believed by some to exceed by far the costs of routine nuclear power plant operations. Kim (1981, 116) reviewed the literature and found that the environmental costs of CFPG has been estimated to exceed those of NPG by a factor ranging from 17 to 61. But most of his coal estimates assume no pollution control, while any new coal plant would in contrast be highly controlled.

Ottinger and coworkers (1990, 31) estimated externality costs for four types of coal plants, including an existing uncontrolled plant, a plant conforming to the U.S. New Source Performance Standards, a plant using advanced fluidized bed combustion, and an integrated gasification combined cycle turbine. See Table 5. The per-unit externality costs assumed by Ottinger are shown in the first column of the table. These costs are much greater than those estimated by, for example, Cifuentes and Lave (1993) and by the forthcoming New York State study. All costs include occupational health costs, which are not properly included in such a table. We include these data because they are generally available and have generated considerable discussion, not because we can endorse them. These data suggest external for an old coal-fired plant of 5.8 cents/kWh, and for a new plant meeting U.S. NSPS of 3.9 cents/kWh. These external costs are well above Ottinger's estimated external nuclear costs of 2.9 cents/kWh, although the differences were not nearly as great as reported by Kim. For two clean technologies, the authors found that a coal-fired plant was slightly less costly than a nuclear plant (2.8 and 2.5 cents/kWh). See Table 5.

Ontario Hydro (1993, 79) offered a much lower estimate of the external cost of fossil generation in Ontario that included all fuels (coal, oil, and gas): 0.395 cents/kWh. See Table 6. This estimate is based on generation from a mix of actual Ontario Hydro plants, simulations of the dispersion of pollutants around the plants, and the population density in the vicinity of the plants. This estimate has the potential for much greater relevance to Ontario because it is based on Ontario data. Table 6 shows the unit value for each outcome of the exposure and the contribution of each element to the estimated total damages, as well as the cost per kWh. Table 7 shows the cost per tonne of external harm from the emission of three pollutants from each of four stations in Ontario. These data show how the damage per tonne varies with location. The damage cost per tonne shown in Table 7 is lowest for the

Table 5
External Costs for Coal-Fired Units

Externality	$/lb [A]	*Existing Boiler (1/2% S)* [B]	*AFBC (1.1% S)* [C]	*IGCC (0.45% S)* [D]	*NSPS* [E]
EMISSIONS (LBS/MMBTU)					
[1] SO_2	$2.03	1.80	0.055	0.48	1.2
[2] NO_x	$0.82	0.607	0.3	0.06	0.006
[3] Particulates	$1.19	0.15	0.01	0.01	0.03
[4] CO_2	$0.0068	209	209	209	209
EXTERNALITY COSTS ($/KWH GENERATED)					
		$0.058	$0.028	$0.025	$0.039

Source: Ottinger (1990, 111).

Notes:

[C]: AFBC = Advanced Fluidized Bed Combustion, an emerging technology for clean burning of coal.

[D]: IGCC = Integrated Gasification Combined Cycle, an emerging technology for converting coal to gas and burning it in a gas turbine.

[E]: NSPS regulations require 1.2 lbs/MMBTU and 90% reduction for plants with emissions greater than 0.6 lb/MMBTU; for plants with emissions less than 0.6 lb/MMBTU, NSPS requires 70% reduction in emissions.

[1]: No SO_2 scrubbers are installed on plants [B], [C], or [D].

[2]: NO_x emissions are uncontrolled in each case.

[3]: Particulates emissions vary widely and depend on the ash content and sulfur content of the coal. NSPS requires 0.03 lb/MMBTU and 90% reduction.

Lennox plant because it is in an area of low population density and highest at Lakeview, which is near Toronto. Ottinger, on the other hand, assumes a cost of $2.03 per pound of SO_2 emissions, almost twice the value at Lakeview and nearly ten times the value at Lennox. Ottinger assumes a cost of $.82 per pound of NO_x, almost seven times the Lakeview estimate and 30 times the Lennox estimate. Ottinger assumes a cost of $1.19 per pound of PM, while Ontario Hydro assumes a cost of $12 to $47 per tonne, about one one-hundredth as great. This range of costs reflects the uncertainty in estimating the harm caused by the emission of conventional air pollutants. We note, however, that the Ontario Hydro figures are much closer to those of Cifuentes and Lave (1993) than are Ottinger's. Cifuentes and Lave's (1993) externality costs as a percentage of Ottinger's costs are: 7%-28% for SO_2; 39%-133% for PM; 6%-18% for NO_2.

Table 6
External Impacts of Fossil Generation in Ontario Based on a Mix of Generation Sites

Receptor	*Pollutants of Concern*	*Physical Impacts at 25 TWh*	*Unit Values ($000)*	*Monetized $M 1992*	*¢/kWh*
Mortality: (Statistical Deaths)	SO_2, SO_4, O_3, NO_3	4.5	4,726	21.40	0.088
Morbidity: (Admissions)	SO_2, SO_4, O_3, NO_3, TSP	1137.2	44	50.83	0.210
Cancer Cases	Trace Metals	23.3	408	9.53	0.039
Crops	O_3	N/A	N/A	8.32	0.034
Bulding Materials	SO_2	N/A	N/A	5.7	0.024
Total				95.79	0.395

Source: Ontario Hydro (1993, 79).

Table 7
External Costs by Station and per Tonne of Pollutant (1992$ Cdn.)

Station/ Pollutant	*Lakeview (1256)*	*Lambton (12)*	*Lambton (34)*	*Lennox*	*Nanticoke*	*Total System*
SO_2	2345	431	535	375	803	734
SO_3	267	87	193	61	214	188
TSP	46	9	12	9	16	15

Source: Ontario Hydro (1993, 81).

Coal-fired power plants, like nuclear power plants, have a finite operating life, at the end of which they need to be decommissioned. There is a paucity of data on the costs of decommissioning a coal-fired plant. In some cases, there may be costs for dealing with hazardous materials that have accumulated on the site, either in containment or in the soil. Williams (1991) compared two decommissioning studies done eight months apart by the same contractor for two U.S. power plants: one nuclear, one coal-fired. Surprisingly, he found that the costs of decommissioning the coal-fired plant was 14% higher than for the nuclear plant, on a dollars-per-megawatt capacity basis.

As with nuclear plants, great uncertainty attends the estimation of the external costs of coal-fired power generation. We note that there has been a steady downward trend in the emissions per kilowatt-hour from coal plants and that emerging clean coal technologies may further reduce these costs for future new plants.

IV. CONCLUSIONS

Despite the uncertainties identified throughout this paper, at least some tentative conclusions can be drawn regarding the environmental costs of nuclear generation.

First, the evidence suggests that the mining stage of the fuel cycle does not present significant external environmental costs when compared with other mining activity in Ontario. While there are, of course, risks to workers in mines in general, we have not found evidence that uranium mining carries significant special risks not found in other mining activities. In addition, we find evidence that mining wages in general reflect compensation for the special risks of mining so that the costs of these risks are incorporated in the price of the fuel and, therefore, are not external costs.

Similarly we find that the risks to workers at Ontario Hydro's nuclear power plants are small and are likely to be well understood by those workers. We believe that these risks are internalized in the workers' wages and in the Workers' Compensation costs to Ontario Hydro.

Second, virtually all of the literature that we have found suggests that the environmental risks associated with routine operation of Ontario Hydro's nuclear reactors are very small, when compared to background risks faced by local residents in the vicinity of the power plant, and when calculated as a percentage of the overall cost of generation. It seems unlikely that incorporating into the price of nuclear power generation the cost of risks arising from these routine operations would discourage nuclear generation in the future. While the estimates are highly uncertain, it seems likely that the external costs of routine nuclear generation are less than those of coal-fired generation, perhaps only a small fraction of those costs.

The risks of major accidents at nuclear generating stations are subject to significant uncertainty. We find the actual experience with CANDU reactors to be reassuring; there have been no accidents causing serious risks during over about a quarter century of commercial operation in Ontario. Ontario Hydro's estimates of the risk of such accidents are very low. Projections of the risks of such accidents at U.S.-style reactors are low and these are consistent with the world's experience with nuclear reactors, including the Chernobyl and Three Mile Island experience. If it is true that CANDU reactors are at least as safe as light-water reactors, and we have not found evidence to the contrary, and if a worst-case outcome is more like that at Three Mile Island than at Chernobyl, which seems plausible, and if Ontario Hydro's estimates are correct, then the risks arising from major accidents at CANDU reactors represent a very small fraction of the average cost of generation. These are, of course, important if's which we cannot confirm with certainty.

A more serious problem, in our view, is the cost of decommissioning, including the cost of permanent disposal of spent fuel. Early estimates indicated that decommissioning nuclear reactors at the end of their lives would be costly. Over time, the estimates of these costs have risen considerably faster than inflation. In Britain the prospect of the sale of nuclear power plants caused a serious appraisal of decommissioning costs, which in turn produced cost estimates much greater than had been previously estimated. The nuclear plants were unsaleable in part because of these costs. We believe that we will not know the costs of decommissioning and disposal of the radioactive wastes with any certainty until at least one of the main commercial reactors has been decommissioned and the wastes safely disposed of. We believe that there is a real risk that the cost will turn out to be far greater than indicated by past estimates, and therefore far greater than the amount that Ontario Hydro has been charging, but not setting aside, for future decommissioning. However, even a five-fold increase in decommissioning costs over those estimated by Ontario Hydro would not greatly increase the average cost of electricity generated by nuclear power stations.

NOTES

1 I would like to thank Nick Sisto for his very able assistance and contribution to researching and writing this paper.

2 1/1.02 = 0.019. Damages are worth about 2% of their undiscounted value.

3 Personal conversation, Hugh Spence, April 21, 1995.

4 R.S.C. 1985, c. N-28. See Trebilcock and Winter (1994) for a discussion of the incentives that this legislation creates regarding the provision of reactor safety.

5 "Particulate Matter Firmly Linked to Deaths," *Environmental Manager* 1 (April 1995):34.

BIBLIOGRAPHY

Atomic Energy Control Board. *Radioactive Release Data from Canadian Nuclear Generating Stations 1972 to 1988*. Ottawa, 1990.

Atomic Energy of Canada Limited. "A Submission to the Ontario Nuclear Safety Review." AECL-9427. 1987.

__. "Summary of the Environmental Impact Statement on the Concept for Disposal of Canada's Nuclear Fuel Waste." ECL-10721, COG-93-11. 1994.

Baumol, William J., and Wallace E. Oates. *The Theory of Environmental Policy*. 2d. ed. New York: Cambridge University Press, 1988.

Beak Consultants Ltd. "The Cost of Decommissioning Uranium Mill Tailings." 1986.

Biewald, B. and S. Bernow. "Confronting Uncertainty: Contingency Planning for Decommissioning." *Energy Journal: Special Nuclear Decommissioning Issue* (1991):233-245.

Charlton, B. "Ontario Hydro's Tritium Problem." *Probe Post* (1988):12-13.

Christy, R.F. "Risks Associated with Nuclear Power." *Environmental Impact Assessment Review* 3, 2-3 (June-September 1982):133-42.

Cifuentes, L.A. and L.B. Lave. "Economic Evaluation of Air Pollution Abatement." *Annual Review of Energy and Environment* 18 (1993):319-42.

Clarke, E.A., J. McLaughlin, and T.W. Anderson. *Childhood Leukaemia Around Nuclear Facilities - Phase II, Final Report.* Ottawa: Atomic Energy Control Board, 1991.

Dewees, Donald N. "Economic Incentives for Controlling Asbestos Disease." *J. Legal Stud.* 15 (1986):289-319.

Energy Research Group. "Calculation of Environmental Externalities for Ontario Hydro's Nuclear Power Plants." Report produced for Ontario Hydro, 1993.

Freeman, A. Myrick, III. *The Measurement of Environmental and Resource Values.* Washington, D.C.: Resources for the Future, 1993.

Ginoza, O.W. "Assessing the Environmental Impacts of the Nuclear Fuel Cycle." *Environmental Impact Assessment Review,* 3, 2-3 (June-September 1982):155-81.

Gofman, J.W. *Radiation and Human Health.* San Francisco: Sierra Club Books, 1981.

Hebert, M. *The Nuclear Liability Act.* Ottawa: Research Branch, Library of Parliament, 1987.

Hill, P.G. "The Social Costs of Electric Power Generation." Royal Commission on Electric Power Planning, Ontario, 1977.

Jayawardene, N.D. and P.D. Stevens-Guille. "Strategy, Planning and Costing for Decommissioning in Canada." *Energy Journal: Special Nuclear Decommissioning Issue* (1991):181-88.

Kim, S.Y. "Measuring of Environmental Costs: A Survey with Special Reference to Coal and Nuclear Generation." Ontario Hydro, 1981.

Lave, Lester B. and Eugene P. Seskin. *Air Pollution and Human Health.* Baltimore: Johns Hopkins Press, 1977.

MacKenzie, J.J. "The Risks of Nuclear Power: A Second Opinion." *Environmental Impact Assessment Review,* 3, 2-3, (June-September 1982):143-54.

MacKerron, G. "Decommissioning Costs and British Nuclear Policy." *Energy Journal: Special Nuclear Decommissioning Issue* (1991):13-28.

Moore, M. and W. Kip Viscusi. *Compensation Mechanisms for Job Risks.* Princeton: Princeton University Press, 1990.

Morrall, P.F. "A Review of the Record." *Regulation* (November/December, 1986):25.

OECD Nuclear Energy Agency. "Decommissioning of Nuclear Facilities: An Analysis of the Variability of Decommissioning Cost Estimates." (1991).

Ontario Hydro. "Decommissioning by Immediate Dismantlement Preliminary Cost Estimate for Pickering NGS." Report 82/0105. 1982.

___. "Darlington NGS Decommissioning Cost Study." Report 88/0090. 1988.

___. "Summary Report on the Social Costs of Emissions from Ontario Hydro." Report 90/0074. 1990.

___. "Full-Cost Accounting for Decision Making." Survey Team #4, Task Force on Sustainable Energy Development. 1993.

Ostro, Bart. "The Effects of Air Pollution on Work Loss and Morbidity." *Journal of Environmental Economics and Management* 10 (1983):371-82.

Ottinger, R.L., D.R. Wooley, N.A. Robinson, D.R. Hodas, and S.E. Babb. *Environmental Costs of Electricity.* Pace University Center for Environmental Legal Studies: Oceana Publications, 1990.

Oxman, A., G. Torrance, W. Garland, and H. Shannon. "Nuclear Safety in Ontario: A Comprehensive Framework for Decision-Making and a Critical Review of Quantitative Analyses." A Workshop on the Safety of Ontario's CANDU Reactors, Ontario Nuclear Safety Review, 1987.

Pohl, R. "Comments and Contents." *Environmental Impact Assessment Review* 3, 2-3 (June-September 1982):125-32.

Resnikoff, Marvin. "Waste Impacts of the Nuclear Fuel Cycle." Report presented to the Ontario Environmental Assessment Board, Ontario Hydro Demand/ Supply Hearings, December 1992.

Roberts, L.E.J., P.S. Liss, and P.A.H. Saunders. *Power Generation and the Environment.* Oxford University Press, 1990.

Torrie, R.D. "Operational Characteristics and Hazards of CANDU Nuclear Power Reactors." Expert Panel Workshop, Greenpeace Reactor Safety Study, Brunel University, London, 1986.

Trebilcock, Michael and Ralph Winter. "Economic Analysis of Nuclear Accident Law." Working Paper, Law and Economics Workshop Series, Faculty of Law, University of Toronto, April 1994.

United States Environmental Protection Agency. "Air Quality Criteria for Particulate Matter and Sulfur Oxides." Vol. 1. Washington D.C.: USEPA, EPA-600/8-82/029a, December 1982.

Williams, D.H. "The Expanding Decommissioning Focus: A Comparison of Coal and Nuclear Costs." *Energy Journal: Special Nuclear Decommissioning Issue.* (1991):295-304.

Comments by

STEPHEN R. ALLEN
KEITH DINNIE
DUANE PENDERGAST

STEPHEN R. ALLEN

In support of its Full Cost Accounting Program, Ontario Hydro retained Energy Research Group Inc. (ERG) in 1993 to perform cradle-to-grave calculations of the environmental externalities associated with the operation of their five nuclear power stations. These externality cost calculations considered full plant and fuel cycles – construction, upstream fuel, routine operations, accidents, decommissioning, intermediate/low-level waste, used fuel – and their associated radiological and fossil fuel pollutant emissions.

A simple and straightforward methodology rooted in plant-specific data or, in their absence, experienced engineering judgment was employed. Calculations that could not be justified by conservative estimates of uncertainty were avoided. Criteria or factors set by the Canadian Atomic Energy Control Board (AECB) or Ministry of Environment and Energy were used whenever possible. Where such values were not available, those established by the U.S. Nuclear Regulatory Commission (NRC) or Environmental Protection Agency (EPA) were used. While having different designs, light water reactors (LWRs) and CANDUs have similar design philosophies, containment systems, and waste disposal practices. Therefore, in several instances, appropriate LWR values were applied to the Ontario Hydro CANDU reactors.

A range (low, nominal, and high) of parameters and externality costs were calculated to account for differing opinions within the scientific community on matters such as radiation health effects and engineering error. Statistical human health effects, for both fatal and non-fatal cancers, were based on factors published by internationally recognized physicians and scientists serving on the United Nations Scientific Committee on the Effects of Atomic Radiation (UNSCEAR), the International Commission on Radiological Protection (ICRP) and the Committee on the Biological Effects of Ionizing Radiations (BEIR V). The ICRP values are used by Ontario Hydro and, therefore, represent the nominal case. USCEAR and BEIR V were the high and low cases, respectively. Subject to discounting, the value of a statistical human life was set at $4,776,209 and at $358,216 for statistical non-fatal injuries. Damage costs for fossil fuel pollutants tailored to the location of each station (NO_x, SO_2 TPM) were established by Ontario Hydro.

The results, on a weighted average basis, showed that the externality costs for nuclear power plants in the Province of Ontario range from 0.002 to 0.05¢/kWh. This very low range of value is not surprising given the robust design and rigorous regulation of nuclear power plants which effectively internalize externality costs (one reason why nuclear power plants cost so much to construct and operate).

ERG believes that this effort represents the most comprehensive calculation of nuclear power plant externalities performed to date. Moreover, the literature review presented by Professor Dewees does not contradict these calculations.

ERG found several areas of agreement and disagreement with Professor Dewees' conclusions with each briefly discussed below.

Areas of Agreement

In general, ERG found agreement with two key areas of methodology in the calculation of externalities and with the conclusions drawn from assessments of routine plant operation and uranium mill tailings.

Externality cost calculations should be site-specific. Characteristics such as plant age, vintage, design, historical performance, size, as well as surrounding population density and land use, have a significant impact on externality cost calculations. For instance, in its Ontario Hydro calculations, ERG found over an order of magnitude difference in the total externality cost for the Bruce and Pickering facilities – primarily as a result of historical performance and population density differentials. It is also important to note that the generic calculations common in the literature tend to overestimate externality costs – relying on values from multiple and often divergent sources and often reflecting the researcher's biases and misunderstanding of engineering principles.

Occupational health efforts are not externalities. Occupational impacts should not be considered an externality because worker risks are compensated through salary and other benefits, and mitigated through life insurance programs. Further recourse is available through the Canadian judicial process. Finally, through union rules, government regulation, and company training programs workers are continually made aware of the risks associated with radiation.

Radiation releases from routine plant operations cause negligible human health impacts. CANDU reactors emit small quantities of tritium, iodine, and other radioactive particulates during routine operation. Ontario Hydro, like all other utilities, regularly computes the resulting dose to the population surrounding its nuclear power plants. These population dose calculations are a function of the emission magnitude, population density, and weather patterns.

ERG regards the Pace Report (Ottinger, 1990, 372-88) value as an appropriate upper-bound estimate with the caveat that it reflects a nuclear facility located in an area with a high surrounding population density. In the analysis for Ontario Hydro's Full Cost Accounting Program, the externality costs for routine plant operations ranged from 0.0002¢/kWh (Bruce-B) to 0.0164¢/kWh (Pickering-A). These low externality costs (which assumed that emissions occur up to 50 years after cessation of station operations to account for the possible buildup of radioisotopes in the biosphere) are consistent with the findings of the recent U.S. National Institute of Health study of the populations surrounding all 111 nuclear plants in the U.S. – that no evidence of elevated cancer incidence existed.

Uranium mill tailings cause negligible public health impacts. The waste from uranium mining – called mill tailings – emits an inert, but radioactive gas called radon that is produced by the decay of Thorium-230, which has a half life of about 80,000 years. Radon emissions can be effectively reduced to background levels by covering the tailings with several feet of compacted overburden. In Canada, uranium mining operations, in compliance with AECB regulations, often bury the tailings in the closed mine or under deep lakes.

Professor Dewees, in reaching his conclusion that radon emissions do not pose a public health problem, assumes that the tailings always remain covered. ERG agrees that the health effects from mill tailings are negligible but reached this conclusion from a more conservative approach, wherein the tailings become fully exposed to the atmosphere after only 100 years. That is the approach that researchers such as Goffman use in arriving at huge environmental externality costs.

In his calculation Goffman (1981, 462) goes on to assume that the released radon becomes evenly distributed throughout the atmosphere of the northern hemisphere with a resulting annual dose of 4×10^{-9} rem per GW to the individual. This exceedingly small dose to an individual (which would typically be considered zero) becomes more significant when the four billion-person population of the northern hemisphere is factored in (16 person-rem per GW). When integrated over what in a mathematical sense is infinity (Th-230 would take over 400,000 years to decay to a trivial level) the population dose becomes 1,840,000 person-rem per GW – with an associated externality cost (for mortality only) of 0.40¢/kWh. ERG believes, however, that climatic changes and medical advancements must be accounted for when considering such long-time horizons. For instance, climatic changes could cause tailings to be covered with snow or water (which significantly reduces radon exhalation from tailings) and medical science could develop a cure or vaccine against cancer. A "medical discount" rate of just 2% would reduce the externality cost of 0.40¢/kWh to a trivial 0.017¢/kWh.

Areas of Disagreement

ERG strongly disagrees with the conclusions that Professor Dewees draws from his literature review of catastrophic reactor accidents and plant decommissioning. In these instances the paper respectively gives credence to positions that contradict engineering principles and considers issues outside the envelope of classic environmental externalities. The two areas of disagreement are discussed further below.

There is a "major source of divergence among reactor risk estimates centred on the probability of a Chernobyl-like accident at a CANDU *reactor."* Indeed there is a divergence of estimates almost exclusively because some researchers use detailed, reactor-specific probabilistic risk assessments while others misapply and judgmentally modify non-plant specific values. Indeed, as a wise man once said – "accident risk values are like political prisoners: if you torture them enough they will say whatever you want."

For its Full Cost Accounting Program, Ontario Hydro provided ERG with the frequency probabilities, population doses, and off-site financial damages for five categories of reactor accidents that result in the release of radioactive materials. These categories cover the full range of design basis and catastrophic accident consequences. (By definition, a catastrophic accident involves an extensive core meltdown, the joint or sequential failure of all plant safety features *and* a bypass or rupture of the plant's containment.) These accident frequencies and consequences were considered to be bounding estimates, pending the publication of the results of a station-specific risk assessment.

ERG found the catastrophic accident probability of approximately 1×10^{-6} per reactor year calculated by Ontario Hydro to be reasonable.

The largest off-site health and damage cost of about $15 billion was postulated to be at the Darlington site. This value is consistent with catastrophic accident consequences (health effects and property damage) published by the General Accounting Office (GAO), the investigative arm of the U.S. Congress, for U.S. LWRs in highly populated areas.

To compute externality costs, ERG essentially consolidated the five accident categories into an annual station health effect and property damage risk. This was done by summing the product of the population dose and off-site damage costs multiplied by the frequency of occurrence. In so doing, ERG increased all probabilities of occurrence by a factor of two to account for external events, such as earthquakes and tornadoes (an engineering "rule of thumb"). These calculations established the annual expected externality cost, which ranged from 0.000011 to 0.027¢/kWh.

One example of divergent risk is a study performed at Pace University (Ottinger 1990). This seriously flawed assessment of catastrophic accidents at nuclear power plants provides a contrasting example to the conservative method adopted for the Ontario Hydro calculations.

First, the Pace report incorrectly regards a core melt accident as being the same as a catastrophic accident. (Again by definition, a catastrophic accident involves an extensive core meltdown, the joint or sequential failure of all plant safety features, and a massive, early rupture of the containment.) In point of fact, a core melt is only one step in a catastrophic accident scenario. For perspective, the 1979 accident at the Three Mile Island (TMI) facility in Pennsylvania involved a partial core meltdown, but no containment failure and only trivial off-site consequences.

Second, consistent with this misconception, the Pace report uses the upper bound core melt frequency of one in 3,300 reactor years presented in the Reactor Safety Study (WASH-1400) as the frequency of a catastrophic accident. Interestingly, the Pace report also quotes an NRC statement that in the next 21 years there is a 45% chance of a core melt in U.S. reactors – which would mean a core melt frequency of one in 10,000 reactor years. Assuming the accident at TMI results in the five cancer mortalities that Roberts (1990) suggests, its externality cost would be between 0.0001 and 0.00004¢/kWh (U.S.) for the above range of probability. These core melt probabilities are incorrect, however, for exclusive application to a catastrophic accident as they do not account for the containment system. A detailed Probabilistic Risk Assessment (PRA) of Ontario Hydro's Darlington facility estimates a large release frequency of about 8×10^{-7}/reactor year – reflecting the plant's redundant safety systems and massive containment.

Third, this incorrect probability value was applied to the off-site consequences of the April 1986 accident at the Chernobyl No. 4 reactor. It is completely inappropriate to compare the consequences of the Chernobyl accident to those postulated for an LWR or CANDU. The Chernobyl RBMK (the Russian acronym for a boiling water reactor) type of reactor is composed of graphite, has no containment structure, and is subject to uncontrollable power excursions. In contrast, LWRs and CANDUs are moderated by water, have massive containment and redundant safety systems, and are not subject to uncontrollable power excursions. The RBMK design led to a larger, quicker and more dispersed release of radioactive material than could be postulated for a catastrophic accident at an LWR or CANDU plant. Indeed, the Pace report admits that "extrapolating from the results of the Chernobyl RBMK reactor for a domestic LWR is somewhat speculative."

Fourth, human health effect costs of $579 billion and property damages of about $58 billion from the Chernobyl accident are used to compute the externality costs for a LWR catastrophic accident. Essentially, the Pace Report appears to misuse the results of the BEIR V report and overestimates worldwide fatalities by a factor of three. In contrast, the GAO estimates that the largest off-site financial consequences due to a catastrophic accident at a U.S. LWR would be no more than $15.6 billion. This value, calculated for the Indian Point 3 reactor near New York City, includes $13.7 billion for crop damage, property loss, decontamination, and evacuation costs as well as $1.9 billion for morbidity and mortality effects.

Even if the Pace Report's off-site financial consequences of $637 billion for a catastrophic accident are applied to Darlington, the externality cost would only be 0.017¢/kWh given the Ontario Hydro PRA results cited earlier. In reality the off-site consequences from a catastrophic accident at Darlington would probably be less than those of the Indian Point 3 facility, which would reduce the externality cost another one or two orders of magnitude.

In short, ERG believes that accident externality costs – while higher than those of other segments in the plant cycle – are quite small when based on plant-specific, state-of-the-art probabilistic risk assessments. This is a conclusion that the Dewees' paper was unable to reach given its limited scope as a literature review, rather than a critical assessment.

Should the owner of the plant be incapable or unwilling to absorb decommissioning costs when due, public funds would need to be used, and decommissioning would in that case constitute an external cost.

Over the past two decades the cost of decommissioning a nuclear power plant in the United States increased in real terms by about 9% per annum. It is reasonable to assume that a similar effect is being realized in Canada for CANDU reactor decommissioning estimates. Thus, at the present time, there can be little doubt that the cost of decommissioning a nuclear power plant is indeed an escalating liability. (In the U.S., federal regulation mitigates this liability by requiring each plant to collect decommissioning funds on a kilowatt hour basis and place them in interest-bearing escrow accounts.

Moreover, the collection rate is periodically adjusted on the basis of revised decommissioning cost estimates.) In Ontario, any degree of failure to collect (and invest) decommissioning funds over an adequate basis is *not* by definition an "environmental externality," which is the topic of Professor Dewees' paper. And, because the taxpayers in Ontario are for all practical purposes the ratepayers as well, there is a strong argument to be made that uncollected decommissioning funds are not even a financial externality cost. In ERG's opinion, prudent regulation and planning requires that nuclear plants in Canada be internalized whether or not Ontario Hydro is privatized.

Finally, it is important to note that decommissioning does in fact have some environmental externality costs through the combustion of fossil fuel and release of dust-bearing radioactive materials in the actual process of dismantlement. For Ontario Hydro's nuclear power plants, decommissioning externality costs are on the order of 0.00004 to 0.005¢/kWh – which can be considered as trivial.

Conclusion

ERG believes that firm conclusions can be drawn regarding the externality costs of nuclear generation. Indeed it is possible to answer confidently the question that Professor Dewees proffers at the outset of his paper – "whether nuclear is bearing a lesser share of its full costs than is coal-fired generation, and if so what is the magnitude and significance of that shortfall?"

Simply stated, the Ontario Hydro Full Cost Accounting Program evaluation shows that when the full plant and fuel cycle is considered, the externality cost of a nuclear power plant is significantly less than coal-fired generation. Interestingly, the analysis also showed that accident externalities – when computed in an objective and methodical manner – are not significant.

Simply put, the robust design, rugged construction, methodical operation, and rigorous oversight of nuclear plants have internalized most environmental impacts. On the other hand, this has led to their high cost – an effect more than any other that has essentially stopped the construction of new nuclear power plants in North America.

BIBLIOGRAPHY

Gofman, J.W. *Radiation and Human Health.* San Francisco: Sierra Club Books, 1981.

Ottinger, R.L., D.R. Wooley, N.A. Robinson, D.R. Hodas, and S.E. Babb. *Environmental Costs of Electricity.* Pace University Center for Environmental Legal Studies: Oceana Publications, 1990.

Roberts, L.E.J., P.S. Liss, and P.A.H. Saunders. *Power Generation and the Environment.* Oxford University Press, 1990.

KEITH DINNIE

Nuclear power externalities can be divided into two types: those for which an adequate experiential base exists and those for which some form of predictive modelling is required to quantify them. In general, it can be stated that those in the first type are found to be very low in comparison to the internal costs of generation (partly because of the large amount of money invested by the industry in safety and emission control), whereas those in the second type, although still small when quantified using best-estimate models and data, involve sufficient uncertainty that they can become significant with respect to the internal costs if extreme assumptions and data are used. Examples of this second type are reactor accident risk, decommissioning, and radioactive waste disposal. This commentary devotes its attention to a brief discussion of the issues involved in their estimation.

Accident Risk

The expectation value of reactor accident risk is obtained from a product of accident frequency and a monetary valuation of accident consequence. The "pervasive information deficiencies" which accompany the estimation of accident risk is a rather back-handed compliment to the generally exemplary safety record of the nuclear industry in Canada and elsewhere.

As discussed in the main study, there are two general ways of estimating the frequency of large accidents; from the experience base or from simulation. The all-too-common practice of dividing the cumulative worldwide operating experience in reactor-years by the number of large accidents ignores a cardinal rule of reliability theory, namely, that all contributors to the experience base should be similar in terms of design and operational conditions. One thing that has been learned over the years is that there are considerable differences in such parameters between reactors in different countries and that the accident frequency is likely to be determined by local factors rather than by global averages. In any case, the inclusion of the accident at Three Mile Island in the data base is inappropriate because, as the paper acknowledges, the release of radioactivity to the environment was negligible from a human health or environmental damage point of view. The systems in place to contain the radioactivity were successful in doing so.

The estimation of accident frequency using probabilistic safety analysis takes into account all the known local factors but suffers from the impossibility of demonstrating that its predictions are based on a complete description of accident possibilities. Those who cite this drawback and claim that the actual risk must be higher than predicted using these methods often overlook the fact that the treatment of the known spectrum of events is usually quite conservative, providing a compensating margin. The uncertainty distribution associated with accident risk spans values both higher and lower than the expectation value.

Uncertainty also confronts the estimation of consequences. Much of the health impact consequences attributed to the Chernobyl accident arise from the practice of integrating microscopically small individual dose increments over large populations and applying the so-called "linear" dose response hypothesis, thereby requiring extrapolation of the dose response function far beyond its scientific base. This approach tends to produce apparently large numbers of statistical fatalities although, even if real, these impacts would be totally undetectable against the natural background radiation exposure and cancer incidence.

For the accident risk externality to be significant in relation to the internal cost (say, 10%), and assuming for this purpose an average external financial consequence of $10 billion, the frequency of a large offsite release of radioactivity would need to be of the order of 1 per 1000 reactor-years. This is far greater than observed or predicted, far greater than could be accepted by utilities from a business risk perspective, and far greater than would be acceptable to the regulator.

Radioactive Waste Disposal

Estimation of externalities from waste disposal must take into consideration the fact that any public radiation exposure and associated health impacts would only be expected on a timescale of many thousands of years into the future. Recognizing that any individual exposures are again likely to be tiny fractions of natural background, it is also necessary to make assumptions regarding the development of society and medical science over this period. Any non-zero discount factor introduced to account for societal development causes the externality to become vanishingly small because of the length of time involved. The alternative, that society's concern for or ability to manage these exposures will not change with time, seems untenable.

Decommissioning

The conclusion of this study is that the cost of decommissioning and waste disposal may represent the most serious problem with respect to the cost of nuclear power. First, this is not primarily an environmental component of cost although, as in construction of the reactor, some portion of the cost arises due to environmental protection considerations in the design.

Cost uncertainty arises not so much because of technological considerations but because of uncertainties in the policy and decision-making processes. In this arena, political positions and perceptions may be more important than the technical issues, which appear to be relatively straightforward.

Summary

The process of estimating external costs of electricity generation often relies heavily on extrapolation of dose response functions and other assumptions for which there is little empirical basis. Although the expectation values have mostly been found to be small, the resulting uncertainties are large enough to fuel an on-going debate as to the appropriate magnitude of the externalities for practical applications.

DUANE PENDERGAST

Professor Dewees has initiated a broad review of the environmental consequences of nuclear energy in comparison with coal-fuelled generation. He has looked at the full range of the fuel cycle from the point where fuel comes from the ground to decommissioning of plants.

The opening tone, even the title, of Professor Dewees' study faintly suggests a presumption that nuclear power is unfairly bearing a lesser share of social and environmental costs than coal-based electricity. However, as he proceeds through his step-by-step review he finds, in fact, that health and environmental costs of nuclear generation are probably very low relative to this competitor.

He has found widely varying information on the costs associated with accidents, largely based on the major difference in estimates of, and actual, consequences of accidents at the Three Mile Island and Chernobyl reactors. He concludes that these costs have little effect on the cost of electricity in Ontario, as design differences between the Chernobyl and CANDU reactors result in much reduced probability and consequence of major CANDU accidents.

Professor Dewees is less convinced that decommissioning and spent fuel disposal costs will be as low as current nuclear industry estimates. He points out that no Canadian reactors have been completely decommissioned. Canada has no spent fuel disposal facility. He believes that lack of practical experience with these activities poses a risk that actual costs will turn out to be greater than estimated. I can add only a little to the comments of Keith Dinnie and Stephen Allen on these issues.

The environmental assessment of the Canadian spent fuel disposal concept is getting underway in 1995. Technical and economic issues have been thoroughly studied. A good measure of practical experience has been gained through the development of research facilities. A huge amount of information is available to the public. I repeat, for the reader, Professor Dewees' reference to a document that provides a window to this information.[1] I speculate that when this concept is fully evaluated in this public forum that the nuclear industry may be accused of spending too much money on a simple problem.

Professor Dewees cites lack of experience with decommissioning as a basis for projections that decommissioning costs will exceed estimates made by the nuclear industry. I believe the reports he has reviewed do not fully reflect Canadian experience with decommissioning of reactors and other nuclear facilities. In addition to the experience with Gentilly 1, cited by Professor Dewees, decommissioning has commenced at the Douglas Point and NPD reactors in Ontario. Additional information on these projects, and others, is available from the literature.[2, 3, 4] In view of this experience base it seems unlikely cost overruns would be as great as the factor of 5 considered by Professor Dewees. Even with such a large cost overrun, Professor Dewees indicates the cost of decommissioning would not greatly increase the cost of electricity generated by nuclear power stations.

Professor Dewees mentions the potential externality effects from carbon dioxide emissions, which are expected by many climate experts to lead to global warming. The information he cites pays scant attention to costs which may be incurred. Erik Haites raised a concern that proliferation of independent power producers might make the control of carbon dioxide emissions more difficult. I would also like to remind participants of this issue and introduce some illustrations which show the contribution nuclear electricity is making to the reduction of carbon dioxide emissions in Canada.

Figure 1 provides information on the basic cycle of life on earth. Our planes, trains, automobiles, and power plants are becoming a significant part of earth's life based on processing and circulation of carbon containing materials. The information presented shows that annual carbon dioxide from fossil fuels is several percent of that estimated to be processed annually by earth's plant life.

Figure 2 shows the strong correlation between our use of fossil fuels and levels of carbon dioxide measured in the atmosphere.

Nuclear energy releases no carbon dioxide to the atmosphere during plant operation. Figure 3 shows the location of 32 CANDU and related reactors operating throughout the world. Eight more are under construction. Replacement of high carbon content fossil generation by one large CANDU can reduce Canada's carbon dioxide generation by as much as 1%.

The solid line of figure 4 shows Canada's total carbon dioxide emissions from 1950 to 1990. The dashed line diverging from the top indicates the extent of emissions had we not followed the nuclear option for electricity generation and had chosen the carbon-based fossil fuel option. This indicates Canada's 22 nuclear plants are reducing carbon dioxide about 15% from what might have been. This is equal to the magnitude of carbon dioxide emission reduction Canada has proposed in response to international concern with respect to global warming.

In closing I should just like to say I am pleased you are contemplating changes of regulations required to ensure competitive, efficient, and safe electricity supply in the 21st century. The need to develop regulations which realistically reflect all environmental costs is emphasized by several of the contributors to this volume. Professor Dewees' preliminary assessment indicates that nuclear energy does not carry with it significant external environmental costs. Should the costs associated with carbon-dioxide-induced global warming turn out to be significant, carbon-dioxide-free nuclear energy will flourish.

NOTES

1 Atomic Energy of Canada, "Summary of the Environmental Impact Statement on the Concept for Disposal of Canada's Nuclear Fuel Waste," AECL 10721 COG-93-11, 1994.

2 Paul Denault and L. de Pabrita, "Gentilly 1 Nears Static State," *Nuclear Engineering International* (August 1985).

3 G. Pratapagirl, et al., "Cost Estimate for Decommissioning a CANDU 6 MK1 Nuclear Generating Station," CNA/CNS Annual Conference, Ottawa, Ontario, June 4-7, 1989.

4 B. Gupta, "Progress of Decommissioning Projects in Canada," Nuclear Decom '92: Decommissioning of Radioactive Facilities, Institute of Mechanical Engineers, London, February 17-19, 1992.

Figure 1
Estimated Global Carbon Cycle Pools and Fluxes (Circa 1990) (Major Components)

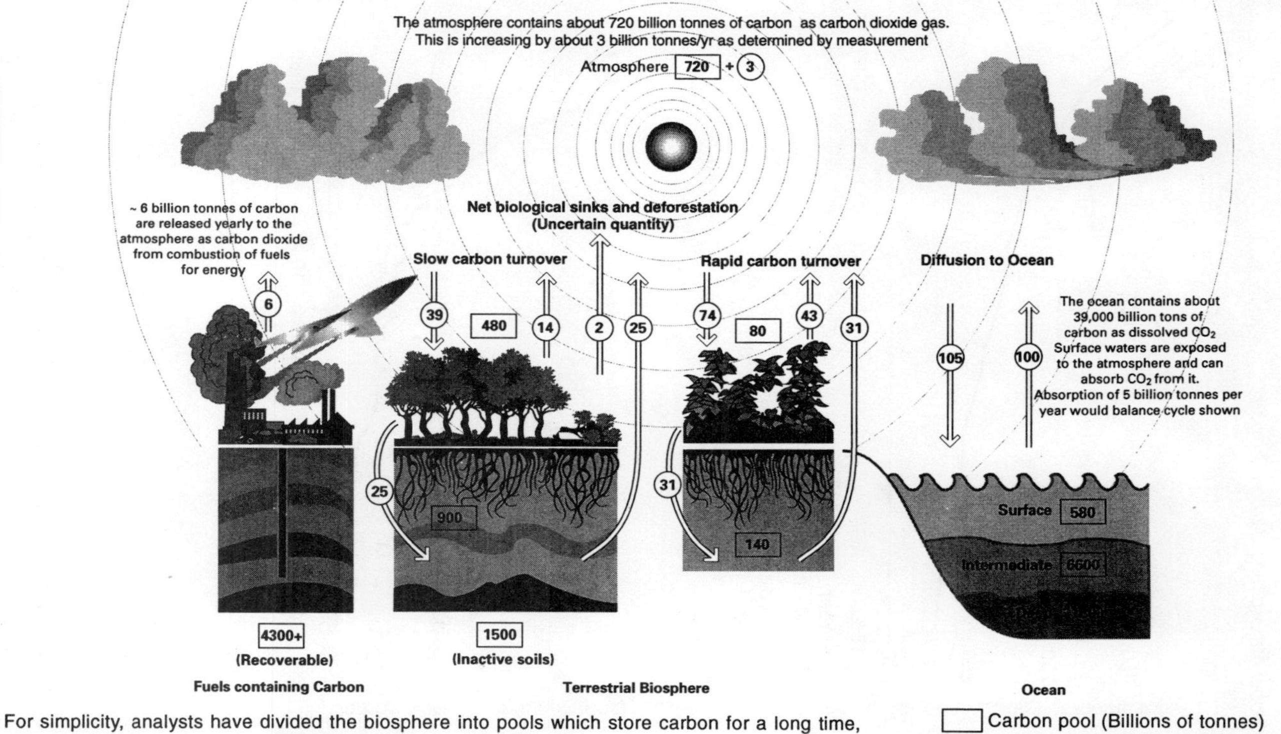

For simplicity, analysts have divided the biosphere into pools which store carbon for a long time, and pools which store and release carbon over a short time

Figure 2
Atmospheric Carbon Dioxide and Emissions

Global Atmospheric Carbon Dioxide

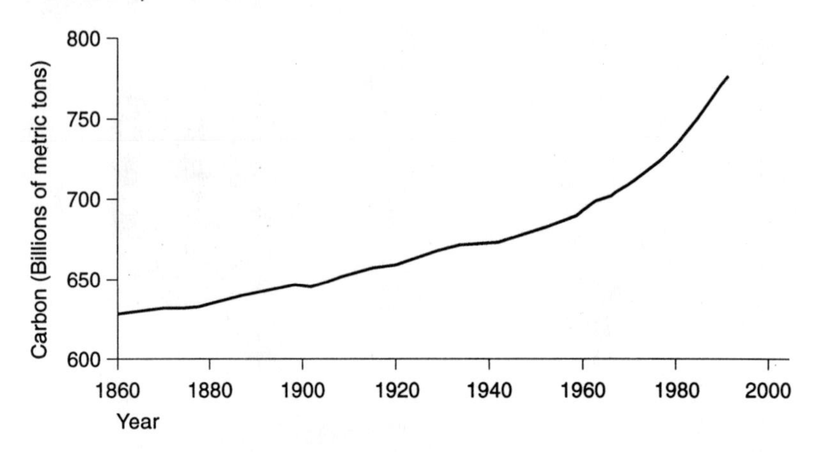

Annual Global Carbon Dioxide Emissions from Fuel

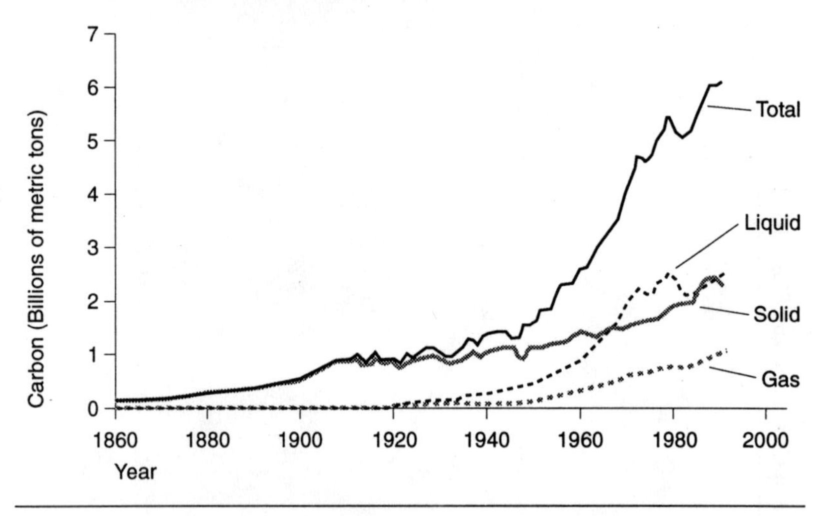

Figure 3
CANDU Technology Contribution to Cutting Global Atmospheric Carbon Dioxide

Electricity generated by conventional alternatives to a typical large CANDU power plant releases 1.6 million tonnes of carbon as carbon dioxide gas each year. This is over 1% of Canada's total carbon dioxide emissions.

The earth's atmosphere is well and quickly mixed. A nuclear power plant anywhere in the world is thus equally effective at reducing carbon dioxide in Canada's share of the atmosphere.

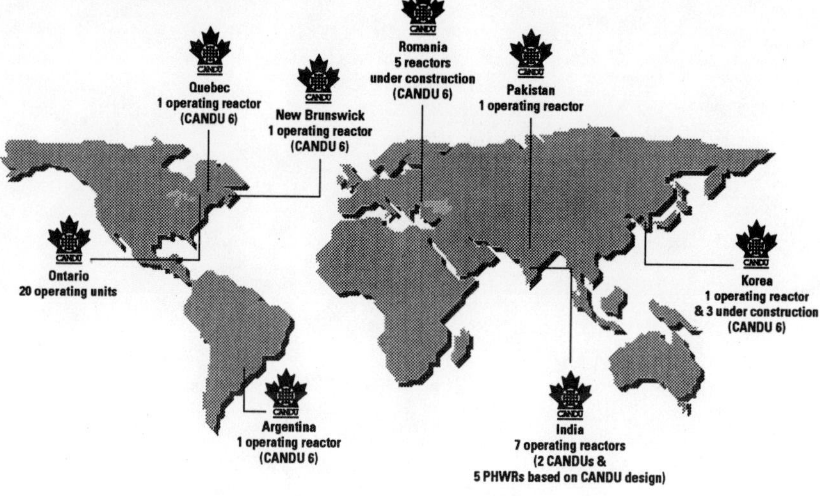

Figure 4
Nuclear Energy's Contribution to Reducing Canada's Carbon Dioxide Emission

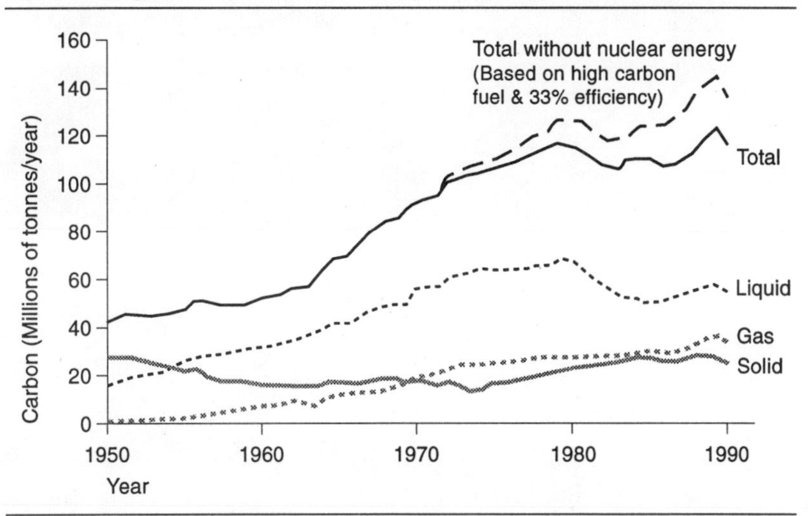

The Distribution of Electricity in Ontario: Restructuring Issues, Costs, and Regulation

ADONIS YATCHEW

I. INTRODUCTION

The purpose of this study is to provide an analysis of the distribution segment of the Ontario electricity industry – its structure, costs, and regulation. The current world-wide debate on the restructuring of electricity industries is focused on introducing competition to the generation segment of electricity industries. While it would appear that a broad range of alternative distributor configurations could co-exist with deregulation and competition in generation, there are certain issues which need to be resolved either a priori or conjointly. Among these are whether any economies of scope will be foregone as a result of restructuring, the obligation to serve in a restructured industry, and the delivery of energy services. Furthermore, for the distribution segment itself, a number of issues need attention, in particular, the degree of integration that is appropriate for it, the regulation of distributors, and whether or not distributing utilities should be privatized.

Historically, the institutional and regulatory structure of the electricity industry as a whole has been driven largely by technological and economic factors. From the beginning of the century to the early 1970s, technological advances resulted in improved fuel efficiencies, increasing scale economies, and a systematic decline in the price of electricity. Thus, producers, public or private, in the U.S. and in Canada, enjoyed a favourable relationship with their regulators and there was little interest in, or need for restructuring of what were typically vertically integrated monopolies.[1]

Since that time, increasing costs and prices, failure to accurately predict declines in demand growth, and the availability of cost effective substitutes have led to pressures to restructure. Indeed, there is widespread belief that restructuring will lead to competition in generation and lower prices through improved capital decision making and enhanced incentives for reducing operating costs; a reduced need for regulation particularly in the generation segment; and, increased accountability and responsiveness to customers. In some jurisdictions, the belief that change in ownership – from public to private – could benefit users and/or raise revenues for the government has provided additional impetus to restructuring.

A number of important technological, economic, and structural features shape the landscape of today's evolving electricity industries:

- maximum scale efficiencies in generation were reached in the early 1970s;
- generation of electricity is no longer a natural monopoly;
- transmission and distribution of electricity continue to be natural monopolies;
- natural gas is not only a cost effective substitute for electricity in many end-uses but also a cost-effective fuel for the generation of electricity;
- technological innovation in information technology has created the possibility of creating wholesale and possibly even retail markets in electricity supply; new energy services are also likely to emerge;
- the industry continues to be capital intensive.

The statistical data that could shed light on the various aspects of the restructuring debate, both abroad and in Ontario, is of varying quality. Nevertheless, the best available empirical evidence and theoretical arguments generally support the following propositions.

First, separation of monopolistic segments of the industry from those amenable to competition is one of the most important steps that can be taken towards institutionalizing competition in generation. Whether competition is to take place at the wholesale level or at the retail level, its implementation is easier if transmission and distribution assets reside in corporations that are separate from those owning generation assets. Any benefits arising from economies of scope in a vertically integrated utility are, in all likelihood, outweighed by the benefits of separation. Such benefits include increased potential for competition and transparency of costs.

Second, private electric utilities with monopoly power are not generally more efficient than public sector monopolies. This is perhaps an unusual conclusion given the fundamentally differing nature of incentives present in private vs. public utilities, but it is supported by a number of empirical studies. Thus, privatization of distribution or transmission assets is unlikely to lead to net benefit to the electricity customer.

Third, the arguments favouring large-scale amalgamation of distributors are weak at best. While there may be efficiency gains to be exploited, these gains are yet to be demonstrated statistically. Furthermore, there are important local accountability benefits that could be lost. Indeed, a decentralized system can confer important benefits: it can be exploited to reduce informational asymmetries and it allows for the possibility of yard-stick competition.

Because of the interrelated nature of the issues involved, an analysis of the distribution segment of the industry cannot be performed without assumptions about how the other segments of the industry would be restructured. Section II therefore briefly sets out industry-wide restructuring issues and principles, as well as certain directions that such restructuring could take. The section also includes a discussion of distribution in several other jurisdictions. Section III provides a discussion of various issues relevant to the distribution sector in Ontario.

II. BACKGROUND AND ANALYTIC FRAMEWORK

Dimensions of Restructuring

The electric utility industry serves four central *functions* – generation, transmission, distribution, and energy supply and services – functions which can be performed by separate entities. Given these functions, there are several dimensions along which choice can take place: the degree of *integration*, both across functions and within functions, the degree of *competition* that is introduced, the mix of public/private *ownership* and the nature and extent of *regulation*.

INTEGRATION

In Ontario the electricity industry is partially integrated with generation, transmission, and some distribution and energy services performed by one firm, and most local distribution and additional energy services performed by municipal utilities. The preponderance of electricity industry assets are in the public sector.Whether or not vertical separation of functions should be effected is driven by two opposing considerations. On the one hand, there may be economies of scope arising from having industry functions incorporated in a single entity. On the other hand, there is considerable risk that an integrated industry will preserve the incumbent's ability to create and sustain barriers to competition. There is relatively little formal econometric investigation of the economies of scope in electricity.

Two studies using U.S. data,[2] find evidence of such economies, though there are difficulties with each study.

More recently, arguments have been raised that demand side management, integrated resource planning (local or system wide), distributed generation decisions and optimal trade-offs between distribution, transmission, and generation investments are best performed within an integrated utility. The essence of the counter-argument is that so long as prices for alternate resources are set correctly, a vertically separated industry will make optimal decisions – acquiring additional supply if that is cheapest, adding to distribution capacity, or investing in demand side management (DSM) if those options are the most desirable.

In addition, it has been argued that some degree of centralized decision making is desirable if environmental issues are to be addressed properly.[3]

On the other hand, the empirical evidence on anti-competitive and potentially inefficient behaviour of vertically integrated monopolies is substantial. The history of the gas and oil pipeline industries is replete with examples of attempts by pipeline owners to favour their own sources of supply. More recently, separation or unbundling of transmission from supply has been the direction taken in the U.S. natural gas industry. Such considerations also formed the basis of divestiture in the U.S. of the Bell operating companies from AT&T. (Local service was seen to create a bottle-neck monopoly.[4]) Indeed the prima facie case that a public or private corporation with monopoly power will act in a fashion so as to preserve and advance its interests, which include retention of monopoly power, is strong.[5,6] Thus, the regulatory blueprint has often involved separation of the portions of the industry that are potentially competitive from those that are natural monopolies.

The logical implications of these arguments for restructuring in Ontario are twofold: first, transmission should be separated from generation. Second, distribution, to the extent that it continues to reside within the main generating company, should also be separated.

The most important advantage flowing from segregation of functions is that this approach separates the interests of the owners of transmission and distribution from the interests of the owners of generation capacity. The alternate approach – legislated access to transmission facilities of vertically integrated utilities – requires constant regulatory supervision to protect against self-dealing.

The creation of separate transmission companies has been undertaken in the U.K., New Zealand, Norway, and Sweden. In each case transmission was in the public sector. On the other hand, in the U.S., where most transmission (and generation) is in the private sector, competition was institutionalized by legislating transmission access in the 1992 Energy Policy Act. To date, no divestiture has been ordered, though the California Public Utilities Commission is investigating the possibility of vertical separation. There is a simple explanation for the differing approaches. Divestiture, in the presence of private property rights, is a difficult, costly and drawn-out process. However, when transmission is in the public sector, separation can be effected relatively more easily. Thus, it is not surprising that the U.S. has taken a different path from most other countries.

COMPETITIVE STRUCTURE

Given that *transmission* and *distribution* continue to be natural monopolies, there is little prospect for the introduction of *direct* competition in these segments. However, the presence of many distributors in Ontario creates the possibility of introducing a surrogate to competition into this segment of the industry through "yardstick competition," which we will discuss further below.

On the *generation* side of the business, a range of models with varying degrees of competition have been discussed at great length though there is insufficient experience to determine which version will prove the most roadworthy. If the grid becomes a common carrier, suppliers and purchasers of electricity could enter into bilateral transactions of their own choosing. Competition would take place at the wholesale level if purchasers were restricted to distribution companies and large industrial customers. In the most ambitious version of this model – retail competition – even residential customers could choose suppliers in much the same way that they now choose long-distance providers. Variants of pool concepts have been proposed to complement or complete such trading arrangements. In this context, a healthy electricity market will undoubtedly require the presence of long-term contracting as well as medium-term and spot markets.

An alternative to transmission access would be a model where the transmission company acquires supply on a competitive basis to meet all contracted demands. (We will refer to this model as "monopsonistic competition.") There are advantages and disadvantages to each approach.

The major argument favouring widespread transmission access is that it likely represents the quickest route to market discipline. However, under the retail competition version of this model, the obligation to supply would effectively disappear. (There would nevertheless be political pressure to try to retain some assurance of supply.) Furthermore, the transmission access approach would also likely introduce greater uncertainty into the system, at least in the short run. Nor could extra-provincial entities be prevented from participating in the market. This, in turn, could complicate an *orderly* transition to competition in generation without the creation of stranded assets and the imposition of unfair burdens on "captive'" customers.[7]

Finally, it seems likely that introduction of transmission access is essentially irreversible. Once a large number of private contracts are in place, it would be difficult to revert to "monopsonistic competition." On the other hand, the transition from "monopsonistic competition" to a model involving transmission access would be much easier to accomplish. As new supply becomes necessary, individual customers, (e.g., distributing utilities and/or large users) could be permitted to acquire their own supply.

Ultimately, there are two potential pitfalls associated with "monopsonistic competition." First, there is a question as to whether adequate market discipline would be present. Second, there is a risk of government influence on grid company choices.[8]

OWNERSHIP

Conventional property rights arguments state that privately owned companies are likely to be more efficiently run than those in the public sector. Public ownership is far too diluted to provide an adequate lever on company performance. In centrally planned economies, the absence of both "exit" and "voice,"[9] results in woefully inadequate performance of state firms.

In mixed economies, there are typically two breeds of companies in industries that are natural monopolies: publicly owned firms and regulated private enterprises. The governance structure of each is subject to its own vulnerabilities, limitations, and ultimately failures. Public corporations do not have the profit motive but often appropriate monopoly rents in other ways, such as to labour or for policy purposes.[10] Given their more direct accountability to political bodies, residential and commercial rates of public enterprises are often lower. Private regulated enterprises, particularly those that are regulated on a rate-of-return basis, are subject to distortions such as the Averch-Johnson effect.

For electrical utilities, the empirical evidence on the impacts of ownership on performance is equivocal.[11] Analysis of cost and price data of U.S. public and private electric utilities suggests no striking cost differences between the two types of entities. (Indeed, publicly owned utilities appear to have a price advantage.) The main reason is that in both cases, suppliers are monopolists and hence are not subject to market discipline.

These statistical conclusions are consistent with the view that monopolies, be they private or public, are used to extract monopoly rents to more or less the same degree, but the benefits are then conferred on different groups. The relevance of these results to the present discussion is that privatization of distribution assets (or for that matter, transmission assets), is unlikely to convey any significant societal benefits.

REGULATION

Historically, Ontario Hydro has been controlled by a changing and ad hoc set of quasi-regulatory instruments. For example, no single regulatory agency has had responsibility for reviewing both capital plans and revenue requirement proposals. Municipal utilities enjoy light-handed regulation by Ontario Hydro but are also responsible to their local constituencies. (In the U.S., most municipal utilities are not regulated but are controlled by municipal bodies.) On the generation side of the business, competition should ultimately obviate the need for regulation of costs and capital plans.[12] Transmission and distribution, on the other hand, would continue to require some form of regulatory oversight.

Given that municipal utilities are accountable to their local constituencies, it can be argued that regulatory authority should also reside at this level. Distribution is a local service, under local control, and like other municipal services, is appropriately within the jurisdiction of municipal authorities. However, asymmetries of information between the would-be regulator (in this case a municipal body) and the regulated (the distribution company) would also be present here. Effective regulatory oversight would be further complicated by the fact that the cost of distribution is currently bundled into the price of electricity.

A potentially useful mechanism for enhancing control of distributor costs while maintaining local control is "yard-stick competition." Under this approach, distributor margins could be compared across jurisdictions, taking local consideration into account, in order to establish cost-minimizing budgets and "best practices." Indeed, such comparisons are already taking place on an informal basis amongst utilities – when one distributor announces a zero rate increase, others face pressure to follow suit. (The pressure comes both from political sources as well as more directly from industrial customers.)

The presence of a large number of distributing utilities in the Province enhances the potential for a sound statistical basis for comparison.[13] Statistical techniques such as production frontier analysis or data envelopment analysis could be applied effectively to these data. (These are the econometric and operations research equivalents to "best practice" analysis.[14])

In summary, the ideal would be to replace regulation with market discipline ("competition where possible, regulation where necessary"). Such may very well be possible within the generation segment of the industry. For the distribution segment, a degree of competitiveness could be injected through the introduction of yard-stick competition. This approach could enhance local control of distribution costs as well as forming part of a scenario where regulation of distribution costs formally devolves to the municipal level. Transmission, however, would need to be regulated at the provincial level.

Finally, it is worth pointing out that the differing scope, degree, and nature of regulation that is appropriate for generation, transmission, and distribution constitutes another argument for vertical separation of the industry.

Distribution in Other Jurisdictions

UNITED KINGDOM

In the United Kingdom, distribution is in the hands of 12 privately owned regional electricity companies (RECs), who jointly own the National Grid Company (NGC) – the transmission company. Generation, transmission, and distribution have been separated.[15] Distribution costs are regulated on a price cap basis relative to the rate of price inflation (RPI). The control applies to income per kilowatt hour distributed to all customers.

Distribution costs comprise about 22% of the final costs of electricity. It should be noted, however, that systems in England and Wales normally receive power at voltages at or below 132 kV. Staff costs comprise about one-third of operating costs with repairs and maintenance a similar proportion.

In the last several years, most profits in the electricity business have come from the distribution segment.[16] As a result, the regulator performed extensive statistical analyses of distributor costs in order to attempt to establish reasonable benchmarks. These analyses took into account such factors as size, customer mix and density, and nature of capital stock. However, because of the small number of data points, considerable judgment needed to be exercised in coming to a conclusion.[17]

As a result of these investigations, distribution charges are being reduced dramatically over the period 1995-2000. Utilities are expected to absorb cost reductions equivalent to reductions of 5.5% to 7.5% *per year* over this period. In addition, distributor capital expenditure plans are being reduced by about 10%.

There is insufficient evidence at this point to determine to what degree the U.K. experiment will be successful and whether there will be a long-term decline in the price of electricity. However, two lessons appear to be emerging. First, on the generation side, competition would have been better served if there were a larger number of generating companies competing against each other. Second, subsequent to privatization, the distribution component of the electricity price has increased to the point where more stringent regulation of this segment of the industry has become necessary.

UNITED STATES

In the United States the preponderance of electricity is provided by vertically integrated investor owned utilities (IOUs). The remainder is distributed by publicly owned utilities (mainly municipal districts) and rural electric cooperatives. Private utility rates are regulated on the basis of a "reasonable rate of return on capital."

Distribution in the U.S. comes in several forms. The vast majority (over 75%) of end-use customers are served by vertically integrated investor owned utilities. About 14% are served by publicly owned systems, many of which are municipal distributing utilities. The remainder are served by rural cooperatives.

As a result of their small size, many municipal utilities are members of Joint Action Agencies. These were formed by public power utilities which were too small to negotiate effectively in bulk power markets or to themselves build generation and transmission projects. Formation of Joint Action Agencies permitted members to pool operating and financial risks and to aggregate demand so as to make feasible the construction of power stations. Over time, these Agencies have grown to provide a broad range of services to member utilities including power supply planning and forecasting, plant operation and maintenance, fuel purchasing, demand side management, financial services, and training (see Table I below). A variety of cost recovery methods are employed depending on the specific nature of the service (e.g., fee for service, recovery through wholesale rates, subscription fees).

While Joint Action Agencies are coming under increasing competitive pressures, their relevance to the present discussion is that small utilities need not necessarily amalgamate in order to avail themselves of many of the services provided in-house within larger utilities.

NORWAY

Distribution of electricity in Norway is performed by over 200 companies owned by municipal or inter-municipal authorities. Private companies are not involved in distribution. Distribution companies receive franchise authorization within a specific geographic area. In effect, they are local natural monopolies which make their own decisions regarding investment, financing, and pricing of electricity. Distributors are locally controlled. Some distributors are vertically integrated but the majority purchase electricity through long-term contracts.

In order to isolate distribution costs, a recent Norwegian study[18] focused on a sample of about 100 Norwegian companies engaging only in distribution. The average number of customers was about 11,500 with the largest serving over 290,000 and the smallest serving 655.

The study found significant returns to increasing the *density* of customers within a fixed distribution area but little in the way of returns to scale. They estimated optimal distributor size to be about 20,000 customers. However, distributors serving 5,000 customers exhibited only slightly

Table 1
Services Provided by U.S. Joint Action Agencies

Power Supply	*Financial Servcices*
backup generation	cost of service studies
control area services (sched. and disp.)	customer billing and collection
load forecasting	insurance
plant operation and maintenance	inventory management
power pooling	investment scenarios
wholesale power contract negotiation	joint purchasing
power pooling	meter reading
	project financing
	rate design
Engineering and Project Management	*Training*
environmental compliance support	computer operator
joint system planning	dispatcher
plant design and construction	lineman
plant operation and maintenance (T&D)	power plant operator
power quality audits	*Other*
renewable energy projects	testing (cable, equipment)
research and development	publications
Fuel Purchasing	legal assistance
Demand Side Management	lobbying
energy conservation programs	event management
integrated resource planning	personnel, administrative
load management	
peak shaving	

Source: American Public Power Association

higher unit costs. Furthermore, there was some evidence of increasing unit costs, i.e., decreasing returns to scale, beyond the 20,000 customer level.

III. CHANGE IN THE DISTRIBUTION SECTOR: EVOLUTIONARY OR REVOLUTIONARY?

Distribution in Ontario

The distribution system in Ontario consists of over 300 municipal utilities which are owned and operated principally by municipal commissions. In most cases, municipal utility boundaries are coincident with the municipal boundaries. About one-third of municipal utilities also administer other services, such as water. Approximately 70% of the power delivered in Ontario is delivered by these municipal utilities.

There are also several private and other companies that distribute power (Gananoque Light and Power Limited, Great Lakes Power Limited, Canadian Niagara Power Company Ltd., Orillia Light and Power) as well as some local private distribution by a few large resource industry companies.

Ontario Hydro is responsible for distribution to the remaining regions of the province. This includes areas where the local municipality has not yet exercised its right to local distribution, as well as the vast rural regions of the province. Rural distribution is subdivided into 13 regional systems. Ontario Hydro also distributes to approximately 100 large direct customers which have demands exceeding 5 MW.

The demarcation line for distribution in Ontario is power delivered below 50 kV. In fact, distributors normally receive service at either 44 kV or 13 kV.

The majority (about 70%) of municipal utility commissioners are elected. Others are appointed by municipal authorities. The authority to create commissions is granted under the Public Utilities Act.

Municipal commissions have the responsibility of overseeing the activities of the municipal utility. Municipal utilities serve 2.8 million customers. Ontario Hydro serves 925,000 retail customers and the above-mentioned 100 or so large direct industrials.

A recent joint study by the Municipal Electric Association and Ontario Hydro (1994) found no significant unexploited returns to scale in the distribution sector.[19,20] Two points should be emphasized. First, the results are preliminary. Second, the analysis was based on 1991 data. Since that time there has been considerable restructuring of the distribution segment within Ontario Hydro and likely efficiency improvements both at Ontario Hydro and within municipal utilities.

A number of issues merit consideration when assessing restructuring within the distribution sector. These include economic efficiency – both static and dynamic – and the regulation of costs; the ability of distributors to fulfil new responsibilities if generation/transmission were to be restructured – in particular, the obligation to procure adequate supply and deliver energy services; and means by which risks to utilities can be mitigated.

Economic Efficiency and Regulation of Costs

Competition has positive effects both in a static sense and in a dynamic sense. At any given point in time, given the state of technology, competitive pressures lead to cost minimization by firms. Over time, such pressures also lead to innovation. Distribution, however, is a local natural monopoly so that direct competition is, at this point, infeasible. Other mechanisms must be relied upon to promote efficiency.

Yard-stick competition, through the use of systematized cost comparisons and benchmarks, is one such alternative. The presence of a large number of utilities of (almost) all size ranges affords important informational advantages – there is a large data base available for estimating benchmarks as well as a large number of natural experiments. This in turn would help to overcome the usual informational asymmetries between the utility and those to which it is accountable, and it would enhance the potential for establishing a sound statistical basis for yard-stick competition.

Province wide regulation, however, would undermine local control of distributing utilities. A plausible alternative would be to have this regulatory responsibility, which currently resides with Ontario Hydro, devolve to municipalities or their municipal commissions. We have argued that functions within the industry should be separated. This would not only allow generators to compete on a more level playing field but it is also desirable from a regulatory point of view. If the validity of separation arguments is to be taken seriously, then separate entities, which would fulfil the current obligations of distributing to the rural sector, would need to be created.

Distributor Responsibilities in a Restructured Industry

If competition is introduced within the generation segment, generators would no longer be responsible for predicting future demand and acquiring adequate supply to meet such demands lest their ability to compete is impaired.

Turning to the menu of choices for competition, if the "monopsonistic competition" model is adopted, then there are two natural candidates for fulfilling the obligation to secure adequate supply: the transmission company and the local distributing utility.

Even in the presence of transmission access, as long as competition remains at the wholesale level, (where distributing utilities and perhaps large industrial customers could choose suppliers), the obligation to secure supply would still remain, but would now fall naturally to distributors. Only if competition propagates to the retail level, would this obligation disappear, or more accurately, devolve to individual customers.

As the industry evolves, it is likely to offer an increasing range of energy services. Statistical analysis of past data cannot be used to determine whether under such circumstances, a new (multi-product) production function will exhibit increasing returns to scale and scope large enough to justify large-scale amalgamation amongst distributing utilities.

However, at such a time, it is precisely the presence of such increasing returns that would create incentives for amalgamation.

In the alternative (and this applies equally to distributor obligations to supply), cooperative efforts amongst utilities, similar to U.S. Joint Action Agencies, could be exploited for delivery of energy services. Indeed, the Municipal Electric Association is a cooperative providing a range of services.

Finally, restructuring of the industry will not only increase the responsibilities of distributors, but will also likely increase the risks they face. While not the focus of this paper, it is recognized that mechanisms exist for mitigating such risks. Among them are the pooling of costs of purchased power through a power pool or through joint ventures by groups of utilities. Contractual ties, both vertical and horizontal, would also be an indispensable tool for risk mitigation.

CONCLUSIONS

Three important themes, and no doubt more, should be kept in mind throughout the restructuring debate. First is the notion that separation of monopolistic segments of the industry from potentially competitive segments is desirable except under the most exceptional circumstances. Such separation not only reduces the risk of cross-subsidy and self-dealing but increases transparency, or conversely reduces the asymmetry of information between the utility and outsiders, be it a regulator, government officials or the public at large.

The second is that private property rights, once created, are difficult to reverse. This is relevant in two important decision areas: in the debate on privatization – e.g., divestiture of transmission would be difficult to accomplish if Ontario Hydro were privatized as a vertically integrated utility; and in the debate on whether to allow transmission access which would result in the proliferation of contractual rights and obligations between buyers and sellers.

The third theme is that decentralization has important benefits associated with it. (One need not belabour the obvious relationship of this proposition to pro-market arguments.) In the present context, a decentralized distribution system reduces informational asymmetries and allows for the possibility of yard-stick competition. Furthermore, accountability of distributors to their municipal governments ensures that distributors are sensitive to local needs. Conversely, centralization of the distribution function within one or a small number of distributing entities would increase informational asymmetries, (who would be the comparison group?) and would reduce local accountability. It would also reduce the flexibility of tying employee remuneration to the local cost of living. Large-scale amalgamation would lead to pressures to increase wages within a distribution company to reflect the highest cost of living centre within its geographic franchise area.[21]

It may be the case that ultimately, small distributors, of which there are many within the province, will not be in a position to negotiate, either individually or cooperatively, the requisite power supply contracts or to provide some future array of energy services. They may choose, as a result, to amalgamate or to contract out such responsibilities. However, allowing such restructuring to *evolve* as a natural response to economic and political forces, has the added advantage that it is more likely to lead to superior alliances – precisely because they are in response to *local* economic and political forces – rather than artificial ones.

ACKNOWLEDGEMENTS

Particular thanks are due to the Municipal Electric Association and to Ontario Hydro for providing access to data and to the latter for direct support of the project. Thanks are also due to the American Public Power Association and to Professor Stephen Littlechild, Director General of Electricity Supply in the United Kingdom.

NOTES

1 For an historical analysis of the relationship between technology and regulation in electricity see R. Hirsh, *Technology and Transformation in the American Electric Utility Industry* (Cambridge University Press, 1989); R. Hirsh, "Regulation and Technology in the Electric Utility Industry: A Historical Analysis of Interdependance and Change," in *Regulation: Economic Theory and History,* ed. J. High (University of Michigan Press, 1991).

2 J. Henderson, "Cost Estimation for Vertically Integrated Firms: The Case of Electricity," in *Analyzing the Impact of Regulatory Change in Public Utilities,* ed. M. Crew (New York, 1985); D.L. Kasserman and J. Mayo, "The Measurement of Vertical Economies and the Efficient Structure of the Electric Utility Industry," *Journal of Industrial Economics* (1991).

3 See e.g., Armand Cohen and S. Kihm, "The Political Economy of Retail Wheeling, or How to Not Re-Fight the Last War," *Electricity Journal* (April 1994):49-61.

4 Robert Crandall, "Halfway Home: U.S. Telecommunications (De)Regulation in the 1970s and 1980s," in *Regulation: Economic Theory and History,* ed. Jack High (University of Michigan Press, 1991).

5 For discussions of the elements of the objective function of private and public firms see e.g., Dave P. Baron and R. Myerson, "Regulating a Monopolist With

Unknown Costs," *Econometrica* 50, 4:911-30; D. Besanko, "On the Use of Regulatory Requirement Information Under Imperfect Information," in *Analyzing the Impact of Regulatory Change in Public Utilities,* ed. M. Crew (Lexington, MA: Lexington Books, 1985); Jean-Jacques Laffont and Jean Tirole, "Using Cost Observations to Regulate Firms," *Journal of Political Economy* 94, 3 (1986):614-41; L. Waverman and A. Yatchew, "Regulation of Electric Power in Canada," forthcoming in *International Comparisons of Electricity Regulation,* eds. R. Gilbert and E. Kahn (Cambridge University Press, 1996).

6 Note that any discussion of the efficiency and benefits of competition assumes that firms will act in their own (or ideally in their owners') interests.

7 With the current excess capacity in Ontario, however, competition is introduced and consideration must be given to the risk of creating stranded assets. (A similar problem arises in many U.S. and other jurisdictions.) The societal inefficiency results from the fact that new generation sources compete against the average costs of electricity from existing generation. On the other hand, the societal (economic) cost of production from existing generation is the marginal cost, which is generally much lower than average costs of production.

8 For economic efficiency reasons, economists generally prefer markets where there are many buyers and sellers. In this case, the presence of multiple players on both sides of the market would reduce the risk of political interference.

9 See A. Hirschman, *Exit Voice and Loyalty: Responses to Decline in Firms, Organizations, and States* (Cambridge, MA: Harvard University Press, 1978) for definitions of these terms.

10 See Waverman and Yatchew.

11 See J. Kwoka, "Pricing in the Electric Power Industry: The Influence of Ownership, Competition and Integration," Harvard Institute of Economic Research Working Paper, 1995; A. Boardman and A. Vining, "Ownership and Performance in Competitive Environments: A Comparison of the Performance of Private, Mixed and State-Owned Enterprises," *Journal of Law and Economics* (1989); S. Atkinson and R. Halvorsen, "The Relative Efficiency of Public and Private Firms in a Regulated Environment: The Case of U.S. Electric Utilities," *Journal of Public Economics* (1985); R. Fare, S. Grosskopf, and J. Logan, "The Relative Performance of Publicly Owned and Privately Owned Electric Utilities," *Journal of Public Economics* (1985); S. Peltzman, "Pricing in Public and Private Enterprises: Electric Utilities in the United States," *Journal of Law and Economics* (1981).

12 While there are important regulatory issues relating to the transition period during which competition in generation would be implemented, these are not the focus of this paper.

13 In contrast, the U.K. system has 12 regional distributors with two additional in Scotland, but the latter remain vertically integrated with generation. A statistical comparison of costs that properly adjusts for differences in local conditions,

such as customer density, age of equipment, customer mix, proportion of below-ground distribution lines, and local wage costs becomes much more difficult.

14 See R. Green and A. Jackson, "Electricity Distribution: A Comparative Efficiency Study of the RECs," manuscript, Department of Applied Economics and Fitzwilliam College, Cambridge (1994).

15 Third party access to the transmission grid is permitted. Generators can sell to the pool or directly to individual large customers. The pool is operated by the NGC and electricity is priced on the basis of bids submitted one day in advance. Statutory responsibilities of the electricity regulator include protection of customers with respect to price, continuity of supply, and quality of service. The regulator must ensure that all reasonable demands for electricity are satisfied, that electricity companies can finance their activities and must promote competition. Other responsibilities relate to efficient use of electricity, environmental protection, rural customers, research and development, health and safety, and the interests of the disabled or those of pensionable age.

16 "Electricity Distribution: Price Control, Reliability and Customer Service," Office of Electricity Regulation, U.K. (October 1993).

17 Green and Jackson also performed regression analyses using production frontier techniques as well as data envelopment analyses on U.K. distributor data. They found significant variation in distributor efficiency. The study did not find increasing returns to scale in distribution, but it should be noted that the smallest distributor serves approximately a million customers.

18 K. Salvanes and S. Tjotta, "Productivity Differences in Multiple Output Industries: An Empirical Application to Electricity Distribution," *Journal of Productivity Analysis* 5 (1994):23-43.

19 Municipal Electric Association/Ontario Hydro, "Joint Study into Retail Electricity Service in Ontario," Interim Report, Toronto (1994).

20 Other studies of interest include L. Christensen and K. Smith, "Economies of Scale in U.S. Electric Power Generation," *Journal of Political Economy* (1976); R. Nelson and W. Primeaux, "The Effects of Competition on Transmission and Distribution Costs in the Municipal Electric Industry," *Land Economics* (1988); W. Primeaux, "An Assessment of X-Efficiency Gained through Competition," *Review of Economics and Statistics* (1977); W. Primeaux, "Estimate of the Price Effect of Competition," *Resources and Energy* (1985); M. Roberts, "Economies of Density and Size in the Production and Delivery of Electric Power," *Land Economics* (1986).

21 Ontario Hydro, like many province-wide corporations, has a uniform wage policy. According to 1993 data, OH employees working at the level of a journeyman power line maintenance worker earned higher wages than any comparable municipal utility employees.

Comments by

MICHAEL GILLESPIE
I.H. (TONY) JENNINGS
ADONIS YATCHEW

MICHAEL GILLESPIE

Dr. Yatchew lays out clearly some important principles underlying the thinking about regulation of the distribution sector of the electricity industry. He proposes that a fundamental starting point should be the separation of the monopolistic segments of the industry from those amenable to competition. He also notes that the ideal target for restructuring and reregulation is to replace regulation with market discipline where possible. He also notes that the regulation of distribution will be affected by how the upstream parts of the industry are configured and deregulated.

The paper spends a considerable amount of its focus on considerations surrounding generation and transmission, including privatization and ownership issues. Given that this session is about distribution and that other components are addressed elsewhere in the conference, I intend to confine my remarks to distribution, except to acknowledge a couple of Dr. Yatchew's points about generation and transmission: specifically that:

1. In introducing competition, there are potential concerns about self-dealing if there is integration of ownership in one utility that spans the generation, transmission, and distribution components.
2. Remedies for self-dealing range from (1) divestment of generation to (2) divestment of transmission and/or system operation to (3) the establishment of regulatory oversight to keep generators and the purchasing transmission entity at fair-dealing distance.

To get the useful focus we need in this session about distribution, I will assume that one of the three above methods are put in place to mitigate concerns about self-dealing upstream.

In the meantime, there is no lack of issues surrounding distribution downstream. One underlying theme I note is that, for all the brave talk about competition and marketplace, what I miss in Dr. Yatchew's argument – and in most discussions about competition in electricity – is the role of the end-use customer in shaping and making a market. The perspective in introducing competition in electricity is very much oriented toward the provider – the exercise of industry specialists designing an optimal industry model based on the two concepts of maximum allocative and economic efficiency, and lowest unit electricity cost. It would be both interesting and relevant to consider what customers mean by and expect of competition.

I should like to suggest that, even before we get to the stage of stakeholdering with customers, we consider customer choice an essential element of real competition in our industry. On that basis, my going-in assumption is that retail access will ultimately be a sine qua non feature of tomorrow's industry.

That is not to say that we might not have to go through some interim steps to get to retail access. But since the distribution end is where the retail interface drives demand, I think it is important to start, as much as we can, with something approximating a real market.

With those assumptions, let me comment on a number of points in the paper:

One is the claim that there is little prospect for the introduction of direct competition in distribution because it is a natural monopoly, and that therefore mechanisms other than competition must be sought to promote efficiency.

This conclusion fails to note adequately the unbundling of services. Only the wires business is a natural monopoly in the sense that only a single set of wires runs down the street. But competition in electricity sales is quite possible and is in fact a distinction that has already been made in natural gas delivery. Furthermore, the municipality or the customers within a utility franchise area could franchise the right to operate, maintain, design, and build wires in any distribution area in a given period of time. A competitive bidding process with performance guarantees is one approach which would disaggregate municipal ownership from the provision of the service.

Even if one pulls up short of a competitive process for getting and maintaining the franchise, the unbundling of services into meter reading, billing, energy audits, consumer information, and conservation measures are all potentially contestable services in a competitive market.

A second point made by Dr. Yatchew is interesting because of its public policy implications. He notes that because municipal utilities are accountable to their local constituencies, it can be argued that regulatory authority should also reside at this level.

The municipal base, all 300-plus of them, is the inherited starting point for defining jurisdictions. One can certainly question whether a political base and a political process of accountability – electing or rejecting commissioners every three years – is the right one for launching a more commercial and competitive provision of services around the distribution wires business. Customers themselves could have some choice in shaping the wires jurisdiction they are served by, for instance, voting on amalgamations and mergers. Such is consistent with the retail access structure, where consumers can select their electricity supplier and/or broker and various providers of energy services.

The author rightly notes that a provincial-wide regulator could enforce consistency at the expense of local control and variation. The public policy decision is whether the electricity industry is to evolve as a public service, such as water and garbage disposal, or as a commercial commodity and service. If the former, then perhaps regulation by municipal council is appropriate. If the latter, then a process might be found to put the lion's share of the decision making about prices and service levels into the hands of customers, with perhaps some regulatory oversight of the basic monopoly wires tariff.

With regard to the regulatory mechanism itself, Dr. Yatchew proposes the use of yard-stick competition, which would compare distributor margins across jurisdictions in order to establish best practices.

I think that benchmarking could be useful in some circumstances and could be used to get some comparative relativities. But I doubt whether benchmarking by itself is sufficient to drive efficiency improvements. In fact, in many jurisdictions, it has not done so to date.

As acknowledged by Dr. Yatchew, Ontario has a system where there are huge variations in size and operating conditions. A more useful approach for regulation would be a price cap mechanism where targets for productivity improvements are part of the performance contract with the municipality or customer jurisdiction area.

The paper states that there is little empirical data to show large efficiency gains through restructuring the municipal system into larger entities.

I find it counter-intuitive that a system with over 200 distributors having fewer than 10 employees is optimal. Different activities, operations versus billing, for example, might have very different economies of scale, and I think we would all like to see more detailed analysis.

In any case, I would suggest that customers themselves are not blind to the potential for efficiency improvements, and could make some call on their own amalgamation. This is not to suggest that utilities themselves might not initiate partnerships, but it makes some sense in a move to create commercial markets to let the users be the choosers, and grant customers some choices over the size and make-up of their local service provider.

Finally, I would like to have seen the paper comment on the scope of regulatory rethinking required as our industry undergoes change. The upstream technological drivers of change that Dr. Yatchew refers to, such as improvements in gas turbine generating technologies that render generation a competitive component, could well be dwarfed by changes downstream.

The emerging notion of electricity as a commodity is useful in the sense that it encompasses many new players such as brokers, marketers, price hedgers, etc. But electricity itself, as Yves Ménard points out, is both product and service, and an increasingly dynamic one.

The model of the electricity industry on which traditional regulatory notions have been based is that of a one-way flow from upstream generation through transmission and distribution to the end-use customers who "consume" the commodity. But accelerating developments in digital control and information technologies will make electric customers selective and pro-active players, designing their own needs, and working with vast stores of information, self-selecting products and services.

The interactive retail interface of a smart house or business for example, could encompass real-time metering, hour-by-hour control feedbacks on pricing including power quality pricing, and interactions with home and office security. Shared or merged services with telecom or cable are possible aspects of these developments.

Convergence with other industries such as natural gas could lead to integration of localized generation and dual-fuel systems.

These potential developments are not decades away: the enabling technology is already here, yet our current regulatory tools and institutions lag behind.

For example, most of our discussion has focused on the traditional notion of electricity rates – consumed kilowatt hours over time. Suppose a service provider at the retail interface wants to provide a customer with an integrated security/lighting/heating/air conditioning package at 10 cents a square foot. Can we ensure that the service provider has the freedom to deliver all of the components on a competitive marketplace basis? If not, are we developing regulatory principles that will at least allow new shaping of traditional services?

It strikes me that while deregulation and competition in the upstream end of the electricity business are vitally important, the greater complexities and challenges may well lie downstream.

I.H. (TONY) JENNINGS

I would like to cover nine areas raised in Dr. Yatchew's presentation. I intend to reinforce some of his conclusions and fill in a few gaps.

I also intend to make references to the long-awaited report of the California Public Utilities Commission, which came out about 10 days ago but I gather has not been discussed. I should point out that the report calls for public comment in 90 days, that it was not a unanimous commission position but was a decision of three out of four commissioners. I am told the majority decision is 500 pages and the minority position is 300 pages; I have not read either but I am working from some articles and press releases. Lastly, in this context I should note that the California legislators have indicated that they plan to have something to say before any decision is made and of course, the U.S. Federal Energy Regulatory Commission (FERC) will also have a chance to comment.

1. Dr. Yatchew started by putting the distribution function in context. He foresaw a separation of generation from transmission and both from distribution; many of you will be aware that the Municipal Electric Association (MEA) has had a Task Force of industry leaders working for some two-and-a-half years looking at industry options which came to similar conclusions in terms of the best form of organization. Our membership (made up of 307 municipal electric utilities in Ontario) supported this broad model almost unanimously at our recent Annual Meeting. I will not take the time to go through the MEA model but I will be referring to it. Suffice it to say here that the model is similar to the structure foreseen by Dr. Yatchew.

   It is interesting to note that the California Commission had apparently come to a similar conclusion, even though, as Dr. Yatchew says, the presence of private capital makes things significantly different. The commission has asked for comments on the actual legal segregation of transmission from distribution and from generation. At a minimum, they are proposing a proxy separation which would have control of all transactions through their transmission grid controlled by an independent third party.

   A missing piece of context is the scale in which we are dealing. For each electricity dollar in Ontario, 70 cents goes to pay for generation, approximately 15 cents goes to pay for transmission, and roughly the final 15 cents is the cost of distribution. Thus, if you were to find a way of achieving a 4% reduction in *distribution* costs, we are really talking about 4% of 15% of the price of electricity. If structural changes are involved in achieving the savings in distribution costs, there will be offsetting implementation costs.

2. Dr. Yatchew notes that separation is the route chosen in public power jurisdictions (and I have noted above that it is even being discussed in jurisdictions where transmission is largely in private-sector hands). He makes the point that changes in structure when private money is involved are significantly more complex. Thus, for Ontario, it is important that no private investment in the electricity system be allowed before any desired restructuring is done.

   Dr. Yatchew also stresses that separation maximizes the benefits for consumers through the full competition that is possible for the generation function. He notes that no competition can be introduced viably for the transmission grid, which is definitely a natural monopoly. He stresses that the current dispersed (horizontally separated) nature of retail level in Ontario allows effective "yard-stick" competition, despite the local natural monopoly involved. This again maximizes benefits for consumers.

   He also discusses the need for horizontal separation. Competition is only effective if there is a sufficient number of competing bodies. In discussions with those involved in implementing the British model, they will generally admit that one of the errors in their system was not breaking up the existing generation into more than a few separate entities.

3. Dr. Yatchew leaves open the question as to whether the transmission system should be operated as a common power pool or as a common carrier with bilateral contracts or open access, but he does note that the move to open access is likely irreversible. Thus, if there are no strong arguments in favour of one over the other – for *all* customers – it makes sense to treat the grid as a power pool, at least in the first instance.

   It is interesting to note that the California Commission has apparently reached the same conclusion. They are proposing that the grid be separated, at least in its operational control. It is unclear whether the transmission grid will be legally separated into "Transco," though this is being increasingly discussed in the U.S. At least operational continuity (supplier, dispatcher, etc.) would be vested with a separate third-party entity. Transactions apparently would be processed at a clearing rate, which would act as an average pricing mechanism, although on a very different basis than here in Ontario. Bilateral contracts or open access would not be allowed for at least two years and then only if certain conditions can be met. It is not clear, but the commission may be saying that retail access may never be necessary since "virtual retail access" is claimed to be likely. "Virtual retail access" is a term used in California to indicate that one can get virtually the full impact of retail access without actually implementing it. Remember, in Ontario we are dealing with a fractional improvement on the 15% share and the claim by some is that the costs of the additional transactions and players (brokers, financiers, etc.) would more than offset any additional gains possible.

One hears a number of people saying that retail access is inevitable, and Michael Gillespie in his comments made this a starting assumption. Judging by the California Commission's reaction, at the moment it is far from inevitable that everyone there will have equal access to alternate suppliers. I will not go into activity in other states.

What is not discussed by Dr. Yatchew is the impact in Ontario. Open access will disturb the current relationships; and in a province of dispersed entities, where currently wholesale rates are identical anywhere in the province, it would seem prudent to try to assess the impact of such a major change before proceeding with it. To date, as far as I am aware, no one has undertaken such a study; certainly Ontario Hydro has indicated it has not and it is my understanding that neither has the Ontario Ministry of Environment and Energy.

4. Ownership of distribution is covered in his discussion, but I am going to argue with Dr. Yatchew's lack of conclusions. Data from the U.S. Department of Energy (1993 data supplied through the American Public Power Association) indicates that the average residential public power customer received power at prices a third lower than those of the investor-owned utilities. Commercial customers of investor-owned utilities paid 16% more for electricity than public power customers, and there is little or no difference between the average rates paid by industrial customers from either type of supplier.

   It is worth noting that some of the moves in the U.S. to transfer franchises from the private sector to municipal public power are being driven by industrial customers at the moment.

   Dr. Yatchew cites John Kwoka's 1995 paper, the most recent dealing with this issue. Kwoka found that while there are special advantages for public power, in terms of advantageous rules for municipal bonds and preference in the supply of low-cost power from the federal power marketing agencies, still he found that ownership itself was the most significant factor accounting for most of the price advantage.

5. Dr. Yatchew indicates that if generation is separated from transmission and is widely enough dispersed, regulation should not be necessary for that function. It would be likely that the transmission system might not require regulation either, so long as the "ownership flows from the customer" – customer representatives being the municipal utilities. Particularly given their non-profit status, the municipal

utilities would have an interest in keeping the cost of transmission minimized.

Regulation of the distribution system is discussed by Dr. Yatchew. In its current form at the time of the conference, the paper is unclear when it notes that "distribution utilities" are normally regulated on rate of return. It is important to understand that in the U.S. most of the municipal electric utilities are not regulated by third-party regulators. The public utilities commissions, which are the regulators in most states, regulate the investor-owned utilities. Virtually all the 2000-plus municipal electric utilities are regulated by their own policy boards, what we in Ontario would call commissions. In this respect, it is worth noting that in the Ontario context, the majority of commissions are directly elected at the local level and thus answerable to their constituencies.

Dr. Yatchew notes that "regulation" in Ontario is carried out by Ontario Hydro and is quite light-handed. It is also administrative regulation without formal hearings. (I would note in passing that elsewhere in the conference there was discussion that there is no appeal/hearing process available. It is my understanding that this is not the case, since the Power Corporation Act allows appeals to the Ontario Hydro Board of Directors in the final instance.) The light-handed regulation is easily understood if one looks at the history of what we now term as regulation.)

As far as I am able to determine, this evolved from the fact that Ontario Hydro (then the Hydro-Electric Power Commission of Ontario) was established not as a crown corporation but as a cooperative of the municipal utilities. It is in fact, a co-op or joint action agency and the regulation has evolved from a responsibility to police the conditions of taking part in that pool. Most of the elements of regulation even today are those which would best protect the members of the pool from "foolishness" or inappropriate action on the part of an individual utility. This is particularly understandable if you recall that Ontario Hydro provides an additional service by being the supplier of last resort. If a municipal utility decides for some reason to "fold-up," the responsibility reverts to Ontario Hydro and can become a pool cost.

6. Let me expand a little on Dr. Yatchew's comments on structure. Many in Ontario and indeed in Canada, consider Ontario's municipal utility structure to be uncommon. While this may be true in Canada, though there are still a few municipal utilities in existence in most provinces and the Québec government is looking at "regionalization," there are many other jurisdictions where municipal utilities are still a strong component. In addition to Norway, which is cited by Dr.

Yatchew, Germany with 800-plus municipal utilities, the U.S. with over 2000, and New Zealand with municipal utilities in most major centres come to mind.

The Norwegian Salvanes and Tjotta (1994) paper was new to me. I find it most interesting that it reaches similar conclusions to the interim report of the retail structure study being undertaken jointly by Ontario Hydro and the Municipal Electric Association. Let me stress that the joint study has only reached interim conclusions and thus cannot speak about its final findings. There is a common view that there are too many municipal electric utilities in this province and that the customer would benefit from "rationalization" of the retail structure. This would appear not to be supported by any of the three studies. In fact, the Norwegian study seems to find a lower point of optimal size (20,000 customers) than was being discussed in Ontario.

This conclusion is seen to be "counter-intuitive." Let me suggest analogies to either the food industry or accommodation industry. Our society has both large grocery stores and small corner suppliers. Similarly, we have multistorey condominiums and rental apartment buildings with gyms and other fancy facilities, and we have small duplexes, triplexes, etc., where the landlords provide minimum service. All suit their respective customers.

There is no reason to assume that the largely rural residential customers of some of our smallest utilities require or desire the same range and level of service that is seen as essential in large urban areas with many large and sophisticated commercial/industrial customers.

I would note too that there are some slight economies of scale cited by Dr. Yatchew, but the potential impact is minimal and may be offset by the cost of making changes. Restructuring boundaries within the electricity system, "separating the systems," changes in metering, etc., are not without cost.

Finally, in this regard I would note that Ontario's legislative framework is clearly designed to support the concept of municipal delivery, and as recently as 1994 the legislature, with the support of all three parties, passed a bill to facilitate further "municipalization." The so-called "boundaries" legislation was the result of several years of joint work by the Ministries of Energy (and Environment), Municipal Affairs, Ontario Hydro, and the Municipal Electric Association. All the major unions involved were consulted at several stages.

In passing, I note that Mr. Gillespie has suggested we "let the users be the choosers." In doing so, we should not be looking at aggregate data across the province but should deal with the locally affected people. My suspicion is that many local users would choose to retain their local utility. Many of you may recall that the Ontario government a few years back promoted the idea of doing away with small municipalities which would have difficulties servicing their constituency. You may also recall that in a number of cases, the local municipal ratepayers fought off amalgamation with their larger neighbours. The small rural residential communities are often satisfied with a different level of service, particularly if it is associated with a different level of taxes.

7. Dr. Yatchew notes that the obligation to serve should theoretically disappear if retail access occurs. I would just like to raise a question in this regard. Given the nature of electricity and its importance in our daily lives and in the economy, I find this conclusion to be "counter-intuitive." It is good economic theory but I cannot believe that any government, regardless of its "colour," would allow a municipality or a significant employing industry to put itself in the position of not having a supply of electricity. An obligation to serve will be laid on somebody's shoulders.

   The current surplus electricity situation in this province makes this a somewhat academic question for the moment. But we are not talking about just this moment. As recently as the end of the 1980s, there were situations where there were serious concerns about the adequacy of electricity supply – at any price. Given regulatory processes from the environmental and land-use planning perspectives which caused great delays and are very costly, a number of major utilities have decided that they will no longer build their own generation. Supply is potentially limited.

8. There are many spokespersons, including Mr. Gillespie, who are assuming that retail wheeling is "inevitable." Let me say that there are a growing number of voices questioning this inevitability. It is important as well to note that continued reference to inevitability may in fact become a self-fulfilling prophecy since it implies a lack of will to resist that change. In the view of the municipal electric utilities, this change is not inevitable. Nor has it been shown to be in the public interest. So far as I am aware, no one has looked at the implications for this province, and the customers of the electricity system, of a change from our common pool and postage stamp rate approach to an open access bilateral trading agreement/common carrier approach. An increasing number of papers in the U.S. are raising questions about vertical separation, despite the initial approach of trying to achieve benefits through open access rather than dealing directly with vertical separation. Private-sector ownership and the attendant rights clearly make separation more difficult in the U.S. It is important to understand that it is transmission access (wholesale bilateral sales) which has been required by the federal government. Retail

access, while it is a cloudy jurisdiction, would appear to fall primarily under state regulation and the conclusions are unclear. It should be noted that the recent California Public Utilities Commission Report does not ensure that retail access will be available to everyone on an equal basis.

9. Lastly, I would just like to react to Dr. Yatchew's suggestion that cooperatives are an effective way to recognize some available economies of scale while maintaining local accountability and responsiveness in such local values as a consistency with local wage rates. Dr. Yatchew recommends cooperatives and implies in his writing that they do not already exist. Let me correct that impression.

   Ontario Hydro was established as a municipally based cooperative in 1906. The Municipal Electric Association acts mainly as a cooperative on behalf of its members and is clearly successfully in competing with other suppliers of a variety of services. In addition, there have been a number of informal and formal local "cooperative" arrangements among municipal electric utilities in Ontario, and new ones are currently being formed.

It should be noted that many opinions have changed since these comments were drafted.

ADONIS YATCHEW

Both Jennings and Gillespie have raised interesting and important points and I am grateful to them for their careful consideration. I would like to respond in two areas. The first dealing with the inevitability of retail competition. The second, with the time sequence of restructuring within the industry.

First, transmission access and even retail competition may be inevitable. However, prior to its introduction it is essential to put in place a mechanism for dealing with stranded assets. Indeed the presence of excess capacity in Ontario argues for a judicious and deliberate move towards competition.

Second, one of the commentators spoke extensively of the need for restructuring within the retail segment of the industry. I would argue that the critical restructuring changes which are *not* likely to occur spontaneously are at the generation and transmission level. Restructuring of the distribution sector is not a necessary *prior* condition for restructuring of transmission and generation and the introduction of competition at the generation level. To give this point some flesh, consider a scenario where tomorrow upon awakening, we find a restructured distribution sector with, for example, 12 regional distributors in Ontario, entirely separate from Ontario Hydro. In such a circumstance, it is unlikely that any changes would result immediately to the transmission and generation segments of the industry that would not have otherwise taken place.

On the other hand, if we wake up tomorrow and find that transmission and generation have been restructured, and competition has been introduced at the generation level, you would find as a result that new obligations and responsibilities had been devolved to distributors. Amongst distributors, you would then likely encounter a few frantic phone calls, ("..where do I get a standard contract for power, how long do I contract for...") but in short order, alliances and cooperatives would form, amalgamations could take place, and if anything the distribution sector would likely rationalize sensibly rather than autocratically because it would be responding to real forces.

In Search of the Cat's Pyjamas: Regulatory Institutions and the Restructured Industry

H. N. JANISCH

INTRODUCTION

This paper has been written with three principal purposes in mind.[1] First, rather than be based on separate speculation as to the future of Ontario Hydro, it is designed to build on Ronald Daniels' and Michael Trebilcock's contribution to the Conference, "Ontario Hydro at the Millennium: Has Monopoly's Moment Passed?" (hereinafter Daniels and Trebilcock). Second, in so doing, the focus will shift onto institutional design issues, particularly the relationship which should exist between the envisioned new regulator, the rest of government, and the legislature. Third, contemporary experience with an analogous sector, telecommunications, will be woven into the analysis in order to highlight the challenges the electrical industry will soon have to face in constructing new regulatory institutions appropriate to a restructured industry. These three purposes will be briefly developed here before an assessment is made of the shortcomings of the current regulatory situation.

Daniels and Trebilcock present a somewhat novel and intriguing challenge to anyone concerned with governmental institutional design in that they have both limited, and ambitious, expectations of regulation. On the one hand, they emphasize that in the shift to a less vertically integrated, more competitive electricity industry in Ontario, they wish regulation to be precisely defined with a mandate "as narrow rather than as broad as possible."[2] Yet on the other hand, they envisage regulation as a seemingly broad-based "agent for change."[3] This would seem to call for a proactive, but narrowly focused regulator, rather different from established reactive, broad-based regulatory agencies.

It has proved to be exceedingly difficult to engraft regulatory agencies on to a parliamentary system of government in which it remains essential to ensure the accountability of regulators to ministers and to the legislature. This is particularly so when such agencies are described as being "independent."[4] It will, as a result, be very important to seek to position any new regulator in our system of government in a manner which provides for an appropriate degree of political accountability but which does not smother the independent initiative necessary for it to be an agent of change, not simply an adjudicator of competing interests.

As Daniels and Trebilcock note, in the electricity industry new technology such as combined cycle gas turbines (CCGTs) are rapidly undermining assumptions about economies of scale and natural monopoly; there is a shift away from centralized to local power generation; distributed generation has greatly enhanced the position of end users, and large customers are threatening to exit by way of self-generation. All in all, there are advanced warnings that the existing, integrated system is unravelling.[5] In telecommunications there have been striking parallels in which digital technology and wireless transmission have undermined natural monopoly; intelligence has been dispersed from the centre of the network to its periphery, thereby greatly increasing the role of end users who are now in a position to threaten to bypass the network itself. Today, there is no longer a single, integrated network but already what may be described as a "network of networks."[6] While forces for change in the two industries appear to be strikingly similar, that which is only being anticipated in the Ontario electrical industry has already largely happened in Canadian telecommunications. Here, indeed, is an industry turned upside down in the 1990s.[7] There is much to be learned about the limits of a regulator as an agent of change and from the terms of the somewhat belated legislative response. Here too is a valuable opportunity for a transfer of experience even though, of course, no industry developments are ever perfectly parallel.

Daniels and Trebilcock are rightly dismissive of the existing regulatory regime. Their standard of evaluation calls for institutional expertise, continuity, and memory for effective external oversight, none of which have been attainable given the diffused, ad hoc, politically driven and crisis-oriented oversight adopted in the past.[8] Thus an appropriate institutional design for regulation should ensure regular, anticipatory legislative oversight and an Ontario Energy Board (OEB) reconstructed in a manner which clarifies its role and responsibilities. And, as is further and rightly insisted, the regulator must *not* be charged with the primary responsibility for fashioning essentially political trade-offs or be allowed to second-guess the strategic and management decisions of firms in the industry "...which would entail the serious risk of undermining any efficiency gains that might be realized by restructuring the Ontario electrical industry along more flexible, dynamic and decentralized lines."[9]

Such then are the general marching orders for the redesign of the legal regulatory regime.

LIMITED, BUT POSITIVE, EXPECTATIONS FOR REGULATION

One of the factors accounting for Ontario Hydro's immediate difficulties was seen as its anachronistic regulatory structure "...characterized by dispersed and fragmented authority that has at times subverted public transparency and fostered government micromanagement."[10] This, Daniels and Trebilcock urge, must be replaced by continuity and expertise, a narrowly focused form of regulation in which political trade-offs have been resolved in the legislation itself and, above all, by a clear set of rules. "It is essential that the regulatory rules of the game be clearly established at the outset of the restructuring process so that private investors can condition their bids for Hydro's assets on a relatively accurate understanding of the future policy environment."[11]

As well as this somewhat predictable general plea for greater clarity and predictability, there is a warning of the trap of multiple and irreconcilable regulatory goals and the importance of structural arrangements as substitutes for regulation. Both need to be expanded on briefly and reinforced with experience from contemporary telecommunications regulation.

Reference is made to the tension between the goals of enhanced efficiency and government revenue raising in the context of privatization.[12] While, no doubt, a very relevant concern in Ontario Hydro's immediate context, the danger of multiple and irreconcilable goals being incorporated into a regulatory scheme should be seen in a somewhat wider context.

For instance, the policy objectives section of the 1993 Telecommunications Act[13] incorporates the monopoly era goal of universal service at affordable rates without reconciling it with the impact of competition on the ability to subsidize the provision of universal service. Similarly, domestic concerns for competitiveness may well require a number of new entrants for its achievement, while international concerns may require that resources be combined in a single national carrier in order to create an effective, world-class competitor.[14] More recently, the Canadian Radio-television and Telecommunications Commission (CRTC) itself issued a report on convergence (between broadcasting and telecommunications) and the information superhighway which envisaged that cultural content concerns could continue to be subsidized in a competitive era. There was simply no recognition that a choice would have to be made between the benefits of monopoly and the benefits of competition.[15]

This exemplifies a politically understandable desire to perpetuate the benefits of monopoly even as there is a move to competition. Similarly, it seems likely that in the context of the electricity industry there will be reluctance to give up on such appealing objectives as power at cost, province building, and the subsidization of rural customers, even as the essential underpinnings of these objectives are knocked out in a move towards decentralized competition. Failure to face up to the need for choices will only mean that the appointed regulator will be left with the principal responsibility for fashioning essentially political trade-offs. This will inevitably lead those who are disappointed with a regulatory outcome to approach the political authorities for its overturn. This, in turn, will undermine the credibility and integrity of the regulatory process, matters to which we will return shortly.

A striking section in the study by Daniels and Trebilcock addresses the importance of structural arrangements as a substitute for regulation. Indeed, as they point out, to the extent that competitive structures can be promoted, to that extent the case for extensive regulation, or in some cases any regulation at all, is obviated.[16] Recent developments in telecommunications confirm the critical importance of getting structure right from the outset. In that industry (in Canada at least) competition has been allowed without any structural change. However, in the Ontario electrical industry it would be possible to get the direction of economic incentives right at the outset and not have to rely on conduct regulation in an effort to achieve outcomes which run counter to structure-induced incentives.

In the United States, a bold decision was made in the mid-1980s to break up AT&T to ensure that the Regional Bell Operating Companies (RBOCs) did not have any incentive to favour a fellow subsidiary long-distance carrier. Divestiture provided for completely separate share ownership of the original long-distance carrier from that of the local companies. As a result, they are indifferent as to whether they interconnect with AT&T or new entrant competitors such as Sprint or MCI.[17]

There is now growing concern in Canada as to the sustainability of competition between long-distance competitors, such as Unitel and Sprint Canada, and the vertically integrated members of Stentor, an alliance of the original telephone companies. Lack of structural reform means that when a non-telephone company competitor seeks access to the local switched network of a Stentor member, it does so as a competitor to the long-distance services provided by that member. In short, crucial bottleneck facilities are controlled by the very companies against which new entrants seek to compete.[18]

Although it is not at all clear that Unitel's present difficulties should be attributed entirely to cost and delays in obtaining access to Stentor-member company local networks, it is apparent that complex costing formulas have not worked satisfactorily. Apart from the immensity of the initial task of allocating costs in as highly integrated an industry as telecommunications (costing here has been well described as "a journey to the centre of the earth"), it is now being belatedly recognized that, in the face of countervailing structure-induced economic incentives, costing is not a game which new entrants (with limited resources) can ever play and win on an ongoing basis.

While the U.S. Congress now faces the daunting task of passing legislation responsive to emerging convergence and multimedia and the collapse of separate industry boundary lines,[19] that country's very dynamic long-distance market stands as a tribute to the benefits of structural solutions over conduct regulation. Similarly in Canada in those sectors, such as cellular, where structural solutions, such as separate subsidiaries and same time out-the-starting gate were adopted, competition is significantly better established than in long distance. While full scale, U.S.-style divestiture would seem unlikely, it should be noted that the Director under the Competition Act is looking into Stentor's cartel-like role and the extent to which it inhibits effective competition.[20]

In all this, one thing is clear. If it is possible to get industry structure right (i.e., competitive) from the outset, this is an opportunity which must not be missed. The immediate significance of pro-competitive structures in the most moderate proposal for reform will be seen in the next section. However, before turning to this issue, it would be well to remind ourselves of the tension apparent in the Daniels and Trebilcock position on regulation and thereby foreshadow some of the institutional design discussion this will provoke.

On the one hand, they appear to have little confidence in regulation. Wherever possible it is to be avoided by pre-emptive structural reforms. Even where they conclude, with but little conviction, that regulation cannot be avoided, they have (as we shall see) little faith in its efficacy. Yet at the same time, there are occasions on which effective regulation will be critical to the success of their grand design. For instance, they rely on regulatory obligations to ensure that a significant portion of the Ontario electricity market is rendered contestable by extraprovincial generators,[21] and envisage that regulation will be able to play a role in limiting inefficient bypass.[22] Moreover, in their conclusion, a reconstituted and reinvigorated OEB is seen as providing "...a neutral external public forum that can act as an agent for change in the industry, given the various existing entrenched interests that have to be moved and monitored in the process of change. This role for the OEB requires a detailed review of its current powers, competencies, and resources."[23]

As noted earlier, this initial selective downgrading of the more traditional role for an economic regulatory agency, later coupled with a far more proactive regulatory role, raises difficult institutional design issues. Indeed, an initial reaction might be to wonder whether such a mixture of roles could ever be undertaken by the same body. At the very least, it turns their last, delightfully understated sentence into a very significant design challenge. However, before addressing this broader concern, it would be well to take a look at some more specific matters.

REGULATORY IMPLICATIONS OF THREE POSSIBLE MODELS

Daniels and Trebilcock put forward three models: Model 1, Competitive Access to Power Pool; Model 2, Wholesale Competition; and Model 3, Retail Competition.[24] While each has somewhat different implications for regulation, in keeping with their overall approach, they start by identifying a very significant structure-incentive issue.

Should the minimalist approach be adopted in which a power pool is created, but Ontario Hydro retains ownership of the grid and continues to provide 90% of the generating capacity in the province, there would be the strongest incentives on the part of the grid owner to favour its own generators. Indeed, they note that any method of creating a competitive market for electricity that leaves intact the present vertically integrated industry will be plagued by disputes raised by allegations of self-dealing.[25]

As we have seen, experience in regulating the telecommunications industry strongly supports the view that conduct regulation alone will not be able to counter structure-created economic incentives. Such a realistic assessment of the limits of regulatory competence leads inevitably to a radical privatization option in which Hydro sells off its generating facilities in four or five bundles of assets. However, Daniels and Trebilcock are not persuaded that nuclear can be privatized, given its cost structure. Nor are they persuaded that there would be any significant benefit in privatizing transmission.

As a result, they create an unnecessary structure-incentive issue for themselves in that they recognize that a government-owned grid will likely favour a government-owned generator or, at least, be seen to be inclined to such favouritism by privately owned competitors.[26] This could be avoided by privatizing transmission. Not only would such a move finesse concerns for self-dealing, but it would also allow for the resolution of the choice of type of regulation of transmission in a manner most consistent with the overall approach of their argument. That approach seeks to have regulation rely on output targets which give maximum latitude to firms in the industry to organize their inputs in such fashion as they judge to be most efficient.[27] However, it is acknowledged that if transmission remains in government ownership, incentive regulation is not likely to be effective.[28] Given their overall preference for structural rather than regulatory solutions and for incentive regulation rather than rate of return regulation, it would seem that Daniels and Trebilcock should move away from their agnosticism with respect to the ownership of transmission.[29]

What, then, are the indispensable regulatory elements in Model 1? On the generating side of things, detailed price regulation is resisted.[30] Reliance is placed on "political voice of one kind or another, much as is the case today" to constrain the publicly owned nuclear generator which currently provides 62% of the total electricity supplied in the province. It is, however, difficult to justify this confidence in the efficacy of today's political constraint in light of yesterday's debacle involving massive rate increases during a recession along with substantial excess capacity, all of which is recorded at the start of their study.[31] Nor would a political voice be likely to expose umbrella pricing (a very likely occurrence) given the cost structure of the publicly owned nuclear generators.

Be that as it may, for our purposes it should be noted that external regulation is at least accepted as a default option in the form of price caps on generation prices if bids into the power pool reflect price increases overtime that are significantly in excess of increases in the CPI. It was also noted that collusion among generators[32] could properly engage the attention of the enforcement authorities under the Competition Act,[33] as could predatory behaviour designed to foreclose entry by more efficient generators by way of temporary low pricing policies.[34]

As for transmission, as a natural monopoly, external regulation was seen as appropriate. Some form of incentive regulation, such as regulatory lag or price caps, was favoured although information and valuation problems were stressed.[35] As for retail rates, as the local distribution companies would remain under municipal ownership in Model 1, accountability similar to that over local taxes and threats of exit were considered adequate to ensure that prices would not be excessive.[36]

It should also be noted that by ignoring stranded costs[37] and payment to "capacity adders,"[38] Daniels and Trebilcock sought again to simplify considerably the range and complexity of the regulatory tasks implicit in their conception of Model 1.

In Model 2, generators would be able to compete directly for wholesale business with local distribution companies and large industrial users by way of bilateral contracts. However, somewhat ironically, as one moves towards more comprehensive competition, the need for regulation increases. Thus in Model 2, at the distributional or retail level, "...there may be a more complex interplay between structural and regulatory reform."[39] As distribution moves from an inefficient, small municipal basis towards consolidation and even privatization, the need for some form of price regulation, preferably price caps, will grow. With a decrease in accountability to local political constituencies, greater efficiency may be achieved, but at the price of undermining political voice as a credible alternative to regulation.[40]

As well, with Model 2 there may be sufficient distributional concerns to warrant additional regulation. For instance, concerns about large industrial user-negotiated bulk discounts contrasted with "captive" residential rates and a perceived threat to the substantial ($100 million p.a.) subsidy to the almost 1 million rural subscribers served directly by Ontario Hydro, might easily be seen as calling for ongoing regulatory intervention.[41]

Whether this is so or not, Daniels and Trebilcock acknowledge that in Model 3, in which generators will be free to compete directly for retail customers through various kinds of intermediaries, there may well be greater need for regulation. Here then is the pervasive irony: the further progress is made towards complete decentralized competition, the greater the need for regulation. As they put it:

> In order to realize the efficiency gains from structural reforms at the retail level, there may be an enhanced need for regulation at this level. This is a case where structural change may not be an effective substitute for regulation, but may indeed intensify the need for new forms of regulation. Ideally, distribution charges would be levied on an unbundled basis (i.e., separate from charges for electricity) and would be subject to some form of price cap or incentive-based regulation to mitigate the dual problems of potential monopoly pricing and potential discrimination between local distribution companies' own customers and other customers.[42]

One is left to wonder why regulation will be capable of overcoming adverse incentives *here,* but apparently not elsewhere. Be that as it may, it would be useful at this stage to cumulate the perceived need and expectations of regulation in order to develop a better sense of an appropriate institutional design response.

Regulation

a. In General:
- ensure that a significant portion of market rendered contestable by extraprovincial generators;
- limit inefficient bypass.

b. In Generation:
- price regulation as a default option;
- Competition Act to guard against collusion and predatory pricing.

c. In Transmission:
- rate of return or price cap if privately owned;
- to guard against discrimination if government owned.

d. In Wholesale:
- price regulation if distributor no longer small municipal electrical utility (MEU).

e. In Retail:
- implement unbundling;
- price caps or other incentive based regulation to guard against monopoly pricing and discrimination.

And then, of course, there is the nagging question of the regulator acting in a "neutral forum" as an "agent of change" moving "entrenched interests" in a "process of change." At the very least, it seems that even with such rigorous pro-competitive structuralist regulatory sceptics as Daniels and Trebilcock, there is still going to be at least some reregulation in the restructured electrical industry. This would seem to warrant a serious look as to how an appropriately crafted regulatory agency will fit into government.

REGULATION IN A PARLIAMENTARY SYSTEM OF GOVERNMENT

Ministerial responsibility lies at the heart of parliamentary government based on the Westminster Model. A government's authority comes from its ability to command a majority in the legislature and its ability to continue to maintain the confidence of the legislature in essential matters. Ministers are responsible for the administration of their respective departments and collectively the cabinet is responsible to the legislature for the policies and behaviour of the government.[43]

This fundamental scheme of things has meant that rather than place regulatory power in fully fledged, independent agencies, as is often done in the United States, Canada has deliberately chosen a half-way position between independence and accountability. Day-to-day regulation often requires full-time, detached professionalism exercised by bodies having a considerable measure of autonomy. However, governments in a parliamentary system are reluctant to see final decision-making authority handed over to non-elected bodies. As a result, provision has often been made for cabinet review of regulatory decisions and, more recently, for binding policy directions.

This form of uneasy compromise between regulatory independence and political control has often led to tension, confusion, and misunderstanding. Consider two examples from the electrical industry drawn from a very long list of regulatory contretemps at both the federal and provincial levels.

In Nova Scotia, a massive rate increase for electrical power was announced by the Board of Commissioners of Public Utilities just before the 1977 provincial election. Premier Reagan announced that the two most important issues in the campaign were unemployment and electricity rates and that an extensive subsidy system for electricity would be brought in *after* the election. He later blamed his defeat, in part, on the fact that the electorate held "the government" responsible for the rate increases, notwithstanding that they had been approved by a regulatory agency whose decisions were not reviewable by the provincial cabinet.[44]

In Ontario in the mid-1970s, as Neil Freeman points out, there was similar evidence of confusion over role and responsibilities.[45] Here the situation was made even more complex and politically volatile by the deployment of a number of different and uncoordinated actors. While the government had sanctioned Hydro's ambitious expansion plans in 1973, it had been obliged to create a royal commission on electric power planning in March 1975. The controversy nevertheless got ahead of the government when Hydro proposed a 29.7% rate increase for 1976, more than double what it had accepted after the first OEB review for 1975 rates. Before the increase could be steered through the OEB rate hearing, it would face political scrutiny in the lead up to the September election. In July, in an attempt to head off a political outcry, the Treasurer used a mini budget to instruct Hydro to reduce its capital expenditures by $1 billion and to reconsider the amount of its rate increase. In response to the new financial constraints, Hydro, with an election looming, revised its increase down to 25% "...but all was for naught as the government was reduced to a minority."[46]

The complexity inherent to the Canadian compromise between political accountability on the one hand, and regulatory independence on the other, is more than amply illustrated in the contemporary regulatory political brouhaha over the introduction of direct-to-home television (DTH).[47] This technology makes direct competition feasible with the cable industry. As the CRTC has used cable as a chosen instrument to support Canadian content in television programming, it was understandably concerned to limit the impact of DTH. As a result, it actively promoted an all-Canadian consortium (which initially even included the cable companies) primarily designed to provide service only in areas presently unserved by cable.

As well, it was proposed that this service would be carried exclusively on a Canadian satellite. Thus the regulator could claim that it was allowing in an alternative delivery mechanism (somewhat misleadingly called competition) and doing so in a manner which would be most compatible with the nationalistic objectives of the Broadcasting Act.[48]

This initiative was implemented by way of an exemption power in that Act.[49] That is to say, the CRTC ruled that if a DTH service used a Canadian satellite and observed Canadian content requirements, it would not have to be licensed. Such an exemption was granted to the Canadian consortium which, however, by then no longer included the cable industry and thus this policy could be described as being somewhat more pro-competitive. At the time of granting this exemption, the Commission knew that the Power Corporation was in the process of linking up with Direct TV in the United States to provide a Canadian DTH service. Rather than await this development, the CRTC pushed ahead with the exemption. What then caused the headline-grabbing row was that, under the Act, regulatory exemptions are not subject to political review by cabinet, although the Power Corporation (once well described as the Liberal Party at work!) was very favourably positioned to exert massive political pressure on behalf of Direct TV.

Rather than supinely accept this regulatory putsch, the government set up an ad hoc rival policy-making body of "Three Wise Men" (all, incidentally, former senior federal bureaucrats). This panel somewhat predictably favoured a broader licensing approach which would include Power Corporation-Direct TV, as DTH would not now have to be carried exclusively on Canadian satellite facilities.[50] Armed with this report, the government indicated to the CRTC that it should get on with a licensing process and that a binding policy directive, as provided for under the Broadcasting Act, would in due course be issued to that effect.[51] The Commission reacted strongly to this political initiative before both the House of Commons and Senate parliamentary committees and urged, inter alia, that not only would this undermine the integrity of its regulatory processes, but that it was, quite possibly, unlawful.[52] The political issues aired concerned favouritism[53] and a perceived threat to the viability of the role assigned to the regulator;[54] the legal issues concerned that slippery concept, retroactivity, and the use of an ostensibly general directive power, apparently to achieve specific ends.[55]

It is evident that the principal cause of the tension and confusion which has all too often surrounded the regulatory process in Canada has been an unwillingness to provide greater clarity in legislation. This has meant, as Daniels and Trebilcock fear, that critical political tradeoffs are left to the regulators which leads inevitably, in a parliamentary system of government, to intermittent, and possibly destabilizing, political intervention.

Two institutional devices have often been employed to provide for political accountability – ex post cabinet appeals and ex ante policy directives.[56]

There are three commonly made arguments in favour of cabinet or political appeals. First, it is said that regulators only take into account the relatively narrow regulatory factors as presented to them by the immediate parties to a specific application. They tend not to take into account broader factors such as enhanced employment opportunities, encouragement of indigenous r. and d., and steps to overcome regional disparities that may well be of legitimate concern to government. Second, it is said that, in practice, cabinet appeals are not disruptive or destructive to the integrity of the regulatory process because they occur but rarely and, in any event are usually not successful. Third, it has been argued that some individual regulatory decisions are of such seminal significance that no government can afford not to be involved. Should it fail to be involved, this would amount to an abdication of political responsibility for major policy making.

Against political appeals it has been argued, first that they come so late in the decision-making process that they are inevitably disruptive and destructive and only serve short-term political ends at the expense of the entire regulatory system. Second, regulation is designed to get certain matters "out of politics" and cabinet appeals provide only for highly selective accountability. Third, abrupt political overrides may have the effect of demoralizing regulators who work hard to hear all sides, define the issues, study contending views, and write carefully reasoned decisions in the light of their interpretation of their statutory mandate and previous statements of government policy.

One widely supported proposal has been to abolish cabinet appeals in favour of a power to issue binding policy directives. It is thought that this would force broad government policy making out into the open before any specific regulatory decision is made, rather than allowing policy makers to "lie back in the weeds" and overrule a politically unpopular decision on a highly selective basis. It has been suggested that policy directives should involve consultation with the regulator and be the product of an open, public process. They should apply only to general policy questions (see the DTH row above) and not be concerned with individual cases.

It has been said that this proposal would focus political accountability where it can be a reality. A public forum for formulating directives would give ministers and their departments an opportunity to advance openly their policy concerns. It would reinforce the principles of responsible government in that the cabinet would, in the end, have its way. At the same time, rather than the closed "cloak and dagger" way in which cabinet appeals are dealt with, accountability to the legislature and the public would be increased as policy directives would be issued following a public report that grew out of a public hearing. This would bring policy making out into the open and eliminate the danger of imputing too much policy significance to individual decisions. It would maintain, it is said, the integrity and worth of regulatory agencies, but not at the expense of ultimate political accountability. As the Economic Council of Canada concluded in a refreshingly unstuffy assessment: "It would be the cat's pyjamas."[57]

However, it is important to note, as has Margot Priest, that although many studies have recommended that appeals or reviews by cabinet be eliminated, all studies done by politicians have favoured the retention of cabinet review.[58] Indeed, at the federal level, the "double whammy" approach of combining appeals with directive powers has prevailed in recent legislation.[59] However, interestingly enough, in British Columbia cabinet appeals have been recently abolished.[60]

REGULATION AS AGENT FOR CHANGE

It is time to pull together a number of threads. As we have seen, for someone concerned with the design of regulatory institutions, the most striking feature of the Daniels and Trebilcock study lies in its rejection of traditional notions of passivity in favour of regulation as an active means of bringing about change. We need to contrast this expectation of regulation with the nature of conventional regulation, as the institutional design of regulatory agencies up to this point has been responsive to such conventional mandates. This will lead us to explore whether conventional regulation is capable of acting as an agent of change. Thereafter, three interrelated but somewhat more specific needs should be addressed: the need for a clear-cut and unequivocal statutory mandate; the need for structural reform wherever possible, and the need for clarification of the role of the Competition Act. As well, two foundational characteristics of regulation, both of which have great significance for the whole question of institutional design, need to be explored. First, can change-seeking regulation ever be taken "out of politics"? Second, once it is recognized that the purpose of regulation is not to balance competing interests but to implement change, will this not require a radical departure from institutions designed for more conventional regulation? This will lead us, finally, to a consideration whether what is needed is a single, committed, and responsible individual rather than a multimember panel, even though such a panel may be required to deal with the more conventional, regulatory issues identified earlier in this paper.

Recall that what is envisaged is a regulator which can actually move existing entrenched interests in a process of change. This is far removed from the conventional concept of regulation, which seeks to reconcile conflicting entrenched interests from time to time in the name of the "public interest." Indeed, as we have seen, many of the contemporary difficulties with regulation flow from an unwillingness of politicians to commit themselves clearly to a particular course of action. This may be "unwillingness" rather than "inability" in view of the politics of purposeful ambiguity.[61] This explains why broad mandates are delegated to be "filled in" by regulators. They, in turn, may be employed as "fall guys" subject to criticism for vagueness by private parties, judicial bodies, and even the government itself, because the legislature knows all too well just how dangerously political regulation really is. This analysis suggests that the legislature may wish to perpetuate ambiguity so as not to offend. If there are to be winners and losers, that unpleasantness can be left to the regulators. While not, perhaps, as powerful an insight as in a parliamentary system at a time of a new majority government, the notion of purposeful ambiguity is still one possessed of significant explanatory power.

Can conventional regulation act as an agent of change? An intuitive answer must be, at most, only to a quite limited extent, especially when that change involves moving away from an entrenched monopoly towards open competition. As Milton Friedman bluntly put it, in keeping with our continuing feline theme, "Cats can't be made to bark."[62] Empirical support for this view may be found in contemporary developments in telecommunications regulation.

Here the CRTC has been exceedingly active, but despite the expenditure of massive amounts of regulatory energy and enthusiasm, many fundamental issues remain unresolved. To start with, as we have seen, the Commission's statutory mandate is (deliberately) internally inconsistent and it hedges what commitment to competition there is with many reservations which protect beneficiaries under the old regime. As well, the benefits and desirability of competition has to be established over and over again in case after case. This has led to a whole series of highly complex and protracted proceedings. This, in turn, has caused delays which means that new issues, such as local competition and convergence, are pressing in on the regulator before it can adequately resolve older issues, such as long-distance competition. As well, an era of "regulated competition" has given all sides opportunities to use regulation to gain strategic advantage, and there has been constant pressure to convert concerns for an even playing field to concerns for equal outcomes. In regulation, concern for equality of opportunity soon gives way to concern for equality of results.

Most importantly, the type of political intervention inherent in the model of conventional regulation in a parliamentary system of government has frustrated the regulator's ability to deal in a definitive manner with the need to place the industry on an economic basis appropriate to competition. The crucial issue here is that of rate rebalancing. Historically, under a monopoly regime a subsidy flowed from long distance, which was priced well above its cost, to local service in order to achieve maximum connectivity (universal service). In rate rebalancing, rates for local service are raised towards their costs, while those for long distance are lowered towards their costs. The CRTC saw this type of rebalancing as a fundamental prerequisite to competition. However, when it sought to undertake even a relatively modest level of rebalancing, the cabinet, in response to complaints from consumer groups, intervened to put its decision on hold.[63]

The underlying cause of this prevarication is, of course, the lack of a clear mandate in the Telecommunications Act and the invitation and opportunities it provides for those who wish to retain the benefits of monopoly or to be protected from the full impact of market forces. The risk here is of sugar-coated competition in which politicians seek to give guarantees that everyone will be winners. If change is to be achieved, what is needed is unadorned competition. Compare this with what Maurice Strong has recently had to say: "The role of the regulator would be to ensure fair competition and provide incentives for efficiency and sustainability."[64] However, notions such as "fair" or "sustainable" competition are invitations to large-scale regulatory intervention which will go a long way to undermine sought after gains in efficiency. The key remains, as Daniels and Trebilcock recognize, to concentrate on getting structures right. Indeed, an ounce of ex ante structural good sense will be worth a hundred weight of ex post regulation.

Rather than have the regulator, with its distributional agenda and lack of particular expertise, deciding on what is or is not "fair" or "sustainable" competition, it would be much more appropriate to involve the Director under the federal Competition Act. Simply put, the prevention of unfair competition is the proper job of the competition authorities, not economic regulatory commissions.[65] This is because no matter how great are the failings of competition law enforcement, competition law has as its central function the preservation of the effectiveness and vitality of the competitive process. Regulation, by contrast, tends to suppress competition in order to protect competitors. However difficult it is to draw the line between the two in practice, economic regulation has an inherent orientation towards preserving competitors; competition law, the competitive process itself.

However, if the Competition Act is to play a significant role, the judicially-created "regulated conduct defence" will need to be addressed.[66] Briefly stated, the courts have said that where there is a modicum of regulation, this may act to exclude the application of competition law. This has created much confusion and uncertainty as to the role of competition law in areas subject to regulation. It would be preferable to deal with this issue directly in legislative reform, rather than await the outcome of judicial reconsideration.

A traditional justification for regulation has been a perceived need to take certain matters "out of politics." However, this has meant that regulation, unlike adjudication, has often been weakened by political isolation rather than strengthened through independence. If regulation is regarded simply as a process by which competing interests are reconciled in a quasi-adjudicative process (the conventional view), then insulation from politics would be appropriate. If, however, regulation is to be an agent for change, this will require it to be integrated into, not isolated from, the political process. Hence the inappropriateness of a "neutral" external forum. As the day-to-day implementer of government policy, such a regulator would have to be subject to government direction, on the understanding, of course, that directions as subordinate legislative instruments will have to conform to the general policy direction adopted in the legislation. As well, in order to maintain momentum, it would be important for the regulator to work closely with the appropriate legislative oversight committee. As the recent Daniels Task Force Report[67] highlighted, it is quite possible for a regulator to drift so far out of the political orbit that it finds itself incapable of even getting the legislature's attention for badly needed legislative change.

This suggestion for the reintegration of regulation into politics should be seen as limited to situations where there has been a major shift away from the conventional regulatory model. If the structure-incentive issues are adequately addressed; if competing interests have been resolved in the legislation; if the regulator's mandate is that of implementing a clearly articulated general policy and not that of reconciling competing interests on an ongoing regulatory basis; and if the regulator is to be an agent of change, then such a massive departure from traditional regulation calls for a similarly radical departure from traditional institutions and relationships.

This may mean that we should be looking to appoint a single, committed individual rather than the conventional relatively neutral, multimember agency. Judicial values such as impartiality and the benefits of multiple decision makers, may simply no longer be relevant. However, earlier in their presentation, Daniels and Trebilcock identified a certain amount of more traditional regulation. This may have to be dealt with by way of a small adjudicative panel attached to the Office of the Electricity Commissioner for Ontario (OECO).

NOTES

1 I regret that illness prevented me from presenting this paper at the Conference itself. I thank both my commentators and the conference organizer for this opportunity to make a belated contribution.

2 Ronald J. Daniels and Michael Trebilcock, "Ontario Hydro at the Millennium: Has Monopoly's Moment Passed?" 25.

3 Ibid., 45-6.

4 See, for example, H.N. Janisch, "Policy Making in Regulation: Towards a New Definition of the Status of Independent Regulatory Agencies in Canada," *Osgoode Hall Law Journal* 46 (1979):17; "Independence of Administrative Tribunals: In Praise of 'Structural Heretics'," *Canadian Journal of Administrative Law & Practice* 1 (1987):1.
5 Daniels and Trebilcock, 4-8.
6 And may well be moving even further. See, for example, Eli Noam, "The Next Future of Telecommunications: From the Network of Networks to the System of Systems," in *The Future of Telecommunications Policy in Canada,* eds. Globerman et al. (Bureau of Applied Research, Faculty of Commerce and Business Administration, University of British Columbia, and Institute for Policy Analysis, University of Toronto, 1995), 385.
7 H.N. Janisch, "Canadian Telecommunications: The World Turned Upside Down," *The Canadian Law Newsletter*, XVIII, Committee on Canadian Law, International Law and Practice Section, American Bar Association, (1993):5.
8 Daniels and Trebilcock, 25.
9 Ibid.
10 Ibid., 6.
11 Ibid., 37.
12 Ibid., 26-7.
13 Telecommunications Act., S.C. 1993, c. 38.
14 H.N. Janisch, "At Last! A New Canadian Telecommunications Act," 17 Telecom Policy, 691; "New Federal Telecommunications Legislation Act and Federal Provincial Arrangements," *The Future of Telecommunications Policy in Canada,* 149.
15 H.N. Janisch, "International Perspectives: Competition and the Regulatory Challenges of the Information Highway," Pacific Telecom Council 1995 Regional Seminar, Kananaskis, Alberta, June 20-21, 1995.
16 Daniels and Trebilcock, 26.
17 For a stimulating re-assessment of divestiture see Richard Vietor, *Contrived Competition, Regulation and Deregulation in America,* (Cambridge, MA: Harvard University Press, 1994).
18 For a description of the breadth of the regulatory challenge this presents, see Michael Ryan, "Regulation in a Newly Competitive Market," in *The Future of Telecommunications Policy in Canada,* 99.
19 It is not yet clear whether comprehensive reform legislation will be enacted. See Daniel Pearl, "Panel Clears Telecom Bill," *Wall Street Journal,* March 24, 1995, B8, col. 4; "Overhaul of Communications Law Passes the Senate by a Convincing Vote of 81-18," ibid., June 16, 1995, A3, col. 1; "Long Distance Companies Get Through to Congress," ibid., July 24, 1995, A14, col. 1.
20 Jill Vardy, "Competition Bureau Adds Staff to Stentor Probe," *The Financial Post,* June 15, 1995, 1, col. 2.
21 Daniels and Trebilcock, 31.
22 Ibid., 26-7.

23 Ibid., 45-6.
24 Ibid., 28-41.
25 Ibid., 29.
26 Ibid., 35.
27 Ibid., 25.
28 Ibid., 35.
29 Ibid., 32.
30 Ibid., 26.
31 Ibid., 4.
32 Ibid., 32-3.
33 Competition Act, R.S.C. 1985, c. C-34.
34 Daniels and Trebilcock, 33.
35 Ibid., 33-5.
36 Ibid., 35.
37 Ibid., 35-8.
38 Ibid., 38.
39 Ibid., 40.
40 Ibid., 40.
41 Ibid., 41.
42 Ibid., 42-3.
43 For a particularly valuable assessment, see Economic Council of Canada, "Accountability, Policy Making and Regulatory Agencies," in *Responsible Regulation (*Ottawa: Ministry of Supply and Services, 1979), c.5
44 H.N. Janisch, "Policy Making in Regulation," 72-3.
45 Neil B. Freeman, *The Politics of Power: Ontario Hydro and Its Government* (University of Toronto Press, forthcoming 1996), manuscript 250-6.
46 Ibid., 252.
47 For a good overview, see Hugh Winsor, "Many Strings Pulled in Satellite-TV Row," *The Globe & Mail,* May 1, 1995, A1, col.2.
48 Broadcasting Act., R.S.C. 1985, c. B-9.
49 Ibid., s. 7.
50 Hugh Winsor, "Competing Home Satellites Urged," *The Globe & Mail,* April 6, 1995, A1, col. 2; ibid., "Panel Backs Opening Market to Competing Satellite Services," April 7, 1995, A4, col. 4; Lawrence Surtees, "Panel Gives Nod to Satellite Competition," ibid., April 7, 1995, B7, col. 2. As *The Globe & Mail* observed editorially: "One end run provokes another." "Corralling the CRTC," April 8, 1995, D6, col.1.
51 Jill Vardy, "Direct-to-Home Push Is On," *The Financial Post,* April 20, 1995, 9, col. 3; Hugh Winsor, "CRTC Considers Ottawa's Proposal on Satellite-TV," *The Globe & Mail,* April 20, 1995, A10, col. 1; ibid., "CRTC Rejects Request to Change Policy," ibid., April 1995, A4, col. 1; Jill Vardy, "CRTC Scrambles Over Changes," *The Financial Post,* April 21, 1995, 7, col. 1; "DTH Licences Snagged in Red Tape," ibid., April 22, 1995, 11, col. 1; "DTH Licence May be Illegal:

CRTC," ibid., April 25, 1995, 12, col. 5; Hugh Winsor, "Cabinet Expected to Overrule CRTC," *The Globe & Mail,* April 25, 1995, C3, col. 4.

52 Hugh Winsor, "Satellite TV Gets Ottawa's Nod," *The Globe & Mail,* April 27, 1995, A1, col. 2; Jill Vardy, "CRTC Forced to Change Rules," *The Financial Post,* April 27, 1995, 3, col. 1; John Urquhart, "Canada Overturns Regulatory Ruling on TV Transmission," *The Wall Street Journal,* April 27, 1995, B8, col. 5; Jill Vardy, "Expressvu May Sue Over Ruling," *The Financial Post,* April 28, 1995, 4, col. 3, "CRTC, Ottawa Battle over Licences Heats Up," ibid., June 7, 1995, 13, col. 1; Harvey Enchin, "Spicer Rips Ottawa for Trodding on CRTC's Turf," *The Globe & Mail,* June 7, 1995, B1, col. 1. In a lead editorial, "Mr. Spicer Protests Too Much," *The Globe & Mail* observed that the battle between the CRTC and government had "become a turf war and a grudge match." June 8, 1995, A22, col. 1. For a response from the CRTC, see "The CRTC and DTH Satellite Distribution," ibid., June 17, 1995.

53 See, for example, Jeffrey Simpson, "Chrétien Should Clear Up His Satellite-TV Signals," *The Globe & Mail,* April 7, 1995, A26, col. 1; Jill Vardy, "Politics Invades Satellite Battle," *The Financial Post,* April 27, 1995, 13, col. 1; Hugh Winsor, "Liberals Accused of Favouring Friends," *The Globe & Mail,* April 28, 1995, A6, col. 1.

54 Bud Sherman, the CRTC's Vice-Chairman, Telecommunications, had earlier perceptively noted the tension which existed between the government and its "independent" regulator. "The current government is much more assertive about its desire to create policy. They have an emphasis on competition. This government has displayed an interest in creating policy, rather than leaving it up to bureaucrats or arm's length agencies." Joanne Chionella and Michael Urlocker, "CRTC Struggles to Stay Relevant," *The Financial Post,* March 4, 1965, 8, col. 1.

In the end, after the directive to the CRTC had been before the parliamentary legislative committees, a formal and binding order was issued to the commission. See Hugh Winsor, "Government Wants CRTC to Change Satellite Decision," *The Globe & Mail,* June 20, 1995, A6, col. 3; Jill Vardy "Ottawa Committees Say CRTC Should License DTH Operators by Sept. 1," *The Financial Post,* June 22, 1995, 3, col. 1; Peter Morton, "Satellite TV Field Opened to Competition," *The Financial Post,* July 7, 1995, 3, col. 1; Barnie McKenna, "Satellite TV Service Set for Fall Start," *The Globe & Mail,* July 8, 1995, A1, col. 1.

55 The government eventually backed away from using the directive power to knock out the exemption earlier granted the Canadian consortium. However, the consortium does not appear to be in a position to take advantage of its favoured position. See, Harvey Enchin, "Expressvu Big Bang Now Just a Whimper," *The Globe & Mail,* July 18, 1995, B1, col. 4. Despite all the talk about competition and choice, as Terence Corcoran has pointed out, since Power Direct TV has agreed to comply with Canadian content requirements similar to those governing cable T.V., the role of competition and choice will

be severely restricted. "Satellite TV Won't Save Consumers Yet," *The Globe & Mail,* July 8, 1995, B2, col.2.

56 *Responsible Regulation,* 63-4.

57 Ibid., 66.

58 Margot Priest, "Structure and Accountability of Administrative Agencies," 1992 Special Lectures of the Law Society of Upper Canada. *Administrative Law: Principles, Practice and Pluralism* (Toronto: Carswell, 1993) 44. In her most valuable lecture, Priest sets out a comprehensive summary of all the various studies that have been made in recent years of the unique relationship which should exist between independent regulatory agencies and the rest of government in a parliamentary system of government.

59 For concern that the "double whammy" will undermine agency integrity, see Sheridan Scott, "The New Broadcasting Act: An Analysis" 1990, 1 Media and Communications Law Rev. 25 at 54-57.

60 Cabinet Appeals Abolition Act., S.B.C. 1993, c. 38. Also Murray Rankin, "The Cabinet and the Courts: Political Tribunals and Judicial Tribunals," *Canadian Journal of Administrative Law and Practice* 3 (1990):301.

61 This notion is somewhat more fully developed in H.N. Janisch, "The Choice of Decisionmaking Method: Adjudication, Policies and Rulemaking," 1992 Special Lectures of the Law Society of Upper Canada, 259 at 293-4.

62 Quoted by Richard Schultz in his presentation on "Regulatory Objectives in a Changing Environment" at the Seminar on Regulatory Frameworks: The Lesson of Experience for Developing Countries, Commonwealth Telecommunications Organization, Kuala Lumpur, Malaysia, May 16-20, 1995.

63 Steven Globerman, Hudson N. Janisch, and W.T. Stanbury, "Analysis of Telecom Decision 94-19, Review of Regulatory Framework," in *Future of Telecommunications Policy,* 417. Also Richard J. Schultz, "Old Whine in New Bottle: The Politics of Cross-Subsidies in Canadian Telecommunications," ibid., 271.

64 As quoted in the editorial, "A Strong Plan for Ontario Hydro," *The Globe & Mail,* July 5, 1995, A 12, col. 1.

65 Richard J. Schultz and Hudson N. Janisch, *Freedom to Compete, Reforming the Canadian Telecommunications Regulatory System* (Ottawa: Bell Canada, 1993), 20-24.

66 See H.N. Janisch, "From Monopoly Towards Competition in Telecommunications: What Role for Competition Law?" *Canadian Business Law Journal* 23 (1994):239.

67 *Responsibility and Responsiveness.* Final Report of the Ontario Task Force on Securities Regulation (Toronto: Queen's Printer, 1994).

Comments by

NEIL B. FREEMAN
I. BRUCE MACODRUM
ANDREW J. ROMAN

NEIL B. FREEMAN

Given the research I did for my forthcoming book, *The Politics of Power: Ontario Hydro and its Government,* (University of Toronto Press, March 1996), which is in essence a history of Hydro's relations with the government from its inception to the present, I thought what I might best add to Professor Janisch's insights on regulation of the electric power industry in Ontario is an explanation of how the current regulatory regime came to be.[1] In other contributions to this volume, this regime has been called "anachronistic," among other less charitable remarks, and this position, which Professor Janisch acknowledges, may well be true. But a couple of key points might be made in its defense as we consider alternatives for the future.

My first point, one which will be a surprise to many who work within the present system, is that a great deal of thought went into the design of the current regulatory arrangements. My second point is that enormous political will had to be mobilized to put the present system in place. Indeed, the story of how third-party regulation of Ontario Hydro came about may serve as a cautionary tale as discussion proceeds on how to replace the regulatory arrangements. My general point, if I may use the "p" word, is that there are politics to regulation.

Before addressing my two points, I think it is essential to outline the regulatory regime that Ontario Hydro has operated under since 1973. In that year the government passed a matrix of three statutes – the Power Corporation Act, the Ministry of Energy Act, and the Ontario

Energy Board Amendment Act – that put the basic structure in place. In addition to recreating the old Hydro-Electric Power Commission of Ontario (HEPC) as Ontario Hydro in the image of a modern corporation, this matrix established the Ministry and Ontario Energy Board (OEB) oversight where there had been no formal regulation.

In brief, the regulatory regime works as follows. When Ontario Hydro wants rate increases it must forward to the Minister of Energy (now Environment and Energy) its rate proposals eight months before the increases are to take effect. The minister, in turn, must refer the rate proposals to the OEB for a formal rate hearing. Four months before the new rates are to take effect, the OEB is required to present its recommendations to the minister. The minister then passes the OEB's findings on to Ontario Hydro for the consideration of its board of directors. Within this diffuse responsibility are the features of the regulatory structure that baffle the best analysts, which leads to the charge that the system is anachronistic.

The first and most obviously baffling feature is that the OEB does not have the quasi-judicial power to set electric rates – it can only review the rate proposals. The second, connected to the first, is that Ontario Hydro's board of directors, under the Power Corporation Act, has the final authority to set rates, which had been the case for the HEPC as well. The third is that the Minister of Environment and Energy does not have the power to arbitrate disputes between the OEB and Ontario Hydro. The fourth, despite the constraint of the third, is that the Ministry of Energy Act leaves the minister responsible for both Ontario Hydro and the OEB, and the Power Corporation Act and the Ontario Energy Board Act both designate the Minister of Energy as the responsible minister. As has been pointed out in other contributions, the minister, by being in a cross-pressured position, is left with conflicting interests. This situation is not thought to be appropriate by contemporary standards of regulation.

Having laid out the present regulatory structure, let me now explain how Ontario arrived at what many find inexplicable – review rather than regulation. I will start by outlining the oversight of Hydro that existed prior to 1973. While no rate regulation existed in statute, regulation nonetheless existed through highly informal private and public arrangements. Privately, the chair of Ontario Hydro, like any other government agency in those less complex times, would inform the premier of Hydro's plans, in this case on rates, and the premier would give his approval. Not unlike the post office foregoing stamp increases in an election year when it was a government department, political considerations would enter into the discussion of Hydro's affairs. Publicly, the Ontario Municipal Electric Association (OMEA), forerunner to the Municipal Electric Association, was the self-proclaimed regulatory "watch dog" of Hydro. Although Hydro's responsibility to the Municipal Electric Utilities (MEUs) was largely a façade, the Hydro chair would address the OMEA's annual meeting, and the Hydro president and other officials would meet with its power-costing committee, or other representatives.

Despite this regulatory arrangement's informality, it nevertheless served the government and Ontario Hydro very well until the mid-1960s. Historically, Ontario's abundance of water power meant costs of generation were low. In addition, Hydro maximized the efficiency of its infrastructure through promotional rates, thereby holding down costs even further. The result was (and an economist could spell this out in more detail than I) that rates were a bargain in Ontario and never much of a political problem. What changed in the 1960s was the onslaught of relatively high rate increases, with the uniform wholesale rate to the MEUs, which began in 1966, rising by 6% in 1966, 1967, 1968, 4.5% in 1969, and 6% again in 1970. The increases were largely attributable to the loss of Ontario's comparative advantage in water power. As a result, the move to fossil and then nuclear-powered generation increased costs and thus rates, with the larger factor being that nuclear power generally made policy-making more complex. With little prospect of a return to the simpler days of old, rates became a political problem, one for which the government was the politically responsible body.

The rapid and dramatic change in Ontario Hydro's rates in the 1960s fuelled pressure for a revamping of the informal regulatory arrangements to lessen the government's increasing political responsibility for Hydro. The issue was forced on the government primarily by the actions of the Niagara Basic Power Users (NBPU), precursor to the Association of Major Power Consumers of Ontario. It had two problems with the existing rate determination process. First, it objected to the favoured status the OMEA had with Hydro. Second, the NBPU objected to uniform rates on the basis that the rate payments of its members, located as they were close to infrastructure without debt, were being used to subsidize expansion elsewhere in the province. When the NBPU did not get satisfaction from Hydro, it made appeals directly to the cabinet. A secondary reason for the rate-making process entering the political arena is that consumers were beginning to make comparison with U.S. jurisdictions that had regulation. Ironically, the pressure the OMEA placed on the government to maintain the status quo also contributed to the forces on the government to address the issue.

In response to these and other pressures it faced on the Hydro front, the government appointed two intersecting commissions in 1971 – Task Force Hydro (TFH) and the Advisory Committee on Energy (ACE). TFH's focus was the organizational structure and policy goals of Ontario Hydro. The ACE's concern was the government's institutional capacity for managing energy policy. Both TFH and the ACE sought to bring an end to the informal, closed-door decision that had characterized rate determination. Both recommended rate review rather than regulation, setting in motion the present system.

On the surface, review of rates, rather than regulation, has had the appearance of a political compromise, a half-way house to full regulation. Review can be classified as a compromise because it satisfied the public demand for openness, Hydro's desire to maintain control over rates, and the government's need to avoid direct responsibility for rates. There are, however, reasons why review, rather than regulation, was recommended and adopted that are not based on political compromise. I say this even though the outcome has had the appearance of a compromise and the actors have recognized it as a compromise.

The logic of review for TFH and the ACE was that review placed the onus on Hydro to justify its increases rather than on the OEB, as the regulator, to provide Hydro with a sufficient increase. This situation avoided the common problem of the regulator becoming sympathetic with the regulated. Hydro would have to assume full responsibility for rates.

Review was also favoured because it was not thought necessary for either the OEB or the ministry to duplicate Hydro's function in determining rates. Although the OEB performed this function through full regulation, for example, of the gas industry, part of the reason review was accepted was that Hydro was a public rather than private monopoly. While it has been suggested that public monopolies are more arbitrary and less accountable than private ones, TFH, the ACE, and the government did not think this was the case. In sum, public openness was the objective, not public, third-party control of Ontario Hydro's rates.

While full regulation was left open as an incremental policy option, it was not only thought to be unnecessary but also to be undesirable. Regulation posed a political problem because it would have left the government, and not just the OEB, responsible for rate increases. Similarly, ministerial arbitration would also have left the government responsible. In adopting the quasi-regulation of review, the government was purposely avoiding the responsibility it had assumed in the 1960s for rate increases. Moreover, the government favoured review because it still permitted the cabinet to approve rates without public scrutiny. By leaving a diffusion of responsibility amongst Ontario Hydro, the Minister of Energy and the OEB, there could be public openness without clear lines of political responsibility.

The government was also happy to accept review as a compromise, while not discounting its logic, because it pleased Ontario Hydro and the OMEA and appeased the NBPU. Gaining the acceptance of these stakeholders for a new regulatory regime, if only the quasi-regulation of review, was an important consideration. The reason was that an extraordinary degree of political will was required for government to establish review. Indeed, before this time, Hydro, for all intents and purposes, had undergone no significant revamping since its inception in 1906. This had not been the case because no one had ever thought to reform Hydro. Reform had often been suggested and proposals forwarded, but what had been missing was the political will to tackle the job, or possibly to take on Hydro. Part of the problem was the fact that the OMEA, claiming that the MEUs were the owners of Hydro, had always mobilized against the revamping of Hydro for fear that it would lead to greater government control.

The question my survey of the politics of Ontario Hydro's current regulatory regime raises is whether the past has any relevance for the establishment of new regulatory structures? Restructuring and privatization issues are, after all, the subject of the present discussion. My assessment of the situation is that any Ontario government, given the adversarial nature of government and opposition in parliamentary democracy that Professor Janisch points out, will seek to avoid responsibility for rates and rate increases. Regulatory options that increase political costs for government will be a more difficult political sell as government considers revamping or replacing the regulatory regime. My educated guess, having examined the politics of the government/Hydro relationship over time, is that this will be true even of options that envision minimally formal regulation of a competitive electricity market, let alone the informality of moral suasion by government to keep rates in line.

The unknown factor, one which I readily acknowledge, is the extent of the political will that exists for privatization – is it strong enough to realize the objective? Back in 1973, there were two factors which drove Hydro reform, including the establishment of OEB rate review. The first was that technological change made the promotion of large-scale nuclear plants, with all their economic spin-offs, an important economic development strategy for government. The second was that this new shape of generation created the imperative to modernize Hydro's structure to take advantage of the new efficiencies. In 1995, there are also two factors driving reform. The first is the technological changes that, for example, make small gas-fired electricity generation attractive. The second, given the inflexibility of a vertically integrated monopoly, is the need for privatization of at least portions of generation to take advantage of the new efficiencies. In my view, the factors driving reform are equally as compelling for the government now as they were in 1973.

As part of a larger restructuring and privatization of the electric industry in Ontario, regulatory reform will depend on the political will of the government. But while the will may emerge, there are politics to regulation that can pose complications and compromise outcomes even for the most strong-willed of governments.

NOTE

Given Professor Janisch's absence from the conference due to illness, the above commentary was originally prepared before his paper was completed.

I. BRUCE MACODRUM

I am stating my own views here and not those of the Toronto Electric Commissioners, except in a couple of areas where the context should make it clear.

I intend to comment on what I call sign post issues: the issues to think about in order to know where in fact we are, so that we can come up with a plan for the future – whether that plan is to leave where we are or to stay where we are.

It is interesting to note in these political days that many of the significant public institutions were the creation of Conservative governments – the CBC, Canadian National Railways, and Ontario Hydro. Ontario Hydro was also the creation of one of the most activist Conservative governments led by Sir James Whitney – which also created the Workers' Compensation Board and established the funding for the University of Toronto as we know it today.

Ontario Hydro was not created for the usual "socialist" reasons for public ownership. It was created for economic development and competitive reasons. The coalition of interests that successfully pushed for it were manufacturers and municipal politicians; their rallying cry was "power at cost." Ontario Hydro was never intended to be a publicly owned business organization with a government-established mandate, but otherwise acting as a private commercial enterprise. From the outset, Ontario Hydro was to be a part of the provincial public sector, delivering an essential service and overseeing and regulating local providers of that essential service.

This leads to my second point. I heard echoes of what I want to say from Mr. Yves Ménard from Hydro-Québec and my second point is really his California bungalow analogy. Ontario Hydro and the municipal electric distribution system are uniquely Ontario creations. Solutions from the United Kingdom, New Zealand, or even the United States may not be appropriate to our circumstances, which are the result of our unique economic and historic development. Also the solutions elsewhere are relatively new – and it is not yet clear how successful they are going to be. The electrical supply system is a large interconnected and interdependent machine. It has been constructed over a long number of years and while it is quite robust in many of its features, it has its own vulnerabilities. Let us remember the electricity supply is fundamentally an engineering matter – economists, lawyers, accountants, and financiers are the support troops – we cannot plan, design, operate or maintain the machine that is the electric power system.

Before you can debate the appropriate form of regulation, you have to understand what it is you are going to regulate. The last speaker referred to the work of Task Force Hydro. The "Role and Place" study defined Ontario Hydro as a "delivery agency of the Provincial Government receiving broad policy advice from the government." And I do not think that definition, although it was in management consultants' jargon, would have troubled Sir James Whitney or Sir Adam Beck.

They sought to create a new public body that would provide benefits to all sectors of the Ontario economy. Electricity was too important to be left to the motives and machinations of private entrepreneurs. It was like schools, hospitals, water works, and other public services. The fundamental issue, which again is Mr. Ménard's issue, is this: is it an essential service that is supplied at cost or is it just another commodity? If it is an essential service, you do not regulate it, you administer the supply of the service and you do it in an efficient and effective manner according to the highest standards of public administration. If a government is dissatisfied with how the service is administered, you fire the administrator and if the public is dissatisfied with the government – they turf out the government.

I remember being told by a very experienced minister when I was with the Ministry of Energy that the two public institutions that the people of Ontario would not tolerate significant political interference with were the O.P.P. and Ontario Hydro.

Therefore, I wonder whether a more practical topic for debate should be what are the appropriate mechanisms of accountability and administration to ensure that Ontario's electricity supply system is responsive to the needs of the Province in the 21st century?

Remember, we are proceeding into the future looking, as McLuhan said, through a rear-view mirror. Our only guide to the future is the past. I wonder if we will not see a period of reaction to this current economic and cultural globalization. People already see their individuality, their community, and their national identities are threatened. The 21st century may see us trying to strengthen and create mechanisms of economic and cultural independence rather than weakening and abandoning them.

However, the topic of this study is regulation. So if you are going to regulate, you have to have an entity to regulate in the private sector such as Bell Canada, Trans Canada Pipelines, or Consumers Gas.

What is the entity that will be regulated? First let us examine the generating assets of Ontario Hydro. There are the five nuclear plants:

Pickering A and B, Bruce A and B, and Darlington. I suggest these assets are not saleable as long as there is the significant contingent liability of what to do with the nuclear wastes.

Next let us look at the fossil assets: they are the coal-fired stations Nanticoke, Lambton, and Lakeview in southern Ontario, Thunder Bay and Atikokan in northern Ontario, and the oil-fired Lennox station.

Will the new owners have the freedom to choose their source of fuel? Remember that the northern Ontario plants and Nanticoke use substantial quantities of western Canadian coal.

I suspect that arch free enterpriser Premier Ralph Klein will be one of the first on a plane to Toronto if there is a possibility that the new owners of Nanticoke are free to buy the lowest-priced coal. Also, any private purchaser of the fossil assets would prudently require assurances regarding So_2, No_x, and CO_2 emission levels. Will governments be prepared to give these assurances?

This leaves the hydroelectric assets. The motto of Ontario Hydro in English is "the benefits of nature belong to the people." I wonder whether a privatization scheme that is anchored on the hydroelectric resources of the Niagara, St. Lawrence, Ottawa, and the northern rivers would have wide public appeal.

In the U.K., the largest beneficiaries of privatization of the regional distribution utilities have been the executives of these companies, who have seen their salaries and benefits increase significantly. So it would be contrary to my interest to be too negative on this option. There are structural problems in the electricity distribution industry and Toronto Hydro is committed to playing a leading and constructive role in addressing them. We accept that the resolution of this problem is beyond the capacity of the industry to address by itself. Government will have to provide some direction, if not leadership, on this issue.

Before the issue of privatization of the distribution sector can be addressed, both Ontario Hydro's power district and the municipal utilities must be rationalized into approximately equal-sized units.

Some are arguing for larger utilities for example, serving the whole Greater Toronto Area. That is Mississauga's Project Phoenix. Others argue the optimum size is about 100,000 customers. After that you start to lose whatever economies of scale are present. That would mean that Toronto Hydro would be split into two utilities.

There has been discussion about segregating the components of the industry vertically, but many utilities in the distribution sector are diversifying horizontally. Already we have quite a degree of that with our public utilities commissions. For example, the Kingston Public Utilities Commission, sells gas and electricity and also runs the bus line. There is great variation throughout the province in the nature of our PUCs. Toronto

Hydro and other utilities are expanding their activities in the communication area. We are installing fibre optic cable in our ducts in the downtown area in the city. We are renting surplus fibres to customers who are interested. We are also looking at diversification in a number of other areas. Possibilities exist in the training area, possibly energy utilization services, garage services, and a multitude of others that relate to our existing business.

My next issue is: once you have privatized, how do you regulate? This issue is not as complex as many suggest. It really comes down to two fundamental options – do you regulate profits or do you regulate prices?

Most North American regulation, so-called rate base/rate of return regulation, is profit regulation. Although the regulator sets prices, these are based on an "allowable return" and regulated companies tend to make frequent visits to their regulators to adjust their prices to ensure they are obtaining their allowable return or to seek a higher allowable return.

Price regulation is the U.K. model. Prices are permitted to move in some relationship to an indicator of general inflation, and whatever return or profit the regulated company can obtain under these prices are for the benefit of their shareholders.

The North American model is fundamentally a cost-based model with shareholder return being considered a cost of equity capital. The criticism of this approach has always been that it rewards capital expenditure without imposing adequate discipline to promote the effective and efficient use of capital. The price regulation model addresses this defect. However, it is somewhat difficult to develop a logical rationale for tying the price of electricity to the general inflation rate, and it is even more difficult to explain the precise factor by which prices should be above or below the general inflation rate. Both models tend to address the issue of competition for large industrial customers by letting the regulated companies negotiate specific large-customer prices within a prescribed range. The other thing that can be noted is that regulation in Ontario in the gas industry and the U.K. model for regulating electricity prices also extend extensively into the standards of service. The various local distribution companies have quite extensive requirements they must meet in standards of service. They report on their success in meeting these standards to the regulator and their customers annually. So the regulator examines not just the price of the commodity but also the standard of electrical service.

Of the issues facing Ontario, the electricity supply system is not in any crisis state that demands urgent public attention compared to many of the other issues on the public agenda, such as the situation with respect to our welfare system, our municipal taxation system, and a host of other things.

That is not to say that changes should not be made. Ontario Hydro and the municipal utilities are struggling to reduce their costs and to be more productive while at the same time improving standards of customer service. We think about electricity as being a homogeneous product, but it is not. For example, the type of service that our customer expects in the downtown area of Toronto in what we called the network or the 13.8 kV system, which is quite different than what we would supply in a residential area or what most other utilities would supply in much less densely populated areas. The amount of redundancy that is built into the downtown system is greater and the reaction by customers to changes in the system and changes in the operation of the system is different. So we provide in effect different products for different markets and a kilowatt hour is not a kilowatt hour in terms of the cost of providing service and the quality of service that is provided. I would argue that before Ontario Hydro can be restructured, the distribution system must be restructured. If we are going to have competition among generating sources and an open access grid, we have to have purchasing companies of a size and sophistication to manage a portfolio of electricity supply. Government will have to play an important role as a catalyst. Once this restructuring of the retail distribution system is undertaken, we can then examine the issues of public versus private ownership and the appropriate form of regulation.

ANDREW J. ROMAN

Other contributors discuss various techniques for regulating Ontario Hydro. A more basic question, however, is whether effective regulation of a large, provincially owned utility such as Ontario Hydro is even possible.

Is Effective Regulation of Ontario Hydro Rates Possible?

Rate regulatory agencies were created to prevent or to reduce the abuse of monopoly power by shareholder-owned public utilities. But a state-owned enterprise like Ontario Hydro has no incentive to gouge consumers in order to enrich shareholders; Hydro has no shareholders. The Averch-Johnson effect, which describes the economic reason for excessive capital expenditures by shareholder-owned utilities, is also not relevant to Ontario Hydro. Hydro's incentives are more political and bureaucratic than economic.

Hydro's real top management is its political master, the Premier, out of whose own office Hydro has usually been steered. Unlike the shareholders of a private business, the Premier does not seek self-enrichment but, rather, political popularity (especially at election time). So, for decades, Hydro has hired thousands of people, all over the Province, and has paid them very high salaries. Hydro has been not so much in the electricity generation and transmission business as in the megaproject construction business. Its construction projects permit government press releases announcing government-created jobs.

It was former Premier Bill Davis, at a dinner speech I attended about 15 years ago, who announced that although high interest rates had caused other governments to cancel their megaprojects, Darlington would be built. By the time the project's costs had to be put into the rate base, and paid for by the consumer, Bill Davis had retired as a hero and Premier Bob Rae was left to announce the bad news of the rate increases to pay for it. Good news Bill created bad news Bob. This is not to blame one Premier or to praise the other. Regardless of the party in power, decisions about Hydro's investment have very long-range impacts. They are not made by shareholders on economic grounds but by Premiers on political grounds. These decisions are made well above the level any regulator is likely to occupy in the political hierarchy.

Hydro dispenses political largesse. It pays for municipal infrastructure as a reward to nuclear reactor host communities. It creates construction and operation jobs in areas of underemployment all over the Province. Will any premier be willing to delegate control over that much economic and political power to the Ontario Energy Board? It is questionable whether either the Premier or the Treasurer of Ontario would be willing to delegate to a tribunal which must, to remain credible, operate at arm's length, the final say in decisions which have such a large impact on the Provincial economy. Neither Ontario Hydro nor any of its prospective spinoffs (such as a separate transmission grid) can truly be controlled by an independent regulatory tribunal until Hydro is first decoupled from political control in the Premier's office.

Contrast Hydro's situation with what conventional public utility rate regulators do in the United States and Canada. How could any rate regulator of Ontario Hydro punish inefficiency in the regulated company by refusing to approve higher rates to pay for inefficient investments (as is suggested by Daniels and Trebilcock, page 33)? There are no shareholders to punish, only customers and taxpayers – essentially the same people. Hydro's notional balance sheet equity comes from customers, not equity investors. In these circumstances, meaningful rate regulation by a regulatory agency, whether employing rate-of-return techniques or price caps, is improbable, if not impossible.

So far we have been looking at whether the present and reasonably foreseeable ownership of Ontario Hydro, and its highly politicized role, permits effective rate regulation. Based on the same political analysis we turn to the examination of the present regulatory mechanisms.

What Kind of Regulation Do We Have Now?

Ontario Hydro is not unregulated. It is subject to normal regulatory laws, such as federal safety standards for nuclear reactors, or environmental assessment legislation. (Nevertheless, it would be well to recall that Ontario Hydro's largest single project, the Darlington nuclear generating station, was exempted from environmental assessment by Premier Davis on the ground that there was insufficient time for the assessment to be conducted before the electricity was needed.) Ontario Hydro's political clout in this Province is still a formidable force to be reckoned with. Moreover, none of this regulation is economic in nature.

The present regulation of electricity rates in Ontario is more cosmetic than real. It serves as no model for the future. This is a criticism of the system, as established by the Legislature perhaps three decades ago, and not a criticism of anyone in office today, and especially not a criticism of the Ontario Energy Board. The purpose of rate regulation is to protect consumers from unfair rates while preserving the financial integrity of the utility. The present regulation of electricity rates in Ontario is not designed to do this.

Most consumers, both residential and industrial, purchase their electricity from a Municipal Electric Utility (MEU). Yet there is no way a customer of an MEU can complain to a rate regulator about the MEU's rates. That is because, as a matter of law, Ontario Hydro is not just the major vendor of electricity to the MEUs but also the regulator of the rates of the MEUs. In a recent complaint I sent to Ontario Hydro about an MEU's rates on behalf of a client of our firm, I was told in a letter from a senior Hydro official that Hydro had no mechanism in place for hearing complaints about its regulatees. The parties would simply have to work it out themselves somehow. This is not the way most people conceive of how a rate regulator works. Yet this is all the regulatory control we have in Ontario over the distribution monopolies.

For its part, Ontario Hydro is not really rate regulated either. The OEB's statutory mandate is limited to making recommendations about Hydro, not decisions. A direct customer of Ontario Hydro cannot initiate a hearing into Hydro's rates because the Ontario Energy Board has no jurisdiction to respond to requests for hearings from customers. The OEB's jurisdiction, as limited as it is, can be invoked only by a reference from the Minister. The scope of such references is usually heavily influenced by Hydro, which has a much larger public relations and legal staff to lobby the government than the government has to control Hydro. (In the 1993-94 round of severe budget cutting, not only did the public relations and legal branches of Hydro not receive cutbacks but, perhaps uniquely, they enjoyed increases.)

As a practical matter, today the Ontario Energy Board can only make recommendations about rates, which the Minister and Ontario Hydro are likely to follow. If the Board makes recommendations which do not please Hydro – and Hydro, not the Minister, makes the ultimate decision regarding rates – Hydro can ignore the OEB. If this happens repeatedly, the members of the OEB are unlikely to be reappointed, because the output of these lengthy and expensive Hydro rate hearings will have been shown to be less than useful in setting Hydro's rates. In short, it would be a career-limiting move for members of the OEB to make recommendations which wander too far from what Ontario Hydro really wants. Underneath all the courtroom trappings and the cross-examination, therefore, Ontario Hydro hearings are, in effect, more like a debating society than a regulatory proceeding.

Even the topic of the debate is determined not by the customers, for whose protection a regulatory agency normally exists, but by the Minister, as influenced by Ontario Hydro. Thus, for all practical purposes, Hydro is self-regulated.

There is, admittedly, some limited regulatory review and some limited political oversight from "head office," the Office of the Premier. The Premier can appoint (or disappoint) the Chair of Hydro and the Hydro Board Members, and can veto expenditure plans. But since few premiers, and few of their political advisors have had any experience in running a major electrical utility, they are largely captured by the information, and the interpretation of that information, filtering through to them from Ontario Hydro. There is not much room in this political structure for an independent regulatory agency.

The rather limited regulatory review just described is not really "regulation" as that term is normally used. The OEB looks, rather than touches; it reviews rather than regulates. These differences in the nature of the regulatory scrutiny are fundamental. Under review, the public is given the opportunity to complain about rate changes and other related matters, without necessarily having any impact on the result. Under rate regulation, there is also usually a public hearing, but the outcome is normally a legally binding rate decision, not merely a recommendation. The purpose of the OEB's public review is catharsis; the purpose of true regulation is control.

Although the public is increasingly dissatisfied with mere catharsis, the transition to regulatory control will not be an easy one.

Ontario Hydro has a long-standing political tradition and corporate culture of autonomy. The imposition of meaningful politically independent regulatory control would be a radical change for the Province. Despite Hydro's praiseworthy sponsorship of various conferences I've attended recently on the subject, the topics canvassed at last summer's OEB Hearing, H.R. 22, and this summer's Hearing, H.R. 23, are very narrow.

Sir Winston Churchill once said: "Jaw jaw is better than war war." I would not attempt to improve upon Churchill, but I will paraphrase him from Hydro's point of view: a conference about competition is better than competition. Or, as recent speeches by Hydro's Chair, Maurice Strong, and its President have indicated, Hydro is all in favour of competition – but not just yet.

Privatization

To paraphrase Mark Twain, the rumours of Hydro's impending privatization are greatly exaggerated. This is so for several reasons.

First, what premier of Ontario will be anxious to sell off "the Crown Jewels," the organization that still spends billions of dollars every year all across the Province, in so many ridings? And how marketable are the nuclear stations, which generate roughly half of Hydro's power?

Second, "Public Power" is more than merely a popular ideology in Ontario. It borders on being a secular religion, a political sacred cow. Public Power is older, in Ontario, than is Medicare. How strong is the incentive to incur the political wrath of the Hydro unions and the many public power supporters for a hideously complex and politically risky privatization, when the benefits may not be noticed in this term of office, if ever? There is as yet no widespread or powerful public demand for privatization of Ontario Hydro. It will take an act of strong and committed political leadership to privatize Hydro over the opposition such an act would engender.

Third, there is the issue of "stranded assets." Even Professors Daniel and Trebilcock (pages 26-8) accepted the bogeyman of stranded assets being created by inefficient bypass. Even the term "bypass" is interesting and revealing. If someone stops purchasing shoes at Eaton's and starts purchasing them at The Bay, no one would call that "bypass"; we would simply say that the customer had found a new supplier. The fancy and pseudo-technical term "bypass" is reserved for public utilities losing business to their competitors. If Eaton's loses money and has stranded assets as a result of competition, we would say that that's too bad, but that's just the way it is. Not so with Hydro.

The unarticulated assumption is that Hydro's stranded assets are an evil to be avoided. They are assumed to be a temporary result of genuine errors in forecasting, which errors will be cured in time. This assumption rests upon the theory that all previous Hydro managements were inept in making supply and demand forecasts but, because the current crop is so much cleverer, such errors will cease from now on. All current and future forecasts will be absolutely correct, so that the stranded assets will be reduced and, over time, ultimately eliminated. What a delightful dream.

Historically, severe supply and demand imbalances have been recurrent. This is to be expected with province-wide central planning. It is not necessary to have a long memory. Let us look at this decade alone. First we had the massive Demand/Supply Plan (DSP) to cure a perceived shortage crisis. Then, within a few months, after a long and expensive hearing, we saw a sudden about face with the capital cutbacks, when it was discovered that the forecasted deficiency in capacity was really a huge surplus. In how many of the last 20 years have supply and demand been perfectly in balance?

Much of the still massive capital expenditure being made today by Hydro will be tomorrow's stranded assets. Stranded assets – like the poor – will always be with us. Indeed, for an organization dominated by governments sensitive to the imperatives of partisan politics rather than economic efficiency, stranded assets are a competitive advantage. That is because the short-run marginal cost of a utility with large excess capacity will always appear lower than that of most potential generating competition. Stranded assets are simply the price a large, capital intensive, publicly owned utility has to pay, with its customer's money, for a quiet, sheltered life – for competition soon, but not just yet.

Today's stranded assets will be paid for, one way or the other, in higher-than-market electricity rates, or in taxes to repay the debt. But the rates will only come down to market levels in Ontario when there is a market in Ontario. Such a market can be created if we want to do so as a province, and if we get our priorities right. We need three things: de-politicization, privatization, and competition, with all three being introduced at once. At the same time, Ontario Hydro must give up any regulatory role it has over MEUs and over itself. After privatization, new stranded assets would no longer be political problems for the government, but what they should be – an economic problem for shareholders.

The government allows Hydro to favour itself in any competition. For its part, the government also helps Hydro directly, as Hydro pays few taxes, has very low short-run marginal costs, and receives numerous political advantages. Today, if a privately owned competitor could make serious inroads on Hydro, the government would be tempted to intervene to help Hydro. We can see examples of this at the federal level, where the Department of Transport provided special protection to Air Canada against all competitors while it was a crown corporation. Even to this day, it is favoured by the Government of Canada as its chosen instrument of national policy. Competition with Ontario Hydro is simply too risky without clear, unequivocal, and irreversible guarantees of de-politicization. Ontario Hydro is, in many ways, a law unto itself in Ontario. De-politicization, therefore, is the first challenge facing Ontario Hydro today.

Since Ontario Hydro's vastly superior resources of knowledge enable it to dominate electricity policy making in the province, the prospect of privatization is a challenge to Ontario Hydro itself. Do its managers want this change enough to make it happen? Do they really want competition soon? We shall see.

Labour Adjustment at Hydro: Costs, Outcomes, and Alternative Strategies

PETER WARRIAN

Ontario Hydro has undergone significant downsizing of its workforce. The success to date has relied heavily on voluntary severance and early retirement programmes, accompanied by extension of conventional collective agreement mechanisms. This approach has high costs and diminishing results, and it does not deal with the high cost of job loss for individual workers. We will discuss here alternative strategies for dealing with labour adjustment through different kinds of labour-management initiatives.

INTRODUCTION: DEALING WITH WORK FORCE REDUCTIONS AT ONTARIO HYDRO

Ontario Hydro is undergoing a huge process of labour adjustment and dislocation. The total reduction in the work force has been approximately 12,000 employees over two years. Most of the reduction has come in the non-unionized staff, outside of the PWU and Society bargaining units, as the table on the following page shows.

To date Hydro and its bargaining agents have relied on traditional collective agreement mechanisms and very expensive buy-out programmes to achieve the targeted work force reductions. We consider that these traditional techniques may have worked for the parties in the past but their costs, and the financial and socio-economic impacts on workers, will show diminishing returns and limited supportability in the future. The parties must look for new ways to reduce both the social costs of job loss for individual workers as well as the overall expense of the work force

Ontario Hydro Total Staff Levels[1] (1992-1994)

Employee Group	*Sept. 1992*	*Feb. 1995*	*Change*	
			Staff	*%*
Executive	708	457	251	35
Non-represented	2275	1297	978	43
Society	7601	5388	2213	29
PWU	18220	14332	3888	21
Other	126	45	81	64
Total-Regular	28930	21519	7411	26
Non-Regular	1331	214	1117	84
Construction	4342	1110	3232	74
Total	34603	22843	11760	34

reduction programmes used to date. Both improved worker outcomes as well as reduced costs involve new approaches to labour adjustment and training in respect to both the internal labour market of the corporation and the external labour market of the broader economy.

Management Initiatives: Retirement and Buyout Offers

Hydro management has sought to achieve its targets for downsizing through a large and generous set of severance, early retirement, and separation programmes. These have been offered in three Phases.

In Phase I, an Early Retirement Allowance (ERA) and Voluntary Separation Allowance were offered to employees. Some 1582 employees took advantage of the options. In Phase II, there was a Special Retirement Program (SRP) and a Voluntary Separation Program (VSP). Some 4,985 employees took advantage of the offers. In Phase III, Special Separation Program (SSP) was offered to achieve an additional 2200 reduction, but only about 1,000 took up the option.

The following table summarizes the outcomes of this combination of programmes:

Summary of Voluntary Reduction Programs[2]

	Early Retirement Allowance (ERA)	*Voluntary Separation Allowance (VSA)*	*Special Retirement Program (SRP)*	*Voluntary Separation Program (VSP)*
Employees	399	183	387	2598
Total Costs	149 M	6 M	433 M	110 M
Cost /Employee	$106,500	$32,800	$181,400	$42,300

The numbers in the table indicate significant successes in the aggregate, but at very high costs. In addition, the outcomes have shown diminishing results. As indicated, the Phase III program only achieved half of its intended target.

Union Initiatives: Traditional Collective Bargaining Approach

The Power Workers' Union has sought to deal with the risks and costs of job losses by resorting to traditional collective bargaining agreement mechanisms. The three key mechanisms are Article 11 of the collective agreement on "Surplus Staff Procedure," Article 13 on "Employment Security Plan," and the "Purchased Services" agreement.

The Surplus Staff Procedure under Article 11 takes up 18 full pages of the collective agreement, with 27 official sub-clauses.3 An indication of the elaborate detail of the provision is section 11.1, which only deals with "Definitions," including definitions for:

Classification
Job Family
Equal Classifications
Seniority
Work Group
LSEOLS
R/D/HO
Site
Occupational Group
Same Classifications
Lower Classifications
Head Office
Exceptions
R/D/B
Location

The provision deals with internal redeployment of employees and arose in the 1985 period when the corporation was undergoing significant internal change, but with relative employment stability overall.

The type of procedure involved is reflected in Article 11.2 of the agreement, which reads:

> Depending on the business need of the location, where a surplus is identified in a classification at any location, either the least senior regular part-time or the least senior regular full-time employee in the surplus classification in the location shall be declared surplus. Where the least senior employee is required to perform work which cannot be performed by a more senior in the location within four weeks, then the next senior employee within that classification shall be declared surplus.
>
> Where there is more than one employee in a surplus classification at the same location, employees in the surplus classification at the location who are not surplus by virtue of their seniority may be given surplus status for vacancy purposes only. When the surplus has been eliminated or when the search/notice period has elapsed such surplus status shall be withdrawn.

The difficulty this has produced in the present situation is that following each step in the procedure takes a long time and often involves "cross-bumping" of junior employees. In the face of the pervasive downsizing and uncertainty about what jobs will be where in the future, inevitably an extensive pool of employees under notice of surplus has accumulated. In the Customer Services business unit, the pool of surplus employees currently totals some 500 individuals.

In addition, an Employment Security clause was agreed to in 1994.

Employment Security[4]

For the period April 1, 1994 to March 31, 1996, no regular Power Workers' Union-represented employees will be involuntarily terminated (does not include termination for cause or retirement at normal retirement age). During this time period, staff who are currently surplus or become surplus shall have their search/notice period extended to the earlier of placement in a vacancy/placement opportunity or March 31, 1996. Staff declared surplus on or after December 11, 1995 shall receive a minimum 16-week or 4-week search notice as described in Article 11.4

To relieve the pressure on payroll costs of the employment security provision, the parties agreed to reduce the Employer's pension fund contribution.[5] The total cost was $200 million.

As a tactical move, reliance on a temporary employment security agreement and use of pension monies is understandable, but it provides no long-range solution. It can easily be viewed by outsiders as trying to put off the inevitable, and doing so on the basis of an overfunded pension fund paid for by the public at large. Further, by definition, the parties can only go back to the pension surplus a limited number of times. The intricate and cumbersome mechanisms of prolonged bumping rights and entitlements under Article 11 do not provide a long-range solution. They at best can buy some time for adjustment and change.

WHY THE PWU POSITION IS ECONOMICALLY RATIONAL: THE COSTS OF UNEMPLOYMENT

The strength of PWU-represented employees and their bargaining agent's resistance to work force reductions, and consequently their defence of conventional job classification and work rules, is economically rational if measured by a very concrete outcome measure that employees intuitively understand: the social and economic costs of job loss. What does the average PWU member risk if he/she loses their job?

Because neither the union nor Hydro management currently track employees after they have left the corporation, only estimates are possible.

Using data from other sources, the author estimates that a back-of-the-envelope answer to the question is:

Estimated Costs of Job Loss for PWU Members (Current $ per Year)[6]

Male Worker	$9084
Female Worker	$11,367

It must be stressed that these are the income losses if workers *find another job*. In addition, the risks of unemployment for a typical PWU member is 27-41% for males and 27-46% for females.

The estimates are based on the Ontario Ministry of Labour Adjustment Model.[7] The Ministry conducted a research project to track 1200 randomly selected workers through three years of post-layoff experience. They concluded that the burden of difficult labour market adjustments is concentrated on those who had a major investment in the jobs they lost, both in terms of long tenure and very specific skills that are not easily marketed to potential employers in today's labour market. The problem is exacerbated when the displaced worker has a limited amount of education. Also, women, all other factors being the same, can expect to face more difficult adjustment.[8] In all these respects, the PWU-represented employees are uniquely vulnerable to excessive economic losses if they lose their Hydro jobs.

MINISTRY OF LABOUR MODEL OF ECONOMIC LOSSES DUE TO JOB LOSS

At displacement, each year of tenure reduces earnings two to three years later by $132 per year in real terms. Put differently, those with long tenure, defined as 15 or more years with the same employer, will earn at least $2,000 a year less. Displaced workers in the broad occupational group "semi-skilled factory production operatives" earned $2,977 less in their last job than the reference group. Compared with the reference group of college and university graduates, those who have less than a Grade 8 education earn $8,693 less, those with full primary school $4,723 less, some high school $4,283 less, high school graduates $3,731 less and those with some college or university $2,643 less. And, after controlling for all the other factors that explain the post-displacement earnings differential – the fact of being a woman costs the displaced worker $2,283.[9]

The Ministry of Labour model is suggestive of the sorts of outcomes that dislocated workers like PWU members may face if they lose their Hydro jobs. Given the importance of such economic risks to how collective bargaining over Hydro restructuring will take place in the coming years, it is in the parties' interests to do a tracking study to find out the

Weighted Factors in MOL Labour Adjustment Model

Factor	*Wage Loss*	*UI*	*Long Term UI*	*Early Retire*
1. Seniority	132			
< 3 yrs		18.1	6	3
4-14		23.1	10	3
15+		26.7	15	10
2. Occupation				
Semi-Skilled	$2,977	28.9	13	5
Millwright	+$4,950			5
Other		15.2	8	
3. Education				
< Gr 8	$8,693	54.8	51	8
Primary	4,723	41.9	24	10
Some High	4,283	24.7	9	6
High	3,731	15.4	5	5
Some College	2,643	15.2	4	4
Coll/Uni		13.5	6	1
4. Gender				
Men		23.1	8	5
Women	$2,283	21.4	13	5
Total	$2,561	22.2	10	5

actual outcomes. Documenting the outcomes should be a first step in changing them.

SECTOR MODEL OF ADJUSTMENT

Part of the underlying issue in dealing with labour adjustment at Hydro to date is whether the parties, both management and labour, feel or take real ownership for the process and outcomes. It is the argument here that the parties will not be able to deal successfully in new and cooperative ways with the internal restructuring of the workplace, including reskilling, work reorganization, career planning, etc. unless they jointly take ownership and co-manage the external labour market outcomes. In the absence of such a joint effort, both sides will be consigned to a war of attrition over the current collective agreement mechanisms in which all parties, including the public, will be subject to sub-optimal outcomes, to say the least.

New sectoral partnerships dealing with labour adjustment and training have emerged in recent years across Canada as a focal point for labour and management to do their business differently and together in the future.[10]

Sectoral councils have been established in most of Canada's major industries including steel, electrical and electronics, automotive parts, automotive repair and service, aviation maintenance, graphic arts, software development, and hospitals and health care. They continue to grow in popularity, and according to one estimate, thirty-five sector groups existed in some stage of formation in 1992. Government's willingness to fund the councils reflects a belief that industry and labour leaders are best able to define the human resource needs of the workplaces they represent, and to design programs for effectively responding to these challenges. Moreover, it is recognized that these councils represent a strategic and coordinated method for developing comprehensive human resource strategies.

After years of academics preaching to practitioners about Japanese or European models on labour-management cooperation, in the late 1980s the parties themselves started to generate a home-grown model of labour-management cooperation and change. Its strategic importance is for labour and management to come together outside of the confines of the traditional collective bargaining relationship, without seeking to end run the normal bargaining process. The primary objective is to have the parties co-manage labour markets in new ways. Inevitably the first and primary focus will be on the retraining and redeployment of employees who may be displaced as industries and sectors restructure. This will not just involve external redeployment from an existing employer to a new employer. It will also involve internal redeployment to different operations or occupations within an existing employer. This means that the parties will need jointly to address the human resource policies and strategies of the current employer. The necessary extension of the process will be for the labour and management sides to address the long-range human resource strategies for their organizations and industries.

The conclusion of this process in the private sector is that labour and management will exert as much energy in their relationships in the future over negotiating the definition and implementation of the future skill set of their workforce as they have in the past on narrower issues of employee compensation. These ideas are new to the public sector but they have become a common place of leading-edge developments in the private sector. What has been said of the new economy is that in an ERA of increased mobility of capital and technology the primary competitive advantage of nations is the skills of their people and the ability to innovate.[11] If the public sector is to contribute what we require for our quality of life and our future economic competitiveness, then these maxims will not only apply to the private firms and industries, they will also apply equally to public institutions.

The parties at Ontario Hydro may learn valuable lessons from the experiences of CSTEC (Canadian Steel Trade and Employment Congress), the first of the Sector Councils which was established in 1986 to deal with the downsizing of the steel industry. CSTEC is a joint venture of the United Steelworkers of America (USWA) and the Canadian steel companies. It is governed by a Board of Directors with Labour and Business Co-Chairs and a membership comprising the CEOs of the six largest steel producers and the six senior officers of the union.

In the period 1988-95, CSTEC provided counselling, training, job search, and placement services to some 12,000 laidoff steelworkers. Adjusting for those who left the workforce through retirement, over 80% of all laidoff steelworkers were placed into new jobs, typically through a period of retraining in community colleges. There are three key factors for the success of the CSTEC/sectoral labour adjustment model:

1. Getting in together as labour-management joint effort, own counsellors, etc.
2. Getting in early, prior to layoff with immediate access to services.
3. One-window shopping for counselling, job search, training referrals, UI.

CSTEC Labout Adjustment Programme Results - October 1994[12]

	Hamilton	*Algoma*
Assisted	781	1,729
Out of Labour Force	128	58
Retired, etc.	73	40
In Training	58	18
In Labour Force	653	1,671
Employed	582(89%)	1,543 (92%)
Unemployed	71 (11.0%)	128(8%)

In 1991, CSTEC took the initiative to extend its mandate in human resource management in the steel industry by expanding its involvement in training to include training of the current work force for future technologies and future skill requirements of those staying within the industry. Through CSTEC, the steel companies and the United Steelworkers have developed a Skill Training Program which is cost-shared with the federal and provincial governments. The proposed funding formula was 1:1, that is one dollar from government for each additional training dollar spent by the industry. In practice, the industry has contributed three dollars in additional training for each one dollar of government training funds.

The objectives of the program are to improve the quality and effectiveness of training and to at least double the amount of training in the industry over three years.

CSTEC Training Programme Objectives

1. At least double the training effort of the industry over three years in terms of broad-based portable skills.
2. Develop industry standards or guidelines in the areas of :
   - training needs analysis
   - training plans and budgets
   - equitable access to training
   - skill training content
   - evaluation.
3. Develop a system of accreditation and certification which will improve the portability and transferability of skills.[13]

Data is now available to begin assessing the achievement of these objectives over the first full year of program operations.

On March 31, 1994, CSTEC completed the first full year of its Skill Training Program. During the year, 14 local Joint Training Committees (JTCs) presented Training Plans. The committees are comprised of local union and management representatives that survey employee and supervisory personnel on training needs, compile and approve training plans for submission to the national Training and Adjustment Committee of CSTEC. The results were some $26 million of CSTEC-eligible training.

Training Expenditures 1993-94

Industry Training	
Established	$10,033,570
Additional	15,914,470
Total	$25,948,040
Government Contributions	
Federal and Provincial	$ 5,289,656

"CSTEC-eligible" training is defined in the Guidelines as training that is broad-based and develops portable, generic skills. As stated above, legislatively mandated items such as health and safety and government-sponsored apprenticeships are specifically excluded. The latter were the kinds of training that formed the bulk of training in the industry up until 1993.

CSTEC Eligible Training[14]

Foundation Skills	Literacy
	Numeracy
	Computer Literacy

Foundation Skills (continued)	Problem Solving Group Skills Communications Skills Training the Trainer
Steel Industry General Skills	Steel Industry Economics Work Re-Organization Facilitator & Supervisor Training Environmental Control Energy Conservation
Steel Industry Technical Skills	Basic Metallurgy Casting Melting Rolling Programmable Logic Controllers Craning Pipe Welding Electrical Resistance Welding Basic Steel Quality Inspection Statistical Process Control
On-the-Job Training	Where it is an integral part of a training plan and includes other methods of formal training
Union-Sponsored Training	Basic Accounting and Financial Analysis Training for Joint Committee Participation Communications Work Organization Ergonomics Technological Change

To provide a base of comparison, the data from the 1991 Human Resources Survey were rerun based on the criteria of CSTEC eligibility.

Comparison of CSTEC-Eligible Training (1990 - 1994)[15]

CSTEC-Eligible Training 1991	$5.7 Million
CSTEC-Eligible 1994	$26 Million

Clearly there was a dramatic increase in the amount of portable, generic skills training being offered in the industry. Over 23,000 steel industry employees received CSTEC training in 1993-94. When combining the 1991 Human Resources Survey data with eligibility, the result was an equally dramatic change in access and distribution of training. In the earlier period, the preponderance of training expenditures, outside of legislatively mandated health and safety training plus apprenticeships, were going to non-bargaining unit technical and supervisory/managerial employees.

Trainees By Occupation (1993-1994)[16]

Occupation	*Number*	%
Managerial	699	3
Supervisory	4,760	20
Production	10,341	44
Clerical	686	3
Trades	4,880	21
Technical	1,979	8
Total	23,345	

Types of Training (1993-1994)[17]

Type	%
Technical Skills	63
Foundation Skills	23
Work Organization	14

The pattern of the first year results of the CSTEC Skill Training Program are clear: there is a much higher level of overall training effort, a greater access to training on the part of hourly employees, and a much greater emphasis on foundational skills and work organization.

The real results of the new training effort in the steel industry provided through CSTEC will only be evident over time, as those skills are put to work on the shop floor. This raises a number of questions about ongoing monitoring and success factors for the CSTEC Program. These include: Will the greater access to training result in utilization of new forms of work organization? Will the emphasis on foundational skills result in a leveraging to higher order skills? Do improved foundational skills enable employees to participate more effectively in work teams and problem solving in the workplace? Will skill enhancement efforts affect the course of collective agreement negotiations in the next major round of industry bargaining in 1996?

The lesson for Hydro from the private Sector Council experience would be to establish a similar body to deal with active labour market policy across Hydro's diverse operations, and ultimately with the Ontario Electric Utility industry as a whole. There are three potential benefits to such a projects:

1. *Economies of Scale and Scope*

Hydro's current efforts have the implicit assumption that the adjustment problem can be handled within its own operations. It currently seeks to offer counselling, transition assistance, and training to employees under surplus notice. In addition it has relied on passive measures like early retirement offers, voluntary severance offers, etc. As the data indicate, this approach, which has had early success, now faces very large cumulated costs and diminishing returns. One of the reasons for this is that PWU members, among others, do not feel that they have any place to go in terms of alternate employment and career opportunities. A natural opportunity lies with other utilities, which is where ex-Hydro staff have often gone in the past. There may also be additional opportunities with Hydro customers. Extending the scope to include broader utilities also offers the prospect of spreading the costs further as well as accessing government funds which are available to assist in deferring the costs.

2. *Joint Management of Worker Transitions*

As discussed above, currently neither Hydro management nor the union take any responsibilities for workers once they have left the corporation. No one knows the true outcomes, beyond personal anecdotes, because there is no tracking data. More fundamentally, however, it should be realized that external labour market outcomes and internal labour market regulation under collective agreements are intimately linked. The reason that PWU members hold to their present positions so strongly is fear of having no alternative. This creates additional pressures to increase internal labour market rigidities through more and more restrictive and expensive contract provisions like the Purchased Services and Employment Security agreements. Hydro's corporate objective of greater internal flexibility and mobility of employees and skills is unlikely to be achieved as long as the outcome for PWU members is a 25% loss of income and 25-40% unemployment rates.

3. *Bridges to Joint Human Resource Practices*

Individuals within Hydro management and the PWU talk at times about achieving a cooperative goal of a high skill, high wage workplace where employees have active participation in their business units and expansive career plans as "knowledge workers." The experience across Canada has been that in order to achieve a new "flexibility-security" deal of this type, some new structures of labour-management cooperation are required in which the management and union come together as equal partners in new ways. The Sector Council Model is the best current example to achieve this goal.

CONCLUSION

We have discussed the current efforts of Ontario Hydro to downsize successfully in a way that meets its operational and financial objectives, while also dealing with employees as fairly and humanely as possible. A series of management retirement and voluntary severance arrangements have been introduce with significant aggregate accomplishments. The PWU, in turn, has sought to safeguard the job security interests of its membership with aggressive new contract guarantees. To deal with the ongoing process of internal and external redeployment of Hydro staff however, will require additional and new approaches. The restrictive job classification and work rule provisions in the collective agreement are there precisely to prevent the huge potential income and social losses that have been estimated for dislocated Hydro workers. To find a new flexible and effective approach for its unionized employees will require moving on both the collective bargaining and active, joint labour adjustment fronts. Labour and management representatives at Hydro would be well-advised to look at an adaptation of the Sector Council model to establish a new partnership in positive cooperation and human resource planning.

The author would like to thank the representatives of Ontario Hydro and the Power Workers' Union who made available their time and insights to assist in the preparation of this study.

NOTES

1 Ontario Hydro, Corporate Compensation and Benefits Division, "Memorandum on VRP Statistics" (November 17, 1994).
2 Ibid., 8.
3 Collective Agreement Between Ontario Hydro and Power Workers' Union, Canadian Union of Public Employees – CLC, Local 1000, April 1, 1994 – March 31, 1996.
4 Memorandum of Agreement, Ontario Hydro Corporation and Power Workers' Union, CUPE Local 1000 (March 30, 1994), 5.
5 Ibid., 14.
6 Estimate based on worker with Grade 12 education, 18 years of employment at Hydro, semi-skilled occupation, and gender.
7 "The Displaced Workers of Ontario: How Do They Fare?" Ontario Ministry of Labour, Economics and Labour Market Research, Toronto (1993).

8 Ibid., xviii.
9 Ibid., 26-30.
10 Carol Haddad, "Sectoral Training Partnerships in Canada: Building Consensus Through Policy and Practice," Eastern Michigan University, Ypsilanti (February 1995).
11 Robert Reich, *The Work of Nations* (New York 1991).
12 CSTEC, *Steel in Our Future News* 1, (Fall 1994), 1.
13 CSTEC, *Steel In Our Future 1995* (Toronto, January 1995), 10.
14 CSTEC, Skill Training Program 1994, (Toronto 1994), 2-4.
15 Recalculation of 1991 HR data by Peter Warrian (March 1995).
16 Training Program, First Year Results 1993-1994, CSTEC Toronto, October 1994, Table 2.
17 Ibid., Table 4.

Comments by

ERIC PRESTON

ERIC PRESTON

Peter Warrian describes many of the events that have occurred at Ontario Hydro over the past two years and presents the actions taken and costs incurred by Hydro in dealing with change. His basic premise is that Hydro has relied on relatively expensive and traditional tools to deal with downsizing, and that these are temporary measures at best. He then argues for a new, jointly managed approach that deals not only with the company's need to downsize but also with the impact on the individual.

Mr. Warrian goes on to emphasize the need for labour and management to address jointly the long-range human resources strategies for their organizations and industries, and cites the experiences of the Canadian Steel Trade and Employment Congress (CSTEC) as an instructive example for the parties at Ontario Hydro. He concludes that Hydro and its unions will have to move to new approaches to internal and external staff redeployment in both collective bargaining and "active, joint labour adjustment fronts."

While I agree with many of the observations made and principles advanced by Mr. Warrian, I have a number of questions about the mechanics of his proposal, and how it would fit within the current Hydro context. I have, therefore, tried to summarize my comments under a series of questions that occurred to me as I read through the material.

Was the large-scale downsizing a success?

Hydro was quite successful in responding to its 1993 financial crisis. It achieved its downsizing not by relying on traditional collective agreement tools, but by implementing large, innovative, and voluntary early retirement and severance programs. Through their voluntary nature, these programs were able to meet the needs of both the company and the individuals electing to participate in the programs. While I agree that the programs appear costly in total, they were determined to have a payback period of about 1.5 years, and were assessed as being less expensive than the cost of implementing the collective agreement alternatives. In addition, these lower staff levels are embedded in the cost base and will continue into the future.

Is downsizing still the issue?

In 1993, Hydro was responding to an urgent financial crisis characterized by high debt, decreasing load, and increasing electricity rates. Although industry restructuring was an emerging issue, our financial condition was "the burning platform." Like many other companies in similar circumstances, Hydro's response was to downsize its workforce and restructure its balance sheet.

While the company has survived the immediate financial challenge, the focus now is on repositioning the company for the new electricity market, and getting ready for competition. This means a change in emphasis from large-scale downsizing to more selective reskilling and redeployment within and outside the company. As Mr. Warrian has noted, this will require new and innovative approaches, but not just to downsizing. The point is that Hydro must focus on becoming a successful company, and it has to determine how it will use its human resources to achieve this. Human resource management must focus on positioning the company for success in a competitive environment, and though it will encompass downsizing, the approach will be much broader in scope.

Where is the balance between downsizing and other HR issues?

Parts of Hydro will continue to be primarily focused on cost and productivity, and some downsizing will be likely to continue. However, it also expects growth opportunities in other parts of the business, and the focus here will be on the more traditional attracting, retaining, and motivating employees. Hydro will also have to investigate new employment concepts and reward systems to be successful in these emerging business areas. This means that the company will have to be concerned both with how it treats employees affected by the continuing or selective downsizing, and also with those employees who stay with the company or who join the company. Issues such as the unbalanced demographics left by downsizing, the need to introduce new graduates into the company, and developing competencies that contribute to competitiveness are immediate and compelling priorities that also have societal implications at least equal to our obligations to departing workers. For both business and societal reasons, we need to pay as much attention to "survivors" and new employees as to the downsized and departing employees.

What's the role of the individual in this process?

In the past, Hydro has often been characterized as paternalistic, with employment provisions that cover employees from cradle to grave. Most literature now indicates that employees cannot expect that concept to continue into the future. Both the company and its employees will have to recognize the need for a new "employment contract" that recognizes that Hydro will not be able to provide the same type of employment certainty in the future as it has in the past, and that employees will have to take a greater measure of responsibility for their own futures independent of the company's. Employees will be expected to develop generic, portable skills that can be moved from company to company or industry to industry. This "continuous learning" and multi-employer career concept represents a major culture change, and it is not clear to me whether such a shift is consistent with a managed labour market approach.

In addition, Hydro employees are generally well educated and highly trained. This should provide them with greater mobility and employability than Mr. Warrian's models suggest. However, this does not mean that some additional training and support is not required to help them make such career moves, and to help employees appreciate that they do have marketable skills.

Can Hydro manage sectoral change?

There is a question of whether Hydro will be able to or want to manage dislocation of employees in an entire industry sector. While I am not very familiar with the conditions that existed in the steel industry at the time CSTEC was created, it seems that there was a common environment or external threat facing the companies and unions in the industry, and that it enabled a joint approach by all the players. At this point, it is not clear exactly what sector Hydro will be operating in, what it will look like, who all the players will be, and whether there will be a commonality of interests among them. Our ability to forecast accurately and manage staffing changes across an industry and over fairly long periods of time is also a major issue.

What will bring the Hydro parties together?

While it is clear that both Hydro and its unions will face a common future, it is not clear that the interests of both parties in that future will be identical. The track record on cooperation between Hydro and its employee representatives is mixed, and without a common set of interests or a precipitating external event, there may not be enough to bring the parties to a joint approach to fundamental change. In addition, the unions may decide to adopt a strategy that is independent of Hydro's business strategy. The recent actions by the PWU to expand to new employers could be seen as an example of union strategies to follow a path that is not completely dependent on Hydro's future success. This does not mean that Hydro's and the union's interests are in conflict, but rather that they may become more separate, as the organizations follow different strategic paths in pursuit of success in the new economic environment.

Can collective bargaining still work?

Mr. Warrian's paper has focused on some of the existing surplus provisions in the PWU collective agreement, and has identified a need for new approaches. However, it should be noted that Hydro has been successful in negotiating more innovative surplus provisions with its other major union, the Society of Professional and Administrative Employees.

Hydro and the Society have in the past few years developed jointly managed processes to achieve downsizing, and have recently established a $6 million Training Account specifically for training and reskilling. These were developed in collective bargaining "to assist employees affected by downsizing...to find employment within Ontario Hydro and to a lesser extent outside Ontario Hydro." While I am not suggesting that identical provisions will work for PWU employees, the opportunity to develop new approaches within collective bargaining should not be missed.

This experience also indicates that collective bargaining may continue to be the appropriate vehicle not only for designing new and innovative approaches to redeployment, but also for balancing the needs of the company, the needs of its continuing employees, and the needs of those employees leaving the company.

Summary

Mr. Warrian has presented a fairly complete picture of Hydro's success to date in downsizing, and accurately assessed a number of the issues that will face us in the future. In particular, his comments on some of our lengthy and cumbersome surplus procedures, the "stop gap" nature of the current employment security provisions, and the perceived vulnerability of employees facing job loss are all "on the money." The advantages of having both Hydro and its unions jointly focus on addressing these issues are obvious, although the concept of full comanagement goes beyond Hydro's current accountability model. However, the issues raised above cause me to wonder if the proposed processes are really appropriate for Ontario Hydro's situation and, at least, raise the issue of timing. Nonetheless, much of Mr. Warrian's analysis is very much "on point" for Hydro's human resource management, and it may help us find a way to address these concerns in the future.